AF376206

F. Hartmann

The Mathematical Foundation
of Structural Mechanics

With 174 Figures

Springer-Verlag
Berlin Heidelberg New York Tokyo 1985

Dr.-Ing. Friedel Hartmann

University of Dortmund,
Department of Civil Engineering
4600 Dortmund 50, FRG

ISBN-13: 978-3-642-82403-6   e-ISBN-13: 978-3-642-82401-2
DOI: 10.1007/978-3-642-82401-2

Library of Congress Cataloging in Publication Data.
Hartmann, Friedel, 1945— The mathematical foundation of structural mechanics. Bibliography: p. Includes index.
1. Mechanics, Applied-Mathematics. I. Title.
TA350.H37   1985   620.1'01'51   84-26721

Softcover reprint of the hardcover 1st edition 1985

Bookbinding: Helm, Berlin

2061/3020-543210

Preface

This book attempts to acquaint engineers who have mastered the essentials of structural mechanics with the mathematical foundation of their science, of structural mechanics of continua. The prerequisites are modest. A good working knowledge of calculus is sufficient. The intent is to develop a consistent and logical framework of theory which will provide a general understanding of how mathematics forms the basis of structural mechanics. Emphasis is placed on a systematic, unifying and rigorous treatment.

Acknowledgements

The author feels indebted to the engineers Prof. D. Gross, Prof. G. Mehlhorn and Prof. H. G. Schäfer (TH Darmstadt) whose financial support allowed him to follow his inclinations and to study mathematics, to Prof. E. Klingbeil and Prof. W. Wendland (TH Darmstadt) for their unceasing effort to achieve the impossible, to teach an engineer mathematics, to the staff of the Department of Civil Engineering at the University of California, Irvine, for their generous hospitality in the academic year 1980–1981, to Prof. R. Szilard (Univ. of Dortmund) for the liberty he granted the author in his daily chores, to Mrs. Thompson (Univ. of Dortmund) and Prof. L. Kollár (Budapest/Univ. of Dortmund) for their help in the preparation of the final draft, to my young colleagues, Dipl.-Ing. S. Pickhardt, Dipl.-Ing. D. Ziesing and Dipl.-Ing. R. Zotemantel for many fruitful discussions, and to cand. ing. P. Schöpp and Frau Middeldorf for their help in the production of the manuscript.

Dortmund, January 1985   Friedel Hartmann

Contents

Notations

$\boldsymbol{a} = \{a_i\}$	vector		$g_i[\boldsymbol{x}]$	198
$A = [a_{ij}]$	matrix		$G_i[\boldsymbol{x}]$	208
A_i, A_e	p. 16			
$\boldsymbol{A} \cdot \boldsymbol{B} = a_{ij} b_{ij}$	10		$\Gamma_\varepsilon(\boldsymbol{x})$	14
$C^n(\Omega)$	13		$\Gamma'_\varepsilon(\boldsymbol{x})$	15
$C^n(\bar{\Omega})$	13		$\Gamma_{N_\varepsilon}(\boldsymbol{x})$	14
$C^n_p(\bar{\Omega})$	13			
$C^\infty_0(\Omega)$	13		$H^k[a,b]$	26
$C^\infty(\Omega)$	13		$H^k(\Omega)$	28
C^n_{loc}	337		$H^k_0(\Omega)$	28
$C^{-1}[\ ]$	37		$H^{-k}(\Omega)$	234
$c(x)$	155			
$C(x)$	160		Λ	70
$C[\boldsymbol{E}]$	34, 37		$-\boldsymbol{Lu}$	38
			$L_2[a,b]$	24
D	69		$[[M_{nt}]]$	18
∂^i	69			
δ_i	177		$N_\varepsilon(\boldsymbol{x})$	14
$\Delta f = f_{,ii}$			$N_u(\hat{u})$	20
$\Delta\Delta f = f_{,iijj}$				
$\nabla f = \{f_{,i}\}$			$\Omega^c = \mathbb{R}^n - \bar{\Omega}$	13
$\nabla\nabla f = [f_{,ij}]$			$\bar{\Omega} = \Omega \cup \Gamma$	13
$\nabla \boldsymbol{u} = [u_{i,j}]$			$\Omega_\varepsilon(\boldsymbol{x})$	14
div $\boldsymbol{u} = u_{i,i}$			$0(g(x))$	12
div $\boldsymbol{S} = \{\sigma_{ij,j}\}$				
div^2 $\boldsymbol{M} = M_{ij,ji}$			$p:A,\ q:B$; if A then B	
$\Delta\varphi_i, \Delta\varphi_e$	15			
$\delta F(u, \hat{u})$	21		$\mathbb{R}^n$	9
$\delta^2 F(u, \hat{u})$	21		$\boldsymbol{R}^k$	82, 85
			R_1	111
$E(u, \hat{u})$	55		R_2	115
(f,g)	68		Σ	245
$[f,g]$	68		$\tau(\boldsymbol{u})$	38
$F(x^k)$	52		$||u||_k$	24
			u_s, w_s	70
$g_0[\boldsymbol{x}]$	141		$V(u, \hat{u})$	104

Introduction

Unlike mathematicians who live happily in the realm of their intellectual creations and must never bring their symbols in contact with the rough outside world, the engineer identifies mathematical symbols with physical objects.

A vector, P, he considers a force, a function, $w(x)$, the elastic line of a beam and a plane bounded domain, Ω, the middle-surface of a plate.

The symbols thus gain a practical meaning which is often much more important to an engineer than questions of (seemingly abstract) mathematical rigour.

It is this preoccupation with "meaning" which so easily leads to confusion among engineers and which makes it so difficult for a mathematician to argue with an engineer, never being sure with what kind of argument the engineer will come up with next—an argument from mechanics or from mathematics?

Consider the equation

$$\int \frac{M\,\delta M}{EI}\,dx = \int p\,\delta w\,dx \tag{1}$$

which expresses the principle of virtual displacements for a beam. If a mathematician suggests that the virtual displacement δw in this equation may be arbitrarily large, it might happen that the engineer objects:

"No, this is not true because in the derivation of the beam equation, $EI\,w^{IV} = p$, we assumed small deflections, that is we neglected the term $(w')^2$ in the expression of the curvature

$$\frac{1}{\rho} = -\frac{w''}{\left(1 + (w')^2\right)^{3/2}}$$

of the elastic line, hence, also the virtual displacement δw must be small".

The engineer quoted concentrates only on the mechanical meaning of the equation and he, thus, overlooks the fact that the equation is a mathematical expression whose truth does not depend on the magnitude of δw. The magnitude of δw does not enter the proof of this equation.

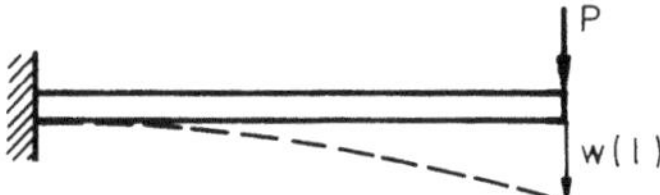

Figure 1

Or consider the following equation

$$\frac{1}{2} P w(l) = \frac{1}{2} \int \frac{M^2}{EI} \, dx$$

which expresses the end deflection of the beam in Fig. 1 in terms of the integral of the bending moment.

In a structural mechanics textbook the proof of this equation might go as follows: The external work is

$$W = \frac{1}{2} P w(l)$$

The internal work is

$$U = \frac{1}{2} \int \frac{M^2}{EI} \, dx,$$

and because the principle of "conservation of energy" requires that the external work is equal to the internal work, $W = U$, we may write

$$\frac{1}{2} P w(l) = \frac{1}{2} \int \frac{M^2}{EI} \, dx \quad (M = -EI \, w'')$$

The proof is done.

But from a mathematical point of view this is not a proof! Why should the integral of the square of the second derivative of a function $w(x)$ be equal to its value at the end of the interval? Because of a law of nature? Is 2 an even number because the world rotates eastwards?

Or think of the constant struggle in the literature with the factor 1/2. One author, starting with the principle of virtual displacements

$$\int (EI \, w'' \, \delta w'' - p \, \delta w) \, dx = 0 \tag{2}$$

obtained via

$$\delta \left(\frac{1}{2} EI \, w''^2 \right) = EI \, w'' \, \delta w'' \tag{3}$$

the equation

$$\delta \left(\frac{1}{2} \int EI \, w''^2 \, dx - \int pw \, dx \right) = 0 \tag{4}$$

To explain the fact that the factor 1/2 appears in front of the first integral but not in front of the second he argued as follows:

"We assume that during the variation the bending moment M changes according to the equation $EI\,w'' = M$, but that the load remains constant. No mechanical reason for these assumptions can be given. They are arbitrary and only justified in that they yield workable results...".

If only the author had dared to be a real engineer, to use his common sense!

The principle of "conservation of energy" and the δ-symbol are, loosely spoken, the jolly jokers of structural mechanics. A talented engineer proves all and everything with these two wild cards. Which sane person would dare to object to an equation as

$$\delta\,U = \delta\,W \qquad ?!$$

Or consider Castigliano's Theorem. In the literature the proof of this theorem starts with the statement that the external work done by a concentrated force is a quadratic form in P

$$W = \frac{1}{2}\,P^2\,\delta_1 = \frac{1}{2}\,P\,\delta \tag{5}$$

(δ_1 is the displacement corresponding to $P = 1$)
and because the internal energy, U, is equal to the external work

$$U = \frac{1}{2}\,P^2\,\delta_1\,,$$

the authors obtain Castigliano's Theorem by differentiating both sides with respect to P

$$\frac{\partial}{\partial P}\,U = P\,\delta_1 = \delta$$

But a mathematician would say this is not a proof either!

Sure, if Eq. (5) holds, then Castigliano's Theorem follows easily. But the problem is Eq. (5). This equation needs a proof. We cannot expect it to be always true.

This equation is, e. g., wrong in the case of elastic plates and elastic bodies because the displacement, δ_1, is infinite as Lord Kelvin found out in 1848, [T2].

Castigliano derived his Theorem in 1875, [C2], for trusses (for which it holds true) and in 1879, [C3], he extended it to elastic bodies by considering such bodies to be made up of a myriad of little truss elements. This extension is not correct.

Castigliano's approach is the typical engineering approach, typical for a Janus-faced science as structural mechanics where mechanical and mathematical reasoning constantly intermingle.

Some colleagues might, at this point, object that if the response of an elastic medium to a concentrated load is an infinite displacement, then the theory is wrong, and we need a more accurate description of physical reality.

But we shall demonstrate in chapter 5 that this seemingly unrealistic result is very natural. If the continuum is of dimension 2 or 3, the order of the governing operator 2 and the degree of the singularity 0 (= concentrated force), then this kind of singularity always occurs.

A more "accurate" description can only be obtained by raising the order of the differential equation from 2 to 4. But this is, we think, not necessary, because the very same differential equation renders excellent results as long as the load is evenly distributed.

Most of our engineering calculations for elastic plates and bodies are based on this equation, so why reject it because it renders (seemingly) strange results in the case of concentrated loads. One cannot eliminate the bad things without eliminating, at the same time, the good things.

To come to a close: we do not cite these examples to poke fun at engineers—the author is himself an engineer (and a mathematician)—but to make the reader understand our point of view, our motivation in publishing this book.

Doing engineering mathematics one easily comes under attack—from two sides.

The engineer does not understand why all this hairsplitting is necessary ("after all the Eiffel tower is still standing"), and the mathematician—never having himself experienced the confusion which so easily arises if the very same symbols once represent abstract entities and once physical objects—does not understand why an engineer sometimes deviates from the straight path of logic, why he needs help in his constant struggle to steer a clear path between abstract rigour and practical vigour.

After working for nearly four years with engineers, a mathematician confessed to the author that he finally began to understand what engineers meant by finite elements... And another mathematician, attending an engineering conference, seemingly in disbelief begged the author to tell him whether engineers really thought of finite elements as little plate elements connected at their edges...

To such mathematicians it might also come as a surprise that engineers even invented their own "functional analysis". While mathematicians speak of "dual pairings", engineers speak of the work of conjugated quantities, of the work of force and displacement. While mathematicians speak of test functions, engineers speak of virtual displacements, etc.

This "functional analysis", a genuine engineering development, is perhaps not so well founded as the real one, but it is dearly loved by engineers. They are very happy with it (as the mathematicians with their version), because it gives them some security, allows them, so to speak, to entertain the thought that they finally found the key to the secrets of mechanics.

Every engineer is well trained in this "functional analysis", and its arguments count in engineering circles at least as much as pure mathematical reasoning. Anyone doing engineering mathematics must take heed of this.

Thus it is not an accumulation of elaborate theories of abstract spaces, what is needed in the first place. More important is it, at the start, to remind the engineer of the simple insight that structural mechanics of continua is applied mathematics, applied infinitesimal calculus.

This insight, somehow, has got lost. Nowadays the mathematical core of structural mechanics is buried under thick layers of "engineering functional analysis" heaved upon by generations of engineers.

Once, at the start, engineers must have known the close connection between calculus and structural mechanics but then in the second generation — after the correct equations were derived, after the hard work was done—engineers slowly shifted the emphasis on meaning, they began to interpret the equations, they created "engineering functional analysis" which, today, has taken over structural mechanics so successfully that what once was (and still is) the basis, is now considered the outgrowth of the latter.

From an educational point of view this engineering functional analysis, certainly, is an ingenious and brilliant summary of the lore of structural mechanics but its defects, also, cannot be neglected: a young student being taught only this kind of structural mechanics will speedily and elegantly solve the old problems (with new numbers) solved already by Bernoulli, Euler, Cauchy et alii, but it is to doubt whether he will manage to solve new problems so effectively as well.

The successful solution of new problems requires, we think, a certain inner freedom, a certain inner detachement which will only prevail if the student has a thorough understanding of structural mechanics and its mathematical basis, that is:

If he knows which logic he really applies when he formulates $\delta U = \delta W$. If he has understood that "meaning" is only a secondary quantity which has nothing whatever to do with the truth of an equation. That an equation is a mathematical expression— open to (important) interpretations—but in its very essence subject only to the laws of mathematics.

The author hopes that this book can contribute to the propagation of this insight.

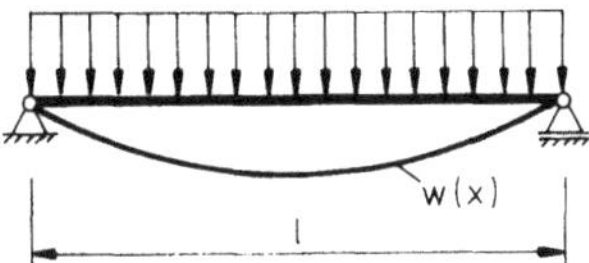

Figure 2

Let us briefly explain which concept we follow in this book.
The equilibrium position, $w(x)$, of a beam, see Fig. 2, can be considered
i) the function $w(x)$ which satisfies the differential equation

$$EI\, w^{IV}(x) = p(x)$$

in the interval between the supports, or
ii) the function $w(x)$ which satisfies the principle of virtual displacements

$$\int \frac{M\,\delta M}{EI}\, dx = \int p\,\delta w\, dx$$

with regard to all admissible virtual displacements δw.

These are the two in-roads to structural mechanics. Equilibrium can be defined in a pointwise sense (differential equation) or in a variational sense.

Sometimes in the literature these two in-roads are taken simultaneously. But this must lead to contradictions because the two condition each other.

If one starts with the differential equation, the principle of virtual displacements is a corollary of the equation $EI\,w^{IV}=p$. Vice versa, if one starts with the principle of virtual displacements, $EI\,w^{IV}=p$ is the Euler equation of the variational principle.

The second approach is more general but also more difficult, and we, therefore, decided to start with equilibrium in a pointwise sense.

The first and decisive step in this approach is the formulation of the differential equation which governs the displacement of the structural element in question. Responsible for this step is the engineer. No mathematician can do this step because it requires some knowledge of mechanics.

But once this step has been done, once the differential equations are established, all that follows is the result of a series of more or less simple mathematical manipulations (mainly integration by parts) which finds its justification not in mechanical laws but in mathematical laws—only.

We essentially only expand the integral

$$\int EI\,w^{IV}\,\delta w\,dx \tag{6}$$

in different directions (or to say it in mathematical terms: we test the differential equation with different functions δw).

In the process of these simple manipulations structural mechanics unfolds itself in all its aspects. No appeal to laws of nature, no conservation of energy principle, no "metaphysics" is necessary.

The credit for this, for the fact that these simple manipulations yield such a rich harvest, goes to the differential equation.

Agreeing upon that $EI\,w^{IV}$ is the governing equation of a beam, means also (as integration by parts reveals) agreeing upon the form of the internal energy,

$$U=\frac{1}{2}\int \frac{M^2}{EI}\,dx$$

upon the form of the force and displacement terms

$$w,\,w',\,M=-EI\,w'',\,V=-EI\,w'''$$

upon the form of the equilibrium conditions,

$$\int EI\,w^{IV}\,r\,dx+[Vr-Mr']_0^l=0 \quad r=a+bx$$

upon the potential energy functional

$$\Pi_1=\frac{1}{2}\int \frac{M^2}{EI}\,dx-\int p\,w\,dx$$

etc.

Negatively spoken this means that in the moment we have decided for a certain differential equation everything is fixed. No options are left.

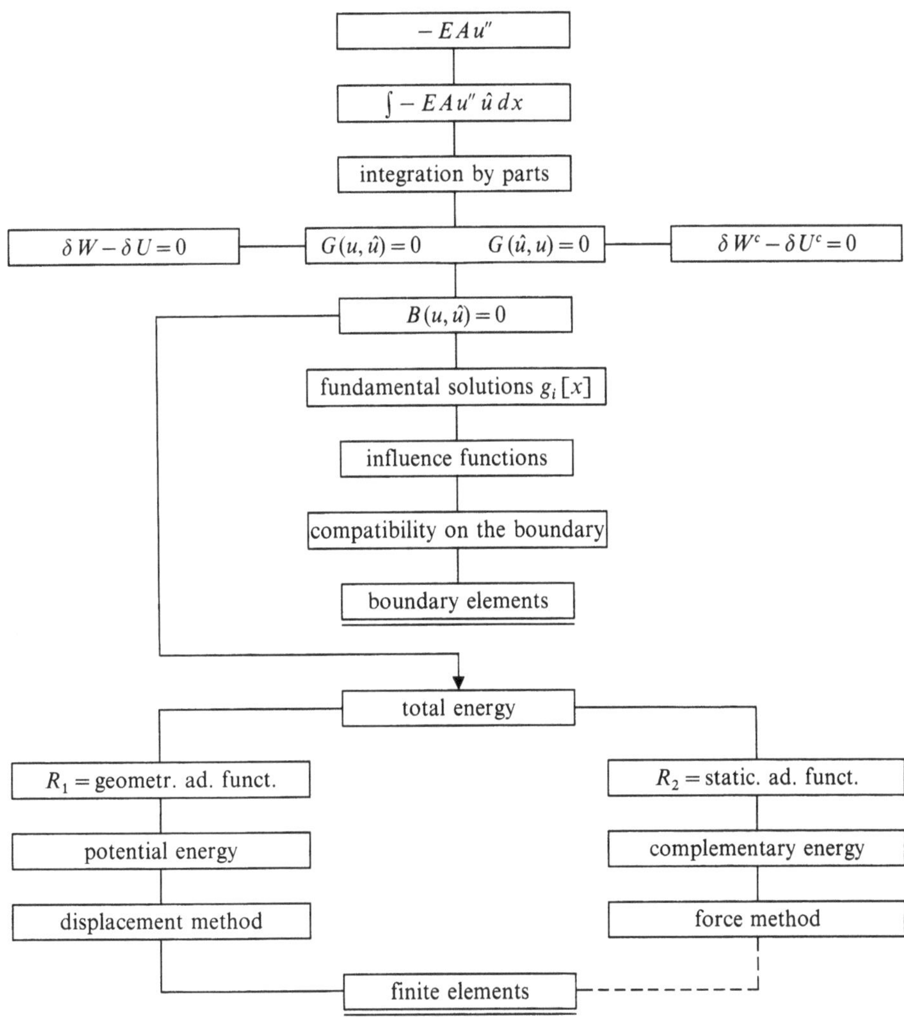

Positively spoken it means that everything is under control. We can be sure that the mathematical properties of the structural element in question are completely and accurately defined by the differential equation.

Any discussion about factors, as 1/2, about the sign of the internal work or about the question whether virtual displacements must be small or not, is superfluous if we trust in logic. Either one can derive it from the differential equation or one can not.

In writing this book we have tried hard to make the book readable, to communicate with the engineer in his language in the hope, that the engineer might, for once, lose his perennial fear of mathematics; though we know, from experience, how much difficulty this causes him.

Hermann Weyl once remarked: *"Men like Einstein or Niels Bohr grope their way in the dark toward their conceptions of general relativity or atomic structure by another type of experience and imagination than those of the mathematician, although no doubt mathematics is an essential ingredient"*, [W 4].

We think we do no one wrong if we modify the beginning and add:

"Men like Einstein and Niels Bohr and engineers in general..."

All this book attempts in the end is to bring some light into this darkness, to strengthen the insight with some logic.

1 Fundamentals

We introduce in this chapter our notations and the principal equations of linear, first-order, structural mechanics.

1.1 Vectors and Matrices

The points of the one-, two- or three-dimensional space of real numbers $\mathbb{R}^n$, $n = 1, 2, 3$, are denoted respectively by

$$
\begin{aligned}
&x && y \\
&x = (x_1, x_2) && y = (y_1, y_2) \\
&x = (x_1, x_2, x_3) && y = (y_1, y_2, y_3)
\end{aligned}
$$

Lower case bold-faced letters are vectors

$$u = \{u_i\}$$

and bold-faced capital letters, usually, matrices

$$A = [a_{ij}]$$

If we use indicial notation then summation over all indices which appear twice is implied. Thus, the scalar product of two vectors reads

$$u \cdot v = \sum_i u_i v_i = u_i v_i$$

and the vector product, $u \times v$, of two vectors u and v of $\mathbb{R}^3$ has the components

$$(u \times v)_i = e_{ijk} u_j v_k$$

where the auxiliary quantity

$$
e_{ijk} = \begin{cases}
1 & \text{if } ijk = 123 \text{ or } 231 \text{ or } 312 \\
-1 & \text{if } ijk = 321 \text{ or } 213 \text{ or } 132 \\
0 & \text{otherwise}
\end{cases}
\tag{1.1}
$$

is the permutation symbol.

The matrix A^T is the transpose of the matrix A. The trace of a matrix A is the sum of the entries on its diagonal

$$\operatorname{tr} A = a_{ii}$$

The (matrix) product of two matrices A $(m \times n)$ and B $(n \times p)$ is a matrix C $(m \times p)$ with the elements

$$c_{ij} = a_{ik} b_{kj}$$

but the scalar product, $A \cdot B$, of two $(n \times n)$ matrices, indicated by a dot, is a number

$$A \cdot B = a_{ij} b_{ij}$$

In some books you might find the notation

$$A \cdot B = \operatorname{tr}(A^T B)$$

Examples:

$$A = \begin{bmatrix} 10 & 2 \\ 2 & 3 \end{bmatrix}, \quad B = \begin{bmatrix} 5 & 7 \\ 6 & 3 \end{bmatrix}, \quad AB = \begin{bmatrix} 62 & 76 \\ 28 & 23 \end{bmatrix} \tag{1.2}$$

$$\operatorname{tr} A = 10 + 3 = 13, \quad \operatorname{tr} B = 5 + 3 = 8$$

$$A \cdot B = a_{11} b_{11} + a_{12} b_{12} + a_{21} b_{21} + a_{22} b_{22} = 85$$

The matrix $I = [\delta_{ij}]$ where

$$\delta_{ij} = \begin{cases} 1 & i = j \\ 0 & i \neq j \end{cases} \quad \text{(Kronecker's delta)}$$

is the unit matrix. Sometimes it is more appropriate, because more suggestive, to express the scalar product of two vectors by a matrix product

$$a \cdot b = a^T b = a_i b_i$$

that is to consider the two vectors a and b $(n \times 1)$ matrices.

Placing the T on the second vector

$$a b^T = C$$

we obtain a quadratic $(n \times n)$ matrix C with the elements

$$c_{ij} = a_i b_j$$

Some books use the notation

$$a b^T = a \otimes b$$

1.2 Differentiation

The partial derivatives of a scalar-valued function are denoted by

$$\frac{\partial f}{\partial x_i} = f_{,x_i} = f_{,i}; \quad \frac{\partial^2 f}{\partial x_i \, \partial x_j} = f_{,x_i x_j} = f_{,ij} \quad \text{etc.}$$

The gradient of a scalar-valued function $f(x)$, $x \in \mathbb{R}^n$, is the vector

$$\nabla f = \operatorname{grad} f = \{f_{,x_1}; f_{,x_2}; \ldots f_{,x_n}\}^T$$

The gradient of a vector-valued (m components) function $f(x) = \{f_i(x)\}$, $x \in \mathbb{R}^n$, is the $(m \times n)$ matrix

$$\nabla f = [f_{i,x_j}]$$

Hence, if we apply the operator ∇ twice to a scalar-valued function then the result is a matrix whose entries are the second derivatives of f

$$\nabla^2 f = \nabla \nabla f = [f_{,x_i x_j}]$$

Example: $f(x) = x_1 x_2 + x_2^3$, $\quad x \in \mathbb{R}^2$

$$\nabla f = \{x_2, x_1 + 3 x_2^2\}^T \quad \nabla^2 f = \begin{bmatrix} 0 & 1 \\ 1 & 6x_2 \end{bmatrix}$$

To apply the Laplace operator Δ to a function means to form the sum of its iterated second derivatives

$$\Delta f = f_{,ii}$$

If Δ is applied twice the result is

$$\Delta \Delta f = f_{,iijj}$$

If $f = \{f_i\}$ is a vector-valued function then the notation Δf means that we apply Δ to every component of f

$$\Delta f = \{\Delta f_1, \Delta f_2, \Delta f_3\}^T$$

The divergence, div A, of a matrix A is a vector whose components are

$$(\operatorname{div} A)_i = a_{ij,j}$$

The divergence, div u, of a vector u is a scalar

$$\operatorname{div} u = u_{i,i}$$

Hence, if we apply the operator div twice to a matrix A we obtain a scalar

$$\operatorname{div}^2 A = \operatorname{div}(\operatorname{div} A) = a_{ij,ji}$$

Example:

$$A(x) = \begin{bmatrix} x_1^2 & x_1 x_2 \\ x_2^3 & x_1 x_2^2 \end{bmatrix} \qquad x \in \mathbb{R}^2$$

$$\operatorname{div} A = \begin{bmatrix} a_{11,1} + a_{12,2} \\ a_{21,1} + a_{22,2} \end{bmatrix} = \begin{bmatrix} 2 x_1 + x_1 \\ 0 + 2 x_1 x_2 \end{bmatrix} = \begin{bmatrix} 3 x_1 \\ 2 x_1 x_2 \end{bmatrix}$$

$$\operatorname{div}^2 A = \frac{\partial}{\partial x_1} 3 x_1 + \frac{\partial}{\partial x_2} 2 x_1 x_2 = 3 + 2 x_1$$

If $f(x)$ and $g(x)$ are two functions defined near x_0 then the expression

$$f(x) = 0(g(x)) \qquad x \to x_0$$

indicates that there exists a positive number M such that

$$|f(x)| < M |g(x)| \quad \text{if } x \to x_0$$

In other words, the function $M |g(x)|$ is an upper bound of $|f(x)|$ if x tends to x_0. The function $g(x) = 1$ is, e.g., such an upper bound for the function $\sin x$ near $x_0 = 0$, $\sin x = 0(1)$ if $x \to 0$.

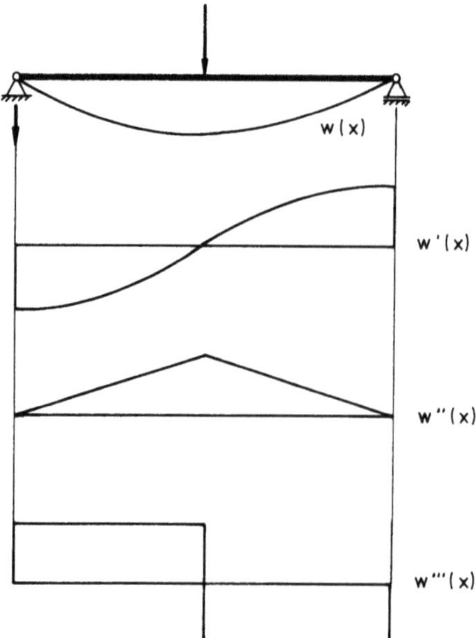

Figure 1.1

We classify functions with respect to their smoothness. A function f is said to belong to $C^n(\bar{\Omega})$ if u and its partial derivatives up to the order n are continuous in $\bar{\Omega}$, the domain Ω plus its boundary Γ. If this is only true in the interior of the domain then we say that u is an element of $C^n(\Omega)$. If the derivatives up to the order $n-1$ are continuous, but the last derivatives, the derivatives of order n, are only piecewise continuous in $\bar{\Omega}$ then we say that u belongs to $C_p^n(\bar{\Omega})$.

If $u = \{u_i\}$ is a vector-valued function then a classification as $u \in C^n(\bar{\Omega})$ means that each component u_i belongs to $C^n(\bar{\Omega})$.

The class $C^0(\bar{\Omega}) = C(\bar{\Omega})$ is the class of all functions which are continuous in $\bar{\Omega}$ and the class $C^\alpha(\bar{\Omega})$ is the class of all functions which are infinitely smooth, that is every derivative is continuous in $\bar{\Omega}$. Polynomials are such functions. Every constant function (the simplest polynomial) belongs to $C^\alpha(\bar{\Omega})$. Finally, $C_0^\infty(\Omega)$ denotes the class of all functions which belong to $C^\alpha(\Omega)$ and which vanish near the boundary Γ. All the derivatives of such functions are zero on the boundary.

The notation $B \subset A$ implies that the set B is contained in the set A, the set B is a subset of A. In the following expression all the classes to the left of a class are subsets of this class.

$$C_0^\alpha(\Omega) \subset C^\gamma(\bar{\Omega}) \subset \ldots C^n(\bar{\Omega}) \subset C^{n-1}(\bar{\Omega}) \ldots C^1(\bar{\Omega}) \subset C(\bar{\Omega}) \subset C_p(\bar{\Omega})$$

Example: The deflection of a beam loaded with a concentrated force, see Fig. 1.1, belongs to $C_p^3[0,l]$ because the shear force $V = -EI\, d^3 w/dx^3$ is discontinuous at $x = 1/2$.

1.3 Domains

The Greek letter Ω denotes a domain and the Greek letter Γ its boundary. The closure of a domain is indicated as

$$\bar{\Omega} = \Omega \cup \Gamma$$

It consists of Ω plus its boundary Γ.

If the domain extends to infinity we say it is unbounded otherwise we call it bounded. The domains considered in this book are always assumed to be bounded.

The complement of an interval $[a, b]$ is the x-axis minus the interval. The complement of a circle is the infinite plane with a hole of the same size as the circle. In general terms the complement Ω^c of an n-dimensional domain Ω is the domain obtained on subtracting the domain $\bar{\Omega}$ from the continuum $\mathbb{R}^n$

$$\Omega^c = \mathbb{R}^n - \bar{\Omega}$$

The orientation on the boundary curve of a plate is positive if the domain lies to the left when the arc-length s increases. The exterior unit normal $n = \{n_i\}$, $|n| = 1$, is the vector which points into the complement, see Fig. 1.2. and forms with the tangent $t = \{t_i\}$, $|t| = 1$, a pair of orthogonal vectors.

$$n_1 = t_2, \qquad n_2 = -t_1$$

The set

$$N_\varepsilon(\boldsymbol{x}_0) = \{\boldsymbol{x} \in \Omega \,|\, |\boldsymbol{x} - \boldsymbol{x}_0| \leqslant \varepsilon\}$$

is called the ε-neighborhood of a point $\boldsymbol{x}_0$. It consists of all those points in a domain Ω whose distance to $\boldsymbol{x}_0$ is smaller than or equal to ε. The boundary of $N_\varepsilon(\boldsymbol{x}_0)$ is denoted by

$$\Gamma_{N_\varepsilon}(\boldsymbol{x}_0) = \{\boldsymbol{x} \in \Omega \,|\, |\boldsymbol{x} - \boldsymbol{x}_0| = \varepsilon\}$$

If $\boldsymbol{x}_0$ is an internal point whose distance to the boundary exceeds ε then $N_\varepsilon(\boldsymbol{x}_0)$ has a symmetric shape, it is a circle or a sphere, see Fig. 1.3a.

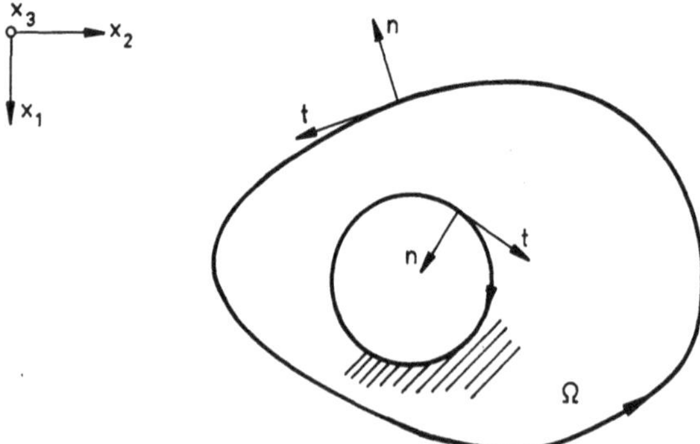

Figure 1.2

The distance of a point from the boundary is, by the way, not a fixed number but depends on the scale employed. Consequently, if $\boldsymbol{x}_0$ is an internal point then it is always possible to choose the scale in such a way that the unit neighborhood $N_1(\boldsymbol{x}_0)$ lies inside Ω. This is not possible if $\boldsymbol{x}_0$ is a boundary point, see Fig. 1.3b.

The analysis of singular integrals (influence functions) requires that we remove small neighborhoods of the source point from the domain.

If $\boldsymbol{x}$ is a point in Ω then we denote by

$$\Omega_\varepsilon(\boldsymbol{x}) = \Omega - N_\varepsilon(\boldsymbol{x})$$

the domain Ω minus the neighborhood $N_\varepsilon(\boldsymbol{x})$. More often we shall simply write Ω_ε instead of $\Omega_\varepsilon(\boldsymbol{x})$.

The boundary of such a domain $\Omega_\varepsilon(\boldsymbol{x})$ we designate by $\Gamma_\varepsilon(\boldsymbol{x})$ or simply by Γ_ε.

If $\boldsymbol{x}$ is an internal point and ε smaller than the distance to the next boundary point then the boundary of the domain Ω_ε consists of the original boundary Γ and the boundary Γ_{N_ε} of the ε-neighborhood, see Fig. 1.3a.

$$\Gamma_\varepsilon(\boldsymbol{x}) = \Gamma \cup \Gamma_{N_\varepsilon}(\boldsymbol{x})$$

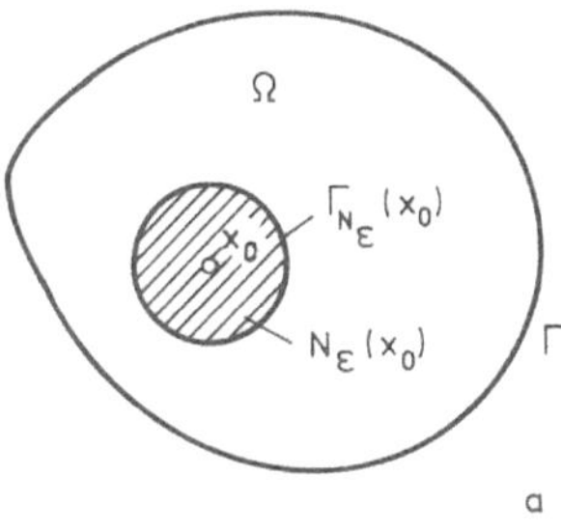
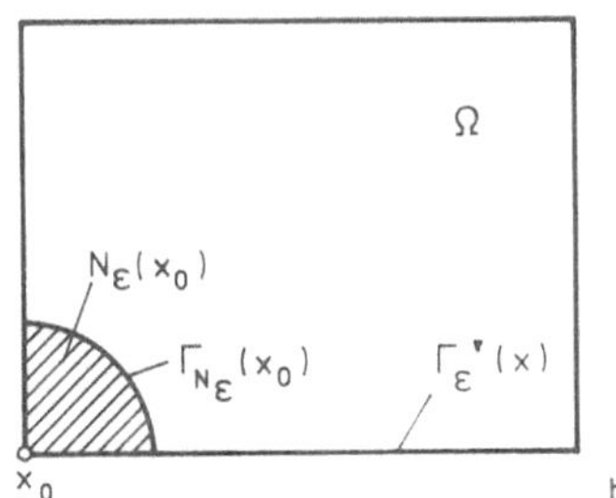

Figure 1.3

If x is a boundary point, see Fig. 1.3b, then $N_\varepsilon(x)$ cuts off a piece of the old boundary and $\Gamma_\varepsilon(x)$ becomes

$$\Gamma_\varepsilon(x) = \Gamma'_\varepsilon(x) \cup \Gamma_{N_\varepsilon}(x)$$

where $\Gamma'_\varepsilon(x)$ denotes the remainder of the original boundary.

Consider the boundary curve Γ of a plane domain Ω. Each point x_0 on Γ possesses two tangents, t_1 and t_2, see Fig. 1.4, (the numbering follows the orientation on the curve) which point away from x_0 and cut the unit circle $\Gamma_{N_1}(x_0)$ at two points and, thus, split the circle in two parts a_i and a_e.

The arc-length of a_i, this is the segment of the circle which for most of its part lies in Ω, is the internal angle of the boundary point x_0. The length of the external arc, a_e, is the external angle of the boundary point x_0.

$$\Delta\varphi_i = \int_{a_i} ds, \qquad \Delta\varphi_e = \int_{a_e} ds, \qquad \Delta\varphi_i + \Delta\varphi_e = 2\pi \tag{1.3}$$

The point x_0 in Fig. 1.3b, e.g., has the internal angle $\Delta\varphi_i = \pi/4$.

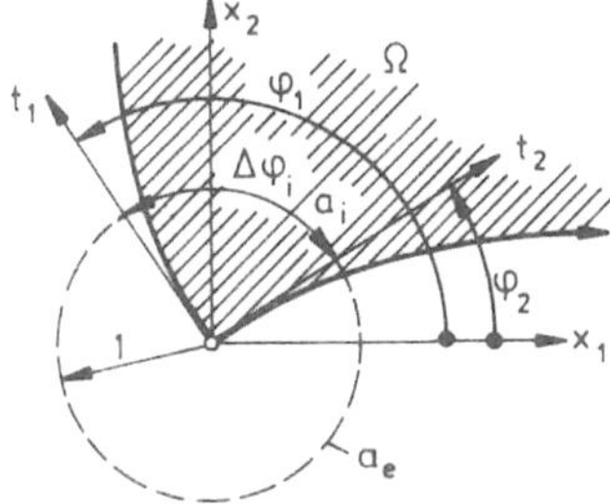

Figure 1.4

The internal and external angles of a point x_0 located on a surface are defined analogously:

The tangents at x_0, that is all the directions which point away from x_0 and which are tangent to the surface form the mantle of a pyramid. This mantle cuts the unit sphere $\Gamma_{N_1}(x_0)$ in a closed curve γ which splits $\Gamma_{N_1}(x_0)$ in two halves,

$$\Gamma_{N_1}(x_0) = A_i \cup A_e$$

A_i is the portion of the unit sphere which for most of its part lies inside the domain Ω. The internal angle $\Phi_i(x_0)$ of the point x_0 is the size of the surface A_i and the external angle $\Phi_e(x_0)$ is the size of the surface A_e,

$$\Phi_i(x_0) = \int_{A_i} ds, \quad \Phi_e(x_0) = \int_{A_e} ds, \quad \Phi_i(x_0) + \Phi_e(x_0) = 4\pi \tag{1.4}$$

The influence a force located at a point x exerts on a distant point y depends on the distance r between the two points and the angle under which the point y is seen from x, see Fig. 1.5.

The distance $r = |y - x|$ (the length of the vector $r = y - x$) is a function of both x and y and possesses, therefore, derivatives with respect to x_i and y_i. A simple calculation shows that

$$r_{,y_i} = \frac{y_i - x_i}{r} = -r_{,x_i} \tag{1.5}$$

Both gradients

$$\operatorname{grad}_y r = \{r_{,y_1}; r_{,y_2}; r_{,y_3}\}^T, \quad \operatorname{grad}_x r = \{r_{,x_1}; r_{,x_2}; r_{,x_3}\}^T$$

have the same length, 1, but point into opposite directions, see Fig. 1.5.

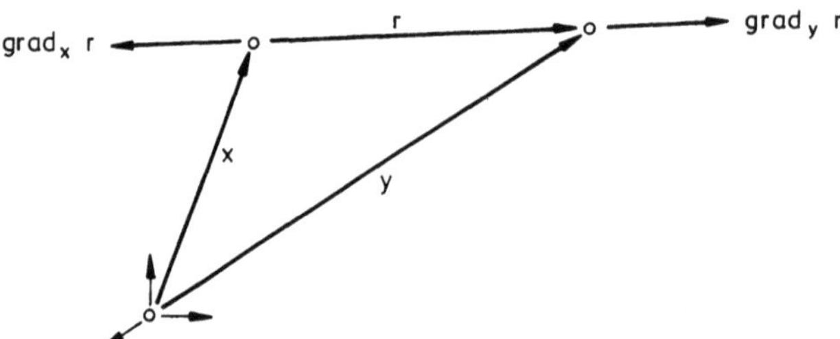

Figure 1.5

Often it is appropriate to choose x as the origin of the coordinate system and to switch to polar coordinates by way of the following equations

$$
\begin{aligned}
y_1 &= r \cos \varphi & y_1 &= r \cos \varphi \sin \vartheta \\
y_2 &= r \sin \varphi & y_2 &= r \sin \varphi \sin \vartheta \\
& & y_3 &= r \cos \vartheta
\end{aligned}
\tag{1.6}
$$

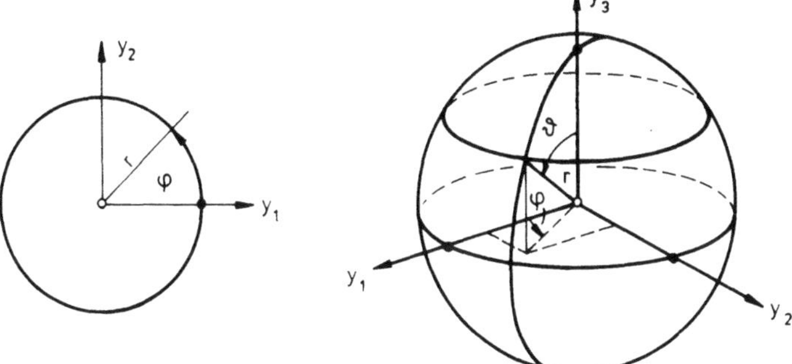

Figure 1.6

In which case the vector r simplifies to

$$r = y - 0 = y$$

and, therefore, Eqs. (1.5) and (1.6) to

$$
\begin{array}{ll}
r,_1 = \cos \varphi & r,_1 = \cos \varphi \sin \vartheta \\
r,_2 = \sin \varphi & r,_2 = \sin \varphi \sin \vartheta \\
 & r,_3 = \cos \vartheta
\end{array}
$$

1.4 Integrals

The expressions

$$\int_\Omega f(x)\, d\Omega, \qquad \int_\Omega g(y, x)\, d\Omega_y$$

are domain integrals. The definition of the volume element $d\Omega$ depends on the dimension of the continuum and the coordinates we use.

$$d\Omega = \begin{cases} dx \\ dx_1\, dx_2 & = r\, dr\, d\varphi \\ dx_1\, dx_2\, dx_3 = \sin \vartheta\, r^2\, dr\, d\varphi\, d\vartheta \end{cases} \tag{1.7}$$

The index y on $d\Omega$ is to indicate that we integrate with respect to y.

A point where a function tends to $+\infty$ or $-\infty$ is called a *singular point*. We say an integral is regular in a domain Ω if the integrand is bounded in Ω and we say it is singular if there are singular points in Ω.

We define the value of a singular integral as the limit of a sequence of regular integrals:

Let $N_\varepsilon(x)$ be a (small) neighborhood of x (Fig. 1.3a) and assume $f(y)$ is singular at x (and only there), then we understand the value of the singular domain integral to be the limit

$$\int_\Omega f(y)\,d\Omega := \lim_{\varepsilon \to 0} \int_{\Omega_\varepsilon(x)} f(y)\,d\Omega \tag{1.8}$$

We call this limit, if it exists, Cauchy's prinicipal value.

Analogously we define as the value of a singular boundary integral (assume $x \in \Gamma$ is a singular point) the limit

$$\int_\Gamma f(y)\,ds := \lim_{\varepsilon \to 0} \int_{\Gamma_{N_\varepsilon}(x)} f(y)\,ds$$

and call it Cauchy's principal value if it exists.

Finally, let Γ be a closed curve, represented by some function $x(s)$, and f a function of the arc-length s. If f is continuous the integral of the derivative df/ds over the closed curve is zero. But if f jumps at n points x^i then the integral is the sum of all these jumps,

$$-\int_\Gamma \frac{d}{ds} f(x(s))\,ds = [[f]] = \sum_i \{f(x^i_+) - f(x^i_-)\} \tag{1.9}$$

The twisting moment M_{nt} of a Kirchhoff plate exhibits such a behaviour. If the boundary is smooth then too is M_{nt} but if Γ has corners then M_{nt} jumps and then $[[M_{nt}]]$ is just the sum of all the corner forces ($=$ jumps).

1.5 Integration by Parts

The integration by parts rule in one dimension is

$$p:\ u, \hat{u} \in C^1[a, b]$$

$$q:\ \int_a^b u'\,\hat{u}\,dx = [u\,\hat{u}]_a^b - \int_a^b u\,\hat{u}'\,dx \tag{1.10}$$

$$[u\,\hat{u}]_a^b := u(b)\,\hat{u}(b) - u(a)\,\hat{u}(a)$$

and in two or three dimensions

$$p:\ u, \hat{u} \in C^1(\bar{\Omega}),\ \Omega \text{ is a regular domain}$$

$$q:\ \int_\Omega u_{,x_i}\,\hat{u}\,d\Omega = \int_\Gamma u\,\hat{u}\,n_i\,ds - \int_\Omega u\,\hat{u}_{,x_i}\,d\Omega \tag{1.11}$$

where in the last equation n_i is the component i of the exterior unit normal n on the boundary Γ of the domain Ω.

Concerning the notion of a regular domain we refer to [G2] p. 13. The domains encountered in practice usually are regular. Edges and vertices are admissible.

An immediate consequence of the last equation, (1.11), are the following results concerning the divergence of vectors (u) and matrices (S).

1. $\qquad p: \boldsymbol{u}, \hat{u} \in C^1(\bar{\Omega})$

$$q: \int_\Omega \operatorname{div} \boldsymbol{u}\,\hat{u}\,d\Omega = \int_\Gamma \boldsymbol{u} \cdot \boldsymbol{n}\,\hat{u}\,ds - \int_\Omega \boldsymbol{u} \cdot \nabla \hat{u}\,d\Omega \tag{1.12}$$

2. $\qquad p: \hat{u}, \boldsymbol{S} \in C^1(\bar{\Omega})$

$$q: \int_\Omega \operatorname{div} \boldsymbol{S} \cdot \hat{u}\,d\Omega = \int_\Gamma \boldsymbol{S}\boldsymbol{n} \cdot \hat{u}\,ds - \int_\Omega \boldsymbol{S} \cdot \nabla \hat{u}\,d\Omega \tag{1.13}$$

Here we used for the first time the notation $p:A$, $q:B$. A notation we shall frequently employ in this book to denote that a statement A implies a statement B.

If, e.g., both functions, $u, \hat{u}$ are from $C^1[a, b]$ then Eq. (1.10) is true. A is the sufficient condition and B is the necessary condition.

With regard to the integration by parts rule our assumption that the two functions $\{u, \hat{u}\}$ are in $C^1(\bar{\Omega})$ is a little bit too strong. In principle it is sufficient that the pair is from $C_p^1(\bar{\Omega})$, that the first derivatives of the functions are only piecewise smooth.

In such a case we would divide Ω into subdomains $\Omega = \cup_i \Omega_i$, where the functions $\{u, \hat{u}\}$ meet the requirements, $\{u, \hat{u}\} \in C^1(\bar{\Omega}_i)$, and we would then perform integration by parts in each subdomain Ω_i and add the single results. The total result would, because the boundary integrals over internal boundaries drop out, be the same result as if we had applied integration by parts in the full domain.

Thus all the integral theorems we formulate in this book for pairs $\{u, \hat{u}\}$ of functions from $C^m \times C^n$ remain valid if the pair $\{u, \hat{u}\}$ belongs to $C_p^m \times C_p^n$. It is only to simplify the notation that we delete the subscript p.

1.6 Gateaux Differentials

Let $f(x)$, $x \in \mathbb{R}^3$, be a scalar-valued function and assume we first evaluate f at x and we then move away from x in a direction $\hat{x}$. The change in f we observe is approximately

$$df = \nabla f(x) \cdot \hat{x} = f_{,x_1}\,\hat{x}_1 + f_{,x_2}\,\hat{x}_2 + f_{,x_3}\,\hat{x}_3 \tag{1.14}$$

The points we pass lie on the straight line $x + \varepsilon\,\hat{x}, 0 \leqslant \varepsilon \leqslant 1$. Along this line the Taylor expansion

$$f(x + \varepsilon\,\hat{x}) = f(x) + \nabla f(x) \cdot \varepsilon\,\hat{x} + 0(\varepsilon^2)$$

is, because both x and $\hat{x}$ are fixed, a function of ε alone.

If we calculate the first derivative of this expression with respect to ε then we obtain the same result as in Eq. (1.14), namely

$$\frac{d}{d\varepsilon}f(x+\varepsilon\hat{x})\bigg|_{\varepsilon=0} = \lim_{\varepsilon\to 0}\left\{\frac{f(x+\varepsilon\hat{x})-f(x)}{\varepsilon}\right\}\bigg|_{\varepsilon=0} = \nabla f(x)\cdot\hat{x} = df$$

Obviously, this ε-algorithm is "equivalent" with standard differentiation. Its advantage is that it still applies in situations where differentiation is not explained as, e. g., in the case of an operator $N(u)$ whose arguments are functions. In such a case an expansion as $dN = N'du$ is meaningless because N' is not explained whereas the ε-algorithm still is.

The expression we derived above we term the Gateaux differential of $f(x)$ at x in the direction of $\hat{x}$ and denote it by $f_x(\hat{x})$

$$f_x(\hat{x}) = \nabla f(x)\cdot\hat{x}$$

In more general terms: if $N(u)$ is an operator then we call the expression

$$\frac{d}{d\varepsilon}N(u+\varepsilon\hat{u})\bigg|_{\varepsilon=0} =: N_u(\hat{u})$$

the Gateaux differential of $N(\ )$ at u in the direction of $\hat{u}$.

The Gateaux differential of the operator $N(u) = (u')^3$ at u in the direction of $\hat{u}$ is, e. g.,

$$N_u(\hat{u}) = \frac{d}{d\varepsilon}(u'+\varepsilon\hat{u}')^3\bigg|_{\varepsilon=0} = 3\,(u')^2\,\hat{u}'$$

The name "differential" is rightly chosen if the operator $N(\ )$ enjoys at u the expansion

$$N(u+\varepsilon\hat{u}) = N(u) + \varepsilon\Delta N + \varepsilon^n R(u,\hat{u},\varepsilon), \qquad n \geqslant 2 \tag{1.15}$$

and if the remainder $\varepsilon^n R(u,\hat{u},\varepsilon)$ has the property

$$\lim_{\varepsilon\to 0}\varepsilon^{n-1}R(u,\hat{u},\varepsilon) = 0$$

(The remainder of $N(u) = (u')^3$ is the expression $\varepsilon^2\left(3\,u'\,(\hat{u}')^2 + \varepsilon\,(\hat{u}')^3\right)$ and, hence, it has this property).

Because, if this expansion, Eq. (1.15), holds then we immediately obtain

$$\frac{d}{d\varepsilon}N(u+\varepsilon\hat{u})\bigg|_{\varepsilon=0} = \lim_{\varepsilon\to 0}\left\{\frac{1}{\varepsilon}\left(N(u+\varepsilon\hat{u})-N(u)\right)\right\}\bigg|_{\varepsilon=0} = \Delta N = N_u(\hat{u})$$

that is the increase ΔN coincides with the Gateaux differential.

If the operator is linear, this implies that $N(u + \varepsilon \hat{u}) = N(u) + \varepsilon N(\hat{u})$, then the Gateaux differential is simply

$$N_u(\hat{u}) = N(\hat{u})$$

and, therefore, does not depend on u just as the increase $df = a\,dx$ of a linear function as $f(x) = a\,x$ depends only on dx and not on x.

1.7 Functionals

A functional is a "function of functions" i.e. an expression whose arguments are functions and whose values are numbers. A very simple functional is the expression $F(u) = u(0)$. The value of this functional at u is the value of the function u at the point $x = 0$.

A linear functional has the property

$$F(a_1 u_1 + a_2 u_2) = a_1 F(u_1) + a_2 F(u_2)$$

A bilinear functional is a functional which depends on two functions u, v and which is linear in both, that is it enjoys the expansion

$$F(a_1 u_1 + a_2 u_2, \hat{a}_1 \hat{u}_1 + \hat{a}_2 \hat{u}_2) = \sum_{i,j=1}^{2} a_i \hat{a}_j F(u_i, \hat{u}_j) \tag{1.16}$$

We call a bilinear functional symmetric if

$$F(u, v) = F(v, u)$$

A quadratic form is a functional

$$F(u) = \frac{1}{2} E(u, u)$$

based on a symmetric bilinear functional $E(u, v)$.

The first variation of a functional $F(u)$ at u in the direction of $\hat{u}$ is the expression

$$\delta F(u, \hat{u}) = \frac{d}{d\varepsilon} F(u + \varepsilon \hat{u}) \Big|_{\varepsilon = 0}$$

and the second variation the expression

$$\delta^2 F(u, \hat{u}) = \frac{d^2}{d\varepsilon^2} F(u + \varepsilon \hat{u}) \Big|_{\varepsilon = 0}$$

Example: The first variation of the functional

$$F(u) = \frac{EA}{2} \int u'^2 \, dx$$

at u in the direction of $\hat{u}$ is

$$\delta F(u, \hat{u}) = \frac{d}{d\varepsilon} \left\{ \frac{EA}{2} \int (u'^2 + 2\varepsilon u' \hat{u}' + \varepsilon^2 \hat{u}'^2) \, dx \right\}\Bigg|_{\varepsilon=0} = EA \int u' \hat{u}' \, dx$$

and its second variation is

$$\delta^2 F(u, \hat{u}) = \frac{d^2}{d\varepsilon^2} \left\{ \ldots \quad \ldots \right\}\Bigg|_{\varepsilon=0} = EA \int \hat{u}'^2 \, dx$$

If the functional depends on two functions, $F = F(u, v)$, then the first variation is the expression

$$\delta F(u, v; \hat{u}, \hat{v}) = \left(\frac{d}{d\varepsilon} F + \frac{d}{d\eta} F \right)\Bigg|_{\varepsilon=n=0}$$

where $F = F(u + \varepsilon \hat{u}, v + \eta \hat{v})$. The second variation is obtained by differentiating twice. The functionals in structural mechanics have the form

$$F(u) = \int_\Omega N(u) \cdot M(u) \, d\Omega$$

that is the integrand is the scalar product (dot!) of two tensors $N(u)$ and $M(u)$ of (equal) degree 0, 1 or 2 (scalars, vectors or matrices).

Particular cases are

$$F(u) = \int N(u) \cdot N(u) \, d\Omega \qquad M(u) = N(u) \tag{1.17}$$

$$F(u) = \int N(u) \cdot u \, d\Omega \qquad M(u) = u \tag{1.18}$$

$$F(u) = \int N(u) \cdot p \, d\Omega \qquad M(u) = p \tag{1.19}$$

The operators $N(\ )$ and $M(\ )$ which appear in these functionals, in general, enjoy expansions as the operator in Eq. (1.15) and consequently the first variation of a functional $F(u)$ in structural mechanics has the form

$$\delta F(u, \hat{u}) = \int \left(N_u(\hat{u}) \cdot M(u) + N(u) \cdot M_u(\hat{u}) \right) d\Omega \qquad \text{"chain-rule"} \tag{1.20}$$

An additional comment deserves the functional in Eq. (1.17), the functional where the same tensor appears twice.

If we perform the first variation of such a functional then we obtain

$$\delta F(u, \hat{u}) = \int \left(N_u(\hat{u}) \cdot N(u) + N(u) \cdot N_u(\hat{u}) \right) d\Omega = 2 \int N_u(\hat{u}) \cdot N(u) \, d\Omega$$

that is the first variation of a "symmetric" functional $(N(u) = M(u))$ is the sum of two identical terms.

These functionals, typically, are preceded by a factor $1/2$ which eliminates the factor 2 which appears after the first variation.

Such a functional is, e.g., the internal energy of a beam

$$F(w) = \frac{1}{2 \, EI} \int M(w) \cdot M(w) \, dx = \frac{1}{2} EI \int w'' \, w'' \, dx$$

where the linear operator $M(w) = - EI \, w''$ appears twice. The first variation is the expression

$$\delta F(w, \hat{w}) = \frac{1}{2 \, EI} \int \left[M(\hat{w}) \cdot M(w) + M(w) \cdot M(\hat{w}) \right] dx = EI \int w'' \, \hat{w}'' \, dx$$

Very popular are expansions such as

$$\Pi(u + \varepsilon \, \hat{u}) = \Pi(u) + \varepsilon \, \delta \Pi(u, \hat{u}) + \frac{\varepsilon^2}{2} \, \delta^2 \Pi(u, \hat{u}) + \ldots \tag{1.21}$$

To understand why the first and second variation appear in such expansions consider the functional

$$F(u) = \int_{\Omega} N(u) \cdot M(u) \, d\Omega$$

If $N(u)$ and $M(u)$ are linear operators then the value of F at $u + \varepsilon \, \hat{u}$ can be expressed as

$$F(u + \varepsilon \, \hat{u}) = \int N(u + \varepsilon \, \hat{u}) \cdot M(u + \varepsilon \, \hat{u}) \, d\Omega = \int N(u) \cdot M(u) \, d\Omega$$

$$+ \varepsilon \int \left(N(\hat{u}) \cdot M(u) + N(u) \cdot M(\hat{u}) \right) d\Omega + \varepsilon^2 \int N(\hat{u}) \cdot M(\hat{u}) \, d\Omega \tag{1.22}$$

The second integral in this expansion is the first variation of $F(u)$ in the direction of $\hat{u}$

$$\delta F(u, \hat{u}) = \int \left(N(\hat{u}) \cdot M(u) + N(u) \cdot M(\hat{u}) \right) d\Omega$$

and the third integral is, up to a factor 2, the second variation of $F(u)$

$$\delta^2 F(u, \hat{u}) = 2 \int N(\hat{u}) \cdot M(\hat{u}) \, d\Omega = 2 \, F(\hat{u}) \tag{1.23}$$

Consequently Eq. (1.22) is identical with

$$F(u + \varepsilon \, \hat{u}) = F(u) + \varepsilon \, \delta \, F(u, \hat{u}) + \frac{\varepsilon^2}{2} \, \delta^2 \, F(u, \hat{u})$$

which is an expression as in Eq. (1.21).

Note that the expansion ends after the second term. This is always so if $N(\ )$ and $M(\ )$ are linear operators, that is in the linear theory of structures.

If $M(u) = p$, i.e. a fixed function, then even the second term in the expansion vanishes as in the following example

$$F(u + \varepsilon \, \hat{u}) = \int (u + \varepsilon \, \hat{u}) \cdot p \, d\Omega = \int u \cdot p \, d\Omega + \varepsilon \int \hat{u} \cdot p \, d\Omega$$

$$= F(u) + \varepsilon \, \delta \, F(u, \hat{u}) \tag{1.24}$$

If the operators $N(\ )$ and $M(\ )$ are nonlinear but possess expansions as the operator in Eq. (1.15) then analogous results as in the linear case are obtained.

1.8 Sobolev Spaces

Let u a function defined on the interval $[a, b]$. The number

$$\|u\|_0 = \left[\int_a^b u^2 \, dx \right]^{1/2}$$

is the L_2-norm (also called H^0-norm) of u. It measures, roughly said, the distance in the mean between u and the function $u = 0$. The smaller $\|u\|_0$ the closer (in terms of area) u is to $u = 0$.

The set of all functions which have a finite L_2-norm ("whose area can be measured") is called $L_2 \, [a, b]$ or $H^0 \, [a, b]$.

In structural mechanics we are not only interested in the distance in displacements but also in the distance in stresses. This is motivation to introduce Sobolev norms of degree $k > 0$,

$$\|u\|_k = [\|u\|_0^2 + \|u'\|_0^2 + \|u''\|_0^2 + \ldots \quad \ldots \|u^{(k)}\|_0^2]^{1/2}$$

$$= [\int (u^2 + u'^2 + u''^2 + \ldots \quad \ldots u^{(k)2}) \, dx]^{1/2}$$

which measure the distance in derivatives ("strains") up to the order k.

These different norms generate different topologies on the set of functions defined on $[a, b]$. The higher the degree the finer the topology.

To judge the distance $d(x, y)$ between two n-dimensional vectors by their first component alone

$$d_1(x, y) = [(x_1 - y_1)^2]^{1/2}$$

would be very crude. Better to compare also the second components

$$d_2(x, y) = [(x_1 - y_1)^2 + (x_2 - y_2)^2]^{1/2}$$

or even the third, etc. The most judicious choice, naturally, is the Euclidean norm

$$d_n(x, y) = [(x_1 - y_1)^2 + \ldots (x_n - y_n)^2]^{1/2}$$

In the same sense the different Sobolev norms $\| \cdot \|_k$ provide different scales to measure the distance between two functions.

The zigzag function u_h in Fig. 1.7 is a good approximation of u in the H^0-norm

$$\|u - u_h\|_0 = \left(\int_0^1 (u - u_h)^2 \, dx \right)^{1/2} = 0.26$$

as the area between u and u_h is small, but a bad approximation in the H^1-norm

$$\|u - u_h\|_1 = \left(\int_0^1 (u - u_h)^2 + (u' - u_h')^2 \, dx \right)^{1/2} = 2.02$$

because the first derivatives $u' = 0$ and $u_h' = $ (up and down) are far apart.

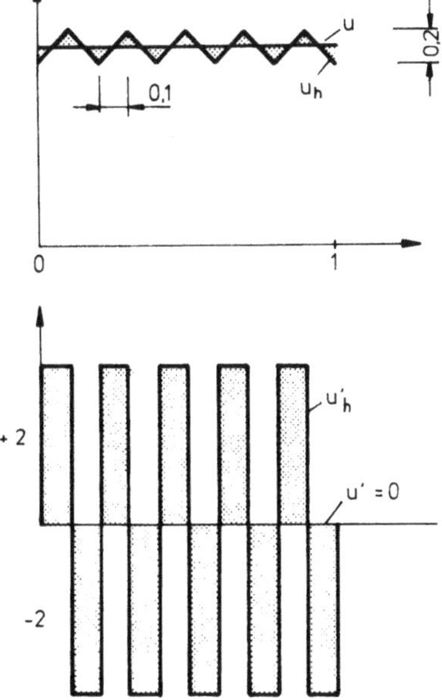

Figure 1.7

Recall that $\mathbb{R}$, the set of all real numbers, is the completion of the set $\mathbb{Q} = \{1/2, 2/3, 3/4 \ldots\}$, the set of all rational numbers, with respect to the Euclidean norm, $|r-q|$. This means: given any real number r then either r is rational, $r = a/b$, or r is the endpoint of a sequence of rational numbers $\{q_n\}$ that converges to r in the Euclidean norm.

$$\lim_{n \to \infty} |r - q_n| = 0$$

The number $r = \sqrt{2}$ is not a rational number but we can construct a sequence of such numbers which converges to $\sqrt{2}$. Hence, $\sqrt{2}$ is a real number. The number $\sqrt{-1}$ is not a rational number and no sequence of rational numbers converges to $\sqrt{-1}$. Hence $\sqrt{-1}$ is not a real number.

The Sobolev space $H^k[a, b]$ is now defined as the completion of the space $C^k[a, b]$ with respect to the norm $\|\cdot\|_k$. This space consists of all those functions which either belong to $C^k[a, b]$ or which are arbitrary close to $C^k[a, b]$, that is given a function $u \in H^k[a, b]$ then either u is in $C^k[a, b]$ or there exists a sequence of C^k-functions $\{v_n\}$ which converges to u in the $\|\cdot\|_k$ norm

$$\lim_{n \to \infty} \|u - v_n\|_k = 0$$

The step function in Fig. 1.8 is not in C^0 but it is the endpoint of the sequence

$$v_n = \frac{4}{\pi} \sum_{j=1}^{n} \frac{1}{2j-1} \sin (2j-1) x \qquad n = 1,2 \ldots$$

of C^0-functions. The single function v_n is the sum of the first n terms of the Fourier series of u which, by definition, converges to u in the L_2-norm,

$$\lim_{n \to \infty} \|u - v_n\|_0 = 0$$

Hence, the step function belongs to $H^0[0,2\pi]$.

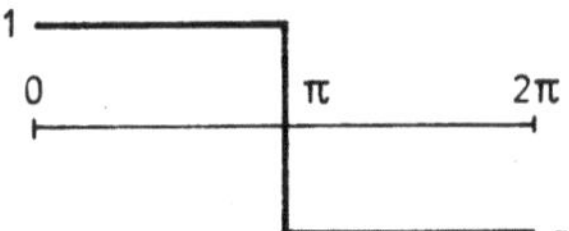

Figure 1.8

The definition of the Sobolev spaces given above can be made more precise, see [F 1] p. 349:

A function u belongs to $H^k[a, b]$ if and only if
(i) there exists a sequence $\{v_n\}$ of C^k-functions such that

$$\lim_{n \to \infty} \|u - v_n\|_0 = 0, \qquad \text{and}$$

(ii) there exists a set of k L_2-functions $\varphi_\varkappa$, $\varkappa = 1,2 \ldots k$, such that

$$\lim_{n \to \infty} \|\varphi_\varkappa - v_n^{(\varkappa)}\|_0 = 0 \qquad \varkappa = 1,2 \ldots k$$

The functions $\varphi_\varkappa$ are called the *strong derivatives* of u. If u is a C^k-function then the sequence is simply the sequence $u, u, u \ldots$ and the strong derivatives the classical derivatives of u.

These strong derivatives should not be confused with the *weak derivatives* of a function u. A function $u^{(\alpha)} \in H^0$ is said to be the weak derivative of u if

$$\int_a^b u^{(\alpha)}\, \varphi\, dx = (-1)^\alpha \int_a^b u\, \varphi^{(\alpha)}\, dx \qquad \forall\, \varphi \in C_0^\infty\, [a, b]$$

that is (in terms of mechanics) if the work done by the load $u^{(\alpha)}$ acting through any virtual displacement φ is equal to the work done by the load $\varphi^{(\alpha)}$ acting through u. If u has a classical derivative of order α then this $u^{(\alpha)}$ is also the strong and the weak derivative of u.

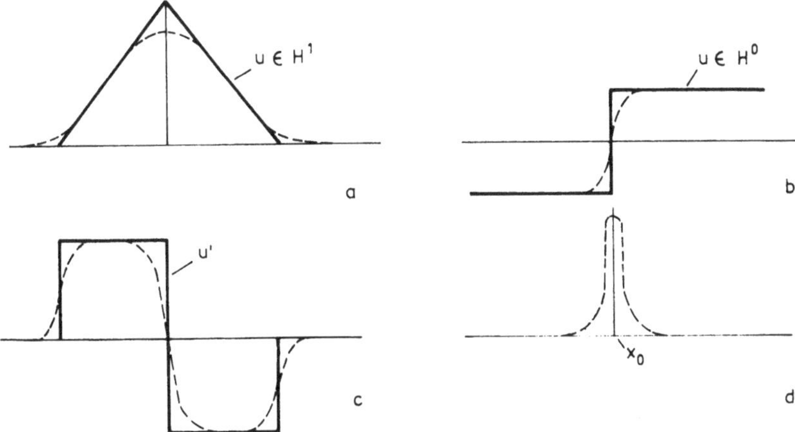

Figure 1.9

The function u in Fig. 1.9a belongs to $H^1\,[0,1]$ while the function in Fig. 1.9b belongs only to $H^0\,[0,1]$.

In the first case there exists a sequence of C^1-functions, see the dashed line, whose zero-th and first derivative converge (in terms of area) to L_2-functions namely u and $u'\,(= \varphi_1)$.

In the second case the derivatives of the functions which approximate u tend to a delta function (a point load concentrated at x_0). Such a delta function does not belong to L_2. Hence, u does not belong to H^1 but only to H^0.

The functions we work with in numerical analysis are usually polynomials or piecewise polynomials, that is functions obtained by piecing together polynomials defined on subsets of the global domain $[a, b]$.

Polynomials are infinitely smooth, they belong to $C^\infty [a, b]$, and, therefore, to $H^k [a, b]$ whatever the value of the index k.

Piecewise polynomials, typically, appear in finite-element applications. They are infinitely smooth locally on the single elements (imagine the interval $[a, b]$ to be partitioned into single elements), but not so with respect to the global domain $[a, b]$. The functions in Fig. 1.9 a, b and c, e. g., are piecewise polynomials.

With respect to this class of functions the space $C_p^k [a, b]$ (the space of all functions whose $k - th$ derivative is at least piecewise continuous in $[a, b]$) and the space $H^k [a, b]$ are equivalent

$$C_p^k [a, b] \cong H^k [a, b]$$

This means: a piecewise polynomial is in $C_p^k [a, b]$ if and only if it is in $H^k [a, b]$.

The extension of Sobolev spaces to functions with several variables poses no additional problems.

Let $u(x)$ a function of points $x = (x_1, x_2, \ldots x_n) \in R^n$. If we denote by $\alpha = (\alpha_1, \alpha_2, \ldots \alpha_n)$ a multi-index (a sequence of natural numbers) with "length"

$$|\alpha| = \sum_{i=1}^{n} \alpha_i$$

and by

$$d^\alpha u(x) = \frac{\partial^{|\alpha|} u(x)}{\partial_{x_1}^{\alpha_1} \partial_{x_2}^{\alpha_2} \ldots \partial_{x_n}^{\alpha_n}}$$

the partial derivatives of u then the H^k-norm of u with respect to a domain $\Omega \subset R^n$ can be written as

$$\|u\|_k = \left(\int_\Omega \sum_{|\alpha| \leq k} (d^\alpha u)^2 \, d\Omega \right)^{1/2}$$

The H^2-norm of the deflection of a Kirchhoff plate is, e. g., the expression

$$\|w\|_2 = \left(\int_\Omega (w^2 + w^2{}_{,1} + w^2{}_{,2} + w^2{}_{,11} + w^2{}_{,12} + w^2{}_{,21} + w^2{}_{,22}) \, d\Omega \right)^{1/2}$$

The space $H_0^k (\Omega)$ is the completion of $C_0^k (\Omega)$ (the functions in $C^k (\Omega)$ whose derivatives of order $\leq k$ vanish on Γ) with respect to the Sobolev norm $\|\cdot\|_k$.

1.9 The Differential Equations

The functions which govern the behaviour of a bar, a beam or a plate etc. can, according to Tonti [T 3], be grouped as follows
   a) *configuration variables (displacements)*
   b) *intermediate variables   (stresses and strains)*
   c) *source variables   (forces)*

As an example consider the displacement $u(x)$ of the bar in Fig. 1.10.

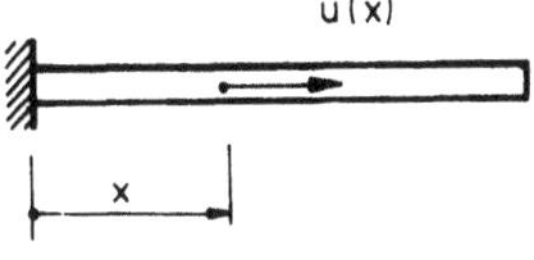

Figure 1.10

We define the strain ε as the first derivative of u,

$$\text{(Def.)} \qquad \varepsilon = \frac{du}{dx}$$

No physical parameter appears in this equation. Only in the next step where we introduce the stress σ as a multiple ($E = $ Young's modulus) of the strain.

$$\text{(Const.)} \qquad \sigma = E\varepsilon$$

The condition that the infinitesimal element in Fig. 1.11 is in equilibrium requires the sum of the stresses, the axial force $N = A\,\sigma$, to satisfy the equation

$$-N + N + dN + p(x)\,dx = 0$$

That is, the axial force N must satisfy the differential equation

$$\text{(Equ.)} \qquad -\frac{dN}{dx} = p$$

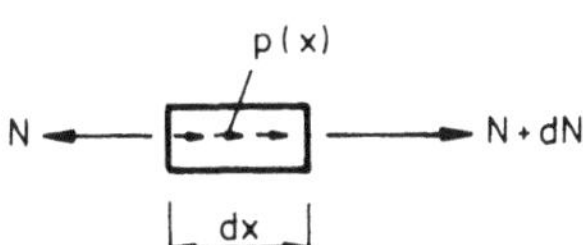

Figure 1.11

The three equations

$$\frac{du}{dx} - \varepsilon = 0$$

$$EA\,\varepsilon - N = 0$$

$$-\frac{dN}{dx} = p$$

so obtained constitute a system of differential equations for three functions u, ε, N.

If we substitute the first equation into the second and then this in turn into the third equation we obtain a differential equation of second degree for $u(x)$ alone

$$-\frac{d}{dx}\left(EA\frac{du}{dx}\right)=p$$

The derivation of the very same equations in the case of beams, plates or elastic bodies would be, more or less, a mere repetition of these steps and we, therefore, list in the following only the resulting equations.

Remark: to have our catalog of equations complete we considered it useful to also include results, namely the equations of the first identities, $G(u, \hat{u})$, which are formulated and proved only in later chapters.

1.9.1 The Straight Slender Frame Element

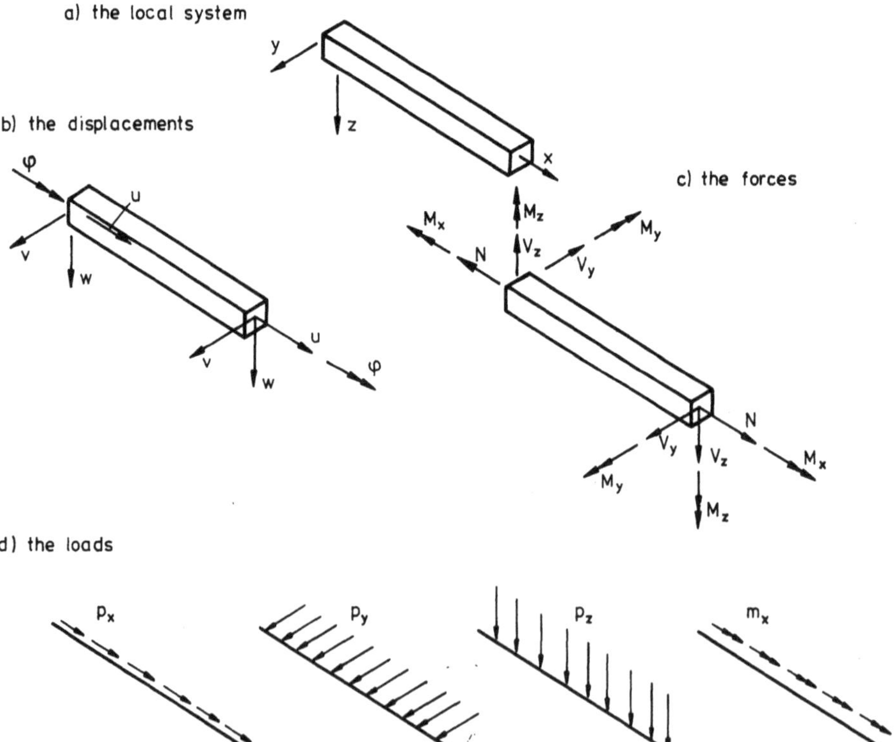

Figure 1.12

1.9.1.1 The Axial Displacement $u(x)$

(Def.) $\qquad\qquad\qquad u' - \varepsilon = 0$ $\qquad\qquad\qquad\qquad$ (1.25a)

(Const.) $\qquad\qquad\qquad EA\varepsilon - N = 0$ $\qquad\qquad\qquad\qquad$ (1.25b)

(Equ.) $\qquad\qquad\qquad -N' = p$ $\qquad\qquad\qquad\qquad$ (1.25c)

Substituting Eq. (a) into Eq. (b) and this in turn into Eq. (c) we obtain

$$-(E A u')' = p$$

where $E =$ Young's modulus, $A(x) =$ cross-sectional area, $p =$ horizontal load.
    The equation of the first identity is

$$G(u, \hat{u}) = \int_a^b -(E A u')' \, \hat{u} \, dx + [N \hat{u}]_a^b - \int_a^b \frac{N \hat{N}}{E A} \, dx = 0$$

If there is a change in temperature of T degrees in the bar then we shall observe, if
the bar can move freely, a displacement u_T which satisfies the differential equation

$$u_T' = \alpha_T T$$

where α_T is the temperature coefficient of the material.

1.9.1.2 The Rotation $\varphi(x)$ of a (circular) bar

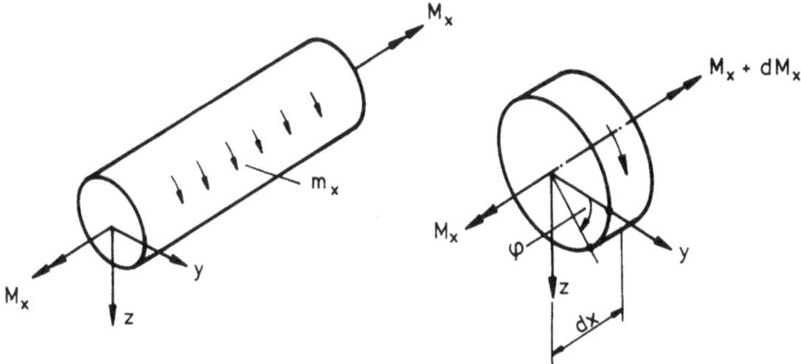

Figure 1.13

(Def.) $$\varphi' - \vartheta = 0$$

(Const.) $$G I_p \vartheta - M_x = 0$$

(Equ.) $$-M_x' = m_x$$

These three equations are equivalent with

$$-(G I_p \varphi')' = m_x$$

where $G = \mu =$ shear modulus, $I_p(x) =$ polar moment of inertia, $m_x =$ torque per unit
length.
    The equation of the first identity is

$$G(\varphi, \hat{\varphi}) = \int_a^b -(G I_p \varphi')' \, \hat{\varphi} \, dx + [M_x \hat{\varphi}]_a^b - \int_a^b \frac{M_x \hat{M}_x}{G I_p} \, dx = 0$$

1.9.1.3 The Lateral and Vertical Deflections v and w

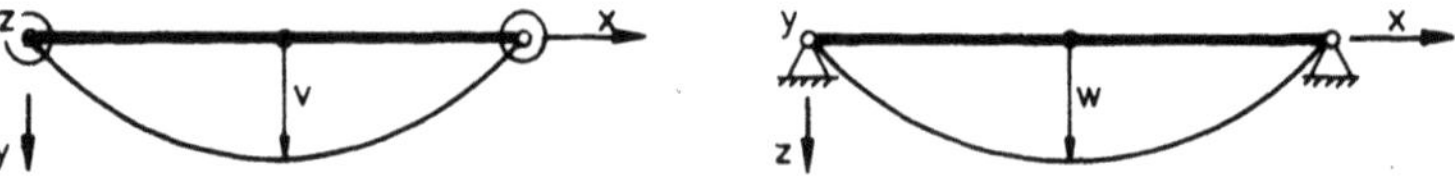

Figure 1.14

(Def.) $\qquad\qquad\qquad \varepsilon - v'' = 0 \qquad\qquad\qquad \varepsilon - w'' = 0$

(Const.) $\qquad\qquad E I_z \, \varepsilon - M_z = 0 \qquad\quad E I_y \, \varepsilon + M_y = 0$

(Equ.) $\qquad\qquad\qquad\quad M_z'' = p_y \qquad\qquad\quad -M_y'' = p_z$

These three equations are equivalent with

$$(E I_z \, v'')'' = p_y \qquad (E I_y \, w'')'' = p_z \qquad\qquad (1.26)$$

where $I_y(x)$, $I_z(x)$ are the moments of inertia of the cross-sectional area with respect to the y- and z-axes; p_y and p_z denote distributed forces in the direction of the y- and z-axes and M_y and M_z are the bending moments.

The shear forces are defined as

$$V_y = -(E I_z \, v'')' \qquad V_z = (E I_y \, w'')'$$

The equations of the first identities are

$$G(v, \hat{v}) = \int_a^b (E I_z \, v'')'' \, \hat{v} \, dx + [\, V_y \, \hat{v} + M_z \, \hat{v}' \,]_a^b - \int_a^b \frac{M_z \, \hat{M}_z}{E I_z} \, dx = 0$$

$$G(w, \hat{w}) = \int_a^b (E I_y \, w'')'' \, \hat{w} \, dx + [\, V_z \, \hat{w} - M_y \, \hat{w}' \,]_a^b - \int_a^b \frac{M_y \, \hat{M}_y}{E I_y} \, dx = 0$$

If there exist non-uniform temperature distributions ΔT_z, ΔT_y, s. Fig. 1.15, and if the beam is free to move then we shall observe displacements v_T and w_T of the axis which satisfy the differential equations

$$v_T'' = \alpha_T \frac{\Delta T_y}{h_y} \qquad w_T'' = \alpha_T \frac{-\Delta T_z}{h_z}$$

where α_T = temperature coefficient of the material, h_y = width of the bar, h_z = height of the bar.

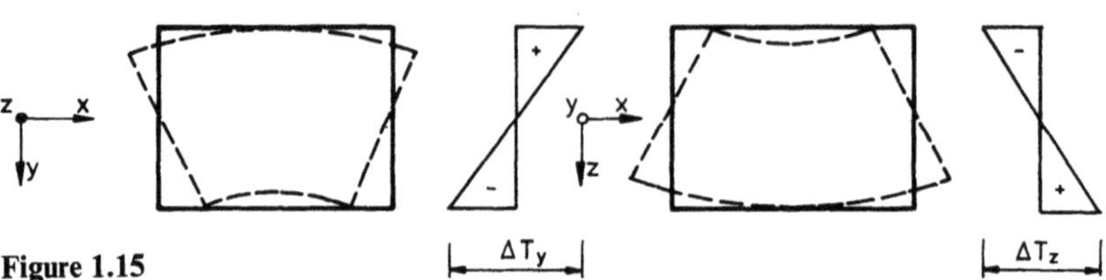

Figure 1.15

1.9.1.4 Shear Deformations

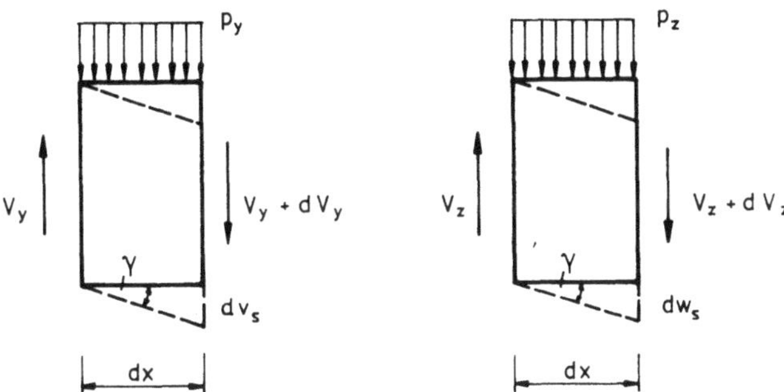

Figure 1.16

It is well known that the solutions v and w of the Bernoulli equations $(E I_z v'')'' = p_y$ and $(E I_y w'')'' = p_z$ do not include shear deformations.

These additional deformations, v_τ and w_τ, satisfy the following systems of equations.

(Def.) $\qquad\qquad \gamma - v_\tau' = 0 \qquad\qquad\qquad \gamma - w_\tau' = 0$

(Const.) $\qquad\qquad \dfrac{1}{\varkappa_y} G A \gamma - V_y = 0 \qquad \dfrac{1}{\varkappa_z} G A \gamma - V_z = 0$

(Equ.) $\qquad\qquad\qquad -V_y' = p \qquad\qquad\qquad -V_z' = p_z$

or equivalently

$$ -\left(\frac{1}{\varkappa_y} G A\, v_\tau'\right)' = p_y \qquad\qquad -\left(\frac{1}{\varkappa_z} G A\, w_\tau'\right)' = p_z $$

where $G = \mu =$ shear modulus, $A(x) =$ cross-sectional area of the beam, $\varkappa_y, \varkappa_z$ = parameters which depend on the shape of the cross section A.

The first identity for v_τ (analogously for w_τ) is

$$ G(v_\tau, \hat{v}_\tau) = \int_a^b -\left(\frac{1}{\varkappa_y} G A\, v_\tau'\right)' \hat{v}_\tau\, dx + [V_y\, \hat{v}_\tau]_a^b - \varkappa_y \int_a^b \frac{V_y\, \hat{V}_y}{G A}\, dx = 0 $$

The shear deformations of a slender beam account, in general, only for about 5 % of the total deflection and, hence, are negligible in all practical computations.

1.9.2 The Kirchhoff Plate

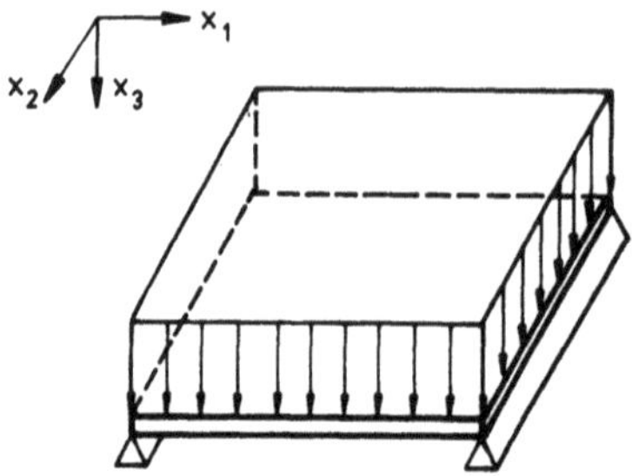

Figure 1.17

(Def.) $\mathbf{E} - \mathbf{E}(w) = \mathbf{0}_{(2 \times 2)}$ (1.27a)

(Const.) $C[\mathbf{E}] + \mathbf{M} = \mathbf{0}_{(2 \times 2)}$ (1.27b)

(Equ.) $-\operatorname{div}^2 \mathbf{M} = p_{(1)}$ (1.27c)

It is $\mathbf{E} = [\varepsilon_{ij}]$ the strain tensor and $\mathbf{E}(\ )$ the operator,

$$\mathbf{E}(w) = \begin{bmatrix} w_{,11} & w_{,12} \\ w_{,21} & w_{,22} \end{bmatrix} = \nabla\nabla\, w$$

$\mathbf{M} = [M_{ij}]$ is the tensor of the bending moments and $C[\ ]$ is the elasticity tensor

$$C[\mathbf{E}] = K\left\{(1 - v)\,\mathbf{E} + v\,(\operatorname{tr} \mathbf{E})\,\mathbf{I}\right\} \cong C^{ijkl}\,\varepsilon_{kl}$$

where $K = Eh^3/12\,(1 - v^2)$ is the constant stiffness of a plate with thickness h.
The elements C^{ijkl} of the elasticity tensor $C[\ ]$ are defined as

$$C^{ijkl} = K\left\{\frac{1 - v}{2}\,(\delta^{ik}\,\delta^{jl} + \delta^{il}\,\delta^{jk}) + v\,\delta^{ij}\,\delta^{kl}\right\}$$ (1.28)

On substituting the Eqs. (a) and (b) consecutively into Eq. (1.27c) we obtain the differential equation

$$K\,\Delta\Delta\, w = p$$ (1.29)

The shear forces are defined as

$$Q_1 = M_{11,1} + M_{12,2}, \qquad Q_2 = M_{21,1} + M_{22,2}$$

At each point of the boundary of a plate the normal $\mathbf{n}$ and the tangent $\mathbf{t}$ form a pair of characteristic directions which determine the magnitude of the bending moment

$$M_n = M_{11}\,n_1^2 + 2\,M_{12}\,n_1\,n_2 + M_{22}\,n_2^2 = M_{ij}\,n_i\,n_j = \mathbf{n}^T\,\mathbf{M}\,\mathbf{n}$$ (1.30)

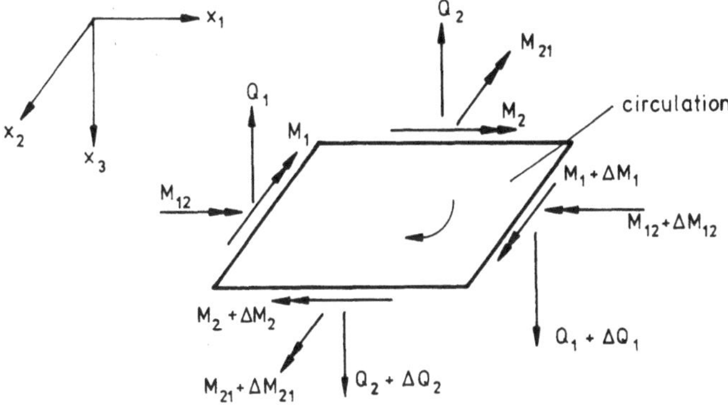

Figure 1.18

the twisting moment

$$M_{nt} = M_{12}(n_1^2 - n_2^2) - (M_{11} - M_{22})n_1 n_2 = M_{ij} n_i t_j = \boldsymbol{n}^T \boldsymbol{M} \boldsymbol{t} \tag{1.31}$$

the shear forces

$$Q_n = Q_1 n_1 + Q_2 n_2 = M_{ij,j} n_i = \operatorname{div} \boldsymbol{M} \cdot \boldsymbol{n} \tag{1.32}$$

and the Kirchhoff-shear

$$V_n = \frac{d}{ds} M_{nt} + Q_n \tag{1.33}$$

on the boundary.

The equation of the first identity is

$$G(w, \hat{w}) = \int_\Omega K \Delta \Delta w \, \hat{w} \, d\Omega + \int_\Gamma \left(V_n \hat{w} - M_n \frac{\partial \hat{w}}{\partial n} \right) ds + [[M_{nt} \, \hat{w}]]$$
$$- E(w, \hat{w}) = 0$$

where $E(w, \hat{w})$ is the symmetric expression

$$E(w, \hat{w}) = K \int_\Omega \left(w_{,11}(\hat{w}_{,11} + v \hat{w}_{,22}) + 2(1-v) w_{,12} \hat{w}_{,12} + w_{,22}(\hat{w}_{,22} \right.$$
$$\left. + v \hat{w}_{,11}) \right) d\Omega$$

and

$$M_n = M_n(w) = -K[(w_{,11} + v w_{,22})n_1^2 + 2(1-v) w_{,12} n_1 n_2$$
$$+ (w_{,22} + v w_{,11}) n_2^2]$$

$$M_{nt} = M_{nt}(w) = -K\left[(w_{,11}(v-1) + w_{,22}(1-v))n_1 n_2\right.$$
$$\left. + (1-v)w_{,12}(n_1^2 - n_2^2)\right]$$

$$V_n = V_n(w) = \frac{d}{ds} M_{nt}(w) + (M_{ij}(w))_{,i} n_j$$

where

$$M_{ij}(w) = -K(1-v)\left(w_{,ij} + \delta_{ij}\frac{v}{1-v}w_{,kk}\right)$$

Let T be a stationary temperature distribution in the plate. If the plate is free to move then the field.

$$T(x_1, x_2, x_3) = T(\mathbf{x}, x_3)$$

causes a deflection $w_T(\mathbf{x})$ of the plate which satisfies the following system of differential equations, see [P1] p. 43–7,

$$E(w_T) = -\alpha_T \theta \mathbf{I}$$

where $\mathbf{I}$ is the unit matrix, (2×2), and θ the function

$$\theta(\mathbf{x}) = \frac{12}{h^3} \int\limits_{-h/2}^{h/2} T(\mathbf{x}, x_3)\, x_3 \, dx_3$$

If the temperature distribution T is constant in the directions x_1 and x_2, zero in the middle plane and linear with respect to x_3 (see Fig. 1.19) then $\theta(\mathbf{x})$ is a constant, $\theta(\mathbf{x}) = \Delta T/2$.

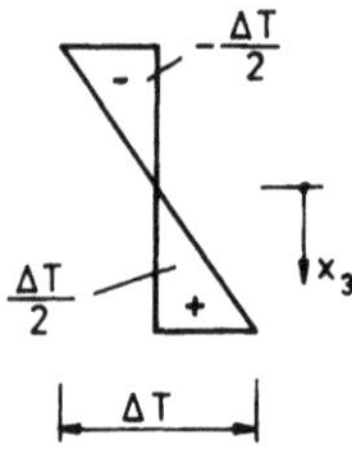

Figure 1.19

1.9.3 The Elastic Body

Let Ω an isotropic elastic body. The displacement vector $\mathbf{u}(\mathbf{x})$ of a material point $\mathbf{x}$ is the vector which points to the new location of the material point $\mathbf{x}$ after the application of the load, see Fig. 1.20.

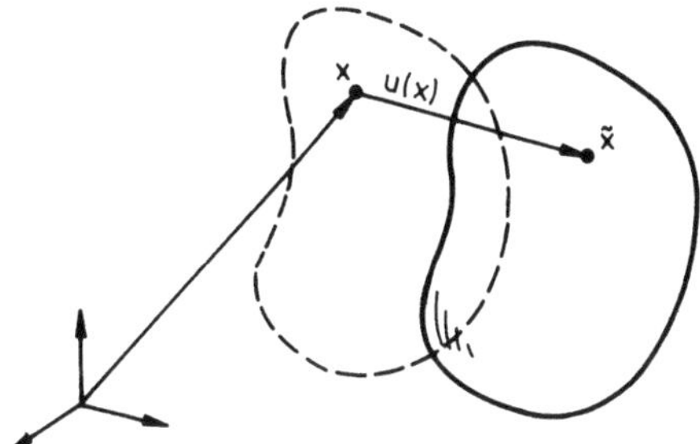

Figure 1.20

The three equations which govern the response of an elastic body are

(Def.) $\qquad E(u) - E = 0_{(3 \times 3)}$ (1.34a)

(Const.) $\qquad C[E] - S = 0_{(3 \times 3)}$ (1.34b)

(Equ.) $\qquad -\operatorname{div} S = p_{(3)}$ (1.34c)

where $E = [\varepsilon_{ij}]$ and $S = [\sigma_{ij}]$ are the strain and the stress tensors, respectively, and $E(\ )$ the operator

$$E(u) = \frac{1}{2}(\nabla u + \nabla u^T) = \frac{1}{2}\begin{bmatrix} 2u_{1,1} & u_{1,2} + u_{2,1} & u_{1,3} + u_{3,1} \\ & 2u_{2,2} & u_{2,3} + u_{3,2} \\ \text{sym.} & & 2u_{3,3} \end{bmatrix} \qquad (1.35)$$

and $C[\ ]$ the elasticity tensor

$$C[E] = 2\mu E + \lambda(\operatorname{tr} E)I \triangleq C^{ijkl}\varepsilon_{kl} \qquad (1.36)$$

with the components

$$C^{ijkl} = \mu\{\delta^{ik}\delta^{jl} + \delta^{il}\delta^{jk} + \lambda\delta^{ij}\delta^{kl}\}, \qquad \lambda = \frac{2\mu v}{1 - 2v} \qquad (1.37)$$

By C^{-1} we denote the inverse of the tensor $C[\ ]$,

$$C^{-1}[C[E]] = E$$

If $C[E] = S$ then the application of this tensor results in

$$C^{-1}[S] = \frac{1}{2\mu}\left\{S - \frac{v}{1+v}(\operatorname{tr} S)I\right\} \triangleq C^{(-1)ijkl}\sigma_{kl}$$

where

$$C^{(-1)ijkl} = \frac{1}{2\mu}\left\{\frac{1}{2}(\delta^{ik}\delta^{jl} + \delta^{il}\delta^{jk}) - \frac{v}{1+v}\delta^{ij}\delta^{kl}\right\}$$

On substituting the Eqs. (a) and (b), consecutively, into Eq. (1.34c) we obtain a differential equation of the second degree for the displacement vector u alone,

$$-Lu(x) = -\left[\mu\Delta u + \frac{\mu}{1-2v}\,\text{grad div } u\right] = p_{(3)} \tag{1.38}$$

which constitutes a system of three equations for the three components u_i

$$-\left\{\mu u_{i,jj} + \frac{\mu}{1-2v}u_{j,ji}\right\} = p_i \qquad i = 1,2,3$$

The traction vector $t = \{t_i\}$ on the surface of the body is the product of the stress tensor and the exterior unit normal

$$Sn = t, \qquad \sigma_{ij}n_j = t_i$$

Because the stress tensor S depends on the displacement field u, see Eqs. (1.34a) and (1.34b), the traction vector t can also be calculated directly from u

$$\tau(u) = 2\mu\frac{\partial u}{\partial n} + \frac{2\mu}{1-2v}n\,\text{div } u + \mu(n\times\text{rot } u) = t \tag{1.39}$$

The components of the traction vector in terms of displacements are then

$$2\mu u_{i,j}n_j + \frac{2\mu v}{1-2v}n_i u_{j,j} + \mu e_{ijk}n_j e_{klm}u_{m,l} = t_i$$

The equation of the first identity is

$$G(u,\hat{u}) = \int_\Omega -Lu\cdot\hat{u}\,d\Omega + \int_\Gamma \tau(u)\cdot\hat{u}\,ds - E(u,\hat{u}) = 0$$

where $E(u,\hat{u})$ is the symmetric expression

$$E(u,\hat{u}) = \int_\Omega \varepsilon_{ij}(u)\,C^{ijkl}\,\varepsilon_{kl}(\hat{u})\,d\Omega = \int_\Omega E(u)\cdot C[E(\hat{u})]\,d\Omega$$

Let T be a stationary temperature distribution within the body. We then observe, if the body is free to move, a deformation of the body. The displacement field u_T of this deformation satisfies the differential equation, see [P1] p. 43–1, (43.1),

$$E(u_T) = \alpha_T T I$$

where I is the unit matrix (3×3) and α_T the temperature coefficient of the material.

The temperature distribution $T(x)$ is itself, by the way, the solution of a differential equation, see [C1] p. 317.

1.9.4 Elastic Plates

A plate is considered an elastic body which is either in a state of plane strain, $\varepsilon_{33} = 0$, or a state of plane stress, $\sigma_{33} = 0$. The equations which apply in a state of plane strain coincide, up to the number of components, with the equations of elastic bodies.

These apply, too, in a state of plane stress if we replace in all equations which govern the behaviour of elastic bodies the constant v by the constant $\bar{v} = v/1 + v$ and replace the functions $u_i, \varepsilon_{ij}, \sigma_{ij}$, etc. by the functions $\bar{u}_i, \bar{\varepsilon}_{ij}, \bar{\sigma}_{ij}$ etc. These latter functions are thickness averages see [G 2] p. 150, as e. g.

$$\bar{u}_i (x_1, x_2) = \frac{1}{h} \int_{-h/2}^{h/2} u_i (x_1, x_2, x_3)\, dx_3$$

1.9.5 The Membrane

Consider a membrane, prestressed by a uniform force N, which deflects under a pressure p, see Fig. 1.21.

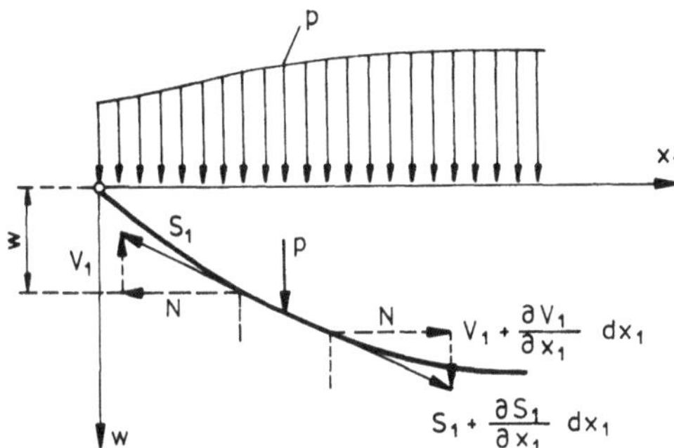

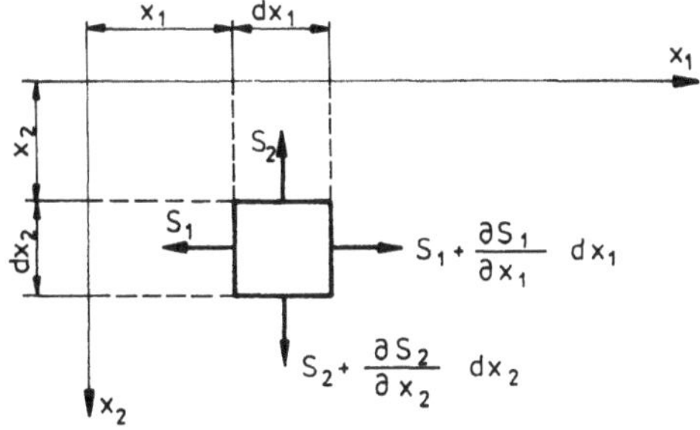

Figure 1.21

Let w be the deflection and

$$\varepsilon = \{\varepsilon_1, \varepsilon_2\} \quad \text{and} \quad v = \{v_1, v_2\}$$

the vector of the strains and the vector of the vertical components of the internal forces, resp. The following equations now apply

(Def.) $\qquad\qquad\qquad \text{grad } w - \varepsilon = 0_{(2)}$

(Const.) $\qquad\qquad\qquad N\varepsilon - v = 0_{(2)}$

(Equ.) $\qquad\qquad\qquad -\text{div } v = p_{(1)}$

or, equivalently,

$$-N\,\Delta w = p$$

The equation of the first identity is

$$G(w, \hat{w}) = \int_{\Omega} -N\,\Delta w\,\hat{w}\,d\Omega + \int_{\Gamma} N\frac{\partial w}{\partial n}\,\hat{w}\,ds - \int_{\Omega} N\,\text{grad } w \cdot \text{grad } \hat{w}\,d\Omega = 0$$

$$(1.40)$$

1.9.6 Reissner's Plate

Consider a plate of uniform thickness h with midplane coordinates x_1, x_2 and thickness coordinates x_3. Loading conditions on the plate faces are

$$(\alpha = 1, 2) \quad \sigma_{\alpha 3} = 0, \quad \sigma_{33} = \pm\frac{p}{2} \quad \text{at } x_3 = \pm\frac{h}{2}$$

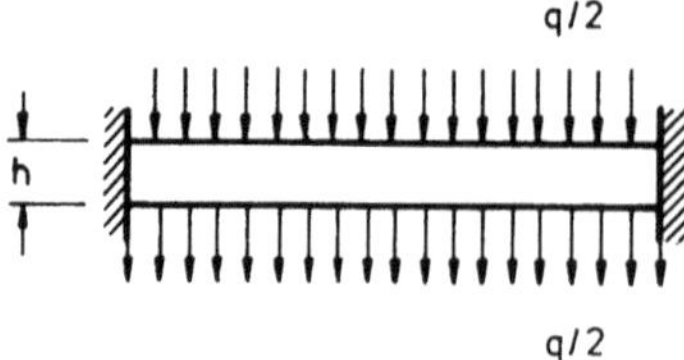

Figure 1.22

The configuration variables are the rotations $\varphi = \{\varphi_1, \varphi_2\}$ and the deflection w defined by

$$\varphi_\alpha = \int_{-h/2}^{h/2} \frac{12\,x_3}{h^3}\,u_\alpha\,dx_3, \quad w = \int_{-h/2}^{h/2} \frac{3}{2h}\left[1 - \left(\frac{2x_3}{h}\right)^2\right]u_3\,dx_3$$

where $u = \{u_i\}$ is the displacement vector of the plate continuum. Let $E = [\varepsilon_{ij}]$ be the (2×2) strain tensor and $M = [M_{ij}]$ be the (2×2) bending-moment tensor and $q = \{q_i\}$ the vector of the two shear forces, q_1 and q_2, then holds, see [W2],

$$\text{(Def.)} \qquad E(\varphi) - E = 0_{(2 \times 2)} \tag{1.41a}$$

$$\varepsilon(\varphi, w) - \varepsilon = 0_{(2)} \tag{1.41b}$$

$$\text{(Const.)} \qquad C[E] - M = 0_{(2 \times 2)} \tag{1.42a}$$

$$a\varepsilon - q = 0_{(2)} \tag{1.42b}$$

$$\text{(Equ.)} \qquad -\operatorname{div} M + q = b \operatorname{grad} p_{(2)} \tag{1.43a}$$

$$-\operatorname{div} q = p_{(1)} \tag{1.43b}$$

where the operators are defined as

$$E(\varphi) = \begin{bmatrix} \varphi_{1,1} & \dfrac{1}{2}(\varphi_{1,2} + \varphi_{2,1}) \\[2mm] \text{sym.} & \varphi_{2,2} \end{bmatrix}; \quad \varepsilon(\varphi, w) = \begin{bmatrix} \varphi_1 + w_{,1} \\[1mm] \varphi_2 + w_{,2} \end{bmatrix}$$

$$C[E] = K(1 - v)E + vK(\operatorname{tr} E)I$$

and the material parameters as

$$K = \frac{Eh^3}{12(1 - v^2)}, \quad a = K\frac{1 - v}{2}\bar{\lambda}^2, \quad b = \frac{v}{1 - v}\frac{1}{\bar{\lambda}^2}, \quad \bar{\lambda}^2 = \frac{10}{h}$$

On substituting the first 2×2 equations, consecutively, into the last two equations we obtain a system of three differential equations of the second degree for the three unknowns $\varphi = \{\varphi_1, \varphi_2\}$ and w

$$-\operatorname{div} C[E(\varphi)] + a\,\varepsilon(\varphi, w) = b \operatorname{grad} p_{(2)}$$

$$-\operatorname{div}\left(a\,\varepsilon(\varphi, w)\right) = p_{(1)}$$

or, at full length,

$$K(1 - v)\left\{-\left(\frac{1}{2}(\varphi_{\alpha,\beta} + \varphi_{\beta,\alpha}) + \frac{v}{1 - v}\varphi_{\gamma,\gamma}\delta_{\alpha\beta}\right)_{,\beta}\right.$$

$$\left. + \bar{\lambda}^2(\varphi_\alpha + w_{,\alpha})\right\} = \frac{v}{1 - v}\frac{1}{\bar{\lambda}^2}p_{,\alpha} \quad \alpha = 1,2$$

$$-\frac{1}{2}K(1 - v)\bar{\lambda}^2(\varphi_\alpha + w_{,\alpha})_{,\alpha} = p$$

The equation of the first identity is

$$
G(\varphi, w; \hat{\varphi}, \hat{w}) = \int_\Omega \big[- \operatorname{div} C[E(\varphi)] \cdot \hat{\varphi} + a\, \varepsilon(\varphi, w) \cdot \hat{\varphi}
$$
$$
- a \operatorname{div} \big(\varepsilon(\varphi, w) \big) \hat{w}\, d\Omega
$$
$$
+ \int_\Gamma C[E(\varphi)]\, n \cdot \hat{\varphi} + a\, \varepsilon(\varphi, w) \cdot n\, \hat{w} \big]\, ds - E(\varphi, w; \hat{\varphi}, \hat{w}) = 0
$$

where $E(\varphi, w; \hat{\varphi}, \hat{w})$ is the symmetric expression

$$
E(\varphi, w; \hat{\varphi}, \hat{w}) = \int_\Omega \big(C[E(\varphi)] \cdot E(\hat{\varphi}) + a\, \varepsilon(\varphi, w) \cdot \varepsilon(\hat{\varphi}, \hat{w}) \big)\, d\Omega
$$

2 Work and Energy

A spring is a very simple elastic element and, therefore, quite appropriate to acquaint us with the principles of structural mechanics.

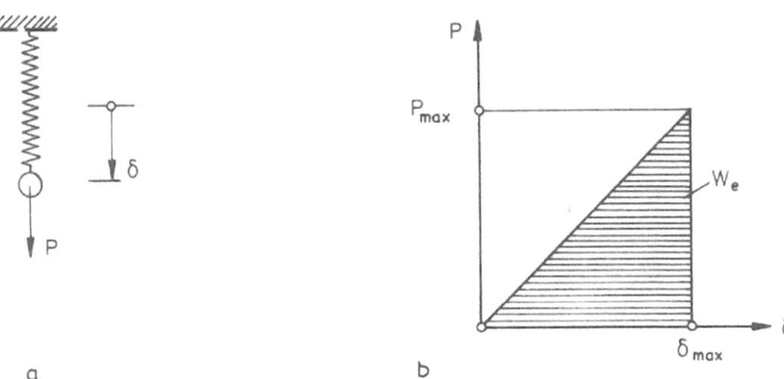

Figure 2.1

If we let a force P act on a spring as in Fig. 2.1a then we observe a displacement δ and we soon recognize that δ is proportional to P

$$P = k\,\delta$$

and that this ratio k, the stiffness of the spring, is different for different materials. Enough about physics.

Now we do mathematics. Let δ, $\hat{\delta}$ be two arbitrary numbers. Then, certainly, the expression

$$\delta\,k\,\hat{\delta} - \delta\,k\,\hat{\delta} = 0$$

is zero or, if we replace the first term $\delta\,k$ by P, also the expression

$$P\hat{\delta} - \delta\,k\,\hat{\delta} = 0$$

We call this (trivial) identity the first identity of the spring and we denote it by

$$G(\delta, \hat{\delta}) = P\,\hat{\delta} - \delta\,k\,\hat{\delta} = 0 \quad \forall\,\delta, \hat{\delta} \quad (\forall = \text{``for all''}) \tag{2.1}$$

Nothing changes if we interchange δ and $\hat{\delta}$

$$G(\hat{\delta}, \delta) = \hat{P}\,\delta - \hat{\delta}\,k\,\delta = 0 \quad \forall\,\hat{\delta}, \delta \tag{2.2}$$

Both equations, (2.1) and (2.2), are zero for all pairs $\{\delta, \hat{\delta}\}$, hence, also their difference

$$B(\delta, \hat{\delta}) = G(\delta, \hat{\delta}) - G(\hat{\delta}, \delta) = P\,\hat{\delta} - \delta\,\hat{P} = 0$$

which we call the second identity of the spring.

Finally we introduce the function

$$F(\delta) = \frac{1}{2}\,\delta\,k\,\delta$$

and call it the internal energy of the spring.

After these preparatory steps it is now a simple matter to formulate the *principle of virtual displacements*

$$G(\delta, \hat{\delta}) = P\,\hat{\delta} - \delta\,k\,\hat{\delta} = 0 \quad \forall\,\hat{\delta}$$

the principle of virtual forces

$$G(\hat{\delta}, \delta) = \hat{P}\,\delta - \hat{\delta}\,k\,\delta = 0 \quad \forall\,\hat{\delta}$$

the principle "eigenwork = int. energy"

$$\frac{1}{2}\,G(\delta, \delta) = \frac{1}{2}\,P\,\delta - \frac{1}{2}\,\delta\,k\,\delta = 0$$

and *Betti's principle*

$$B(\delta, \hat{\delta}) = P\,\hat{\delta} - \delta\,\hat{P} = 0 \quad \forall\,\delta, \hat{\delta}$$

No proof is necessary. These principles are evidently true.

Consider, e. g., the first principle, the principle of virtual displacements. If a spring loaded with a force P is in equilibrium then holds

$$k\,\delta = P \tag{2.3}$$

and, hence, the virtual internal and external work

$$\hat{\delta}\,k\,\delta = \hat{\delta}\,P$$

are the same, whatever the virtual displacements $\hat{\delta}$. Or else, $G(\delta, \hat{\delta}) = 0, \forall\,\hat{\delta}$.

Perhaps we should only explain why we have multiplied the first identity on the diagonal, that is in case $\delta = \hat{\delta}$, with the factor $1/2$ (principle "eigenwork = int. energy").

Assume we load the spring with a (slowly) increasing force P

$$0 \leqslant P \leqslant P_{max}$$

see Fig. 2.1 b.

The area between the straight line and the δ-axis is the work, W_e, done by the external forces

$$W_e = \frac{1}{2} \delta_{max} P_{max}$$

According to Eq. (2.3)

$$P_{max} = k \, \delta_{max}$$

and, hence, it follows

$$W_e = \frac{1}{2} \delta_{max} P_{max} = \frac{1}{2} \delta_{max} k \, \delta_{max} = F(\delta_{max})$$

that is the external work done by the force acting through its own displacements is equal to the internal energy, or simply stated: "eigenwork = int. energy".

We, thus, have learnt that the trivial identity $G(\delta, \hat{\delta}) = \delta \, k \, \hat{\delta} - \delta \, k \, \hat{\delta} = 0$ is the mathematical basis of four principles of mechanics.

But we are not yet at the end. The identity $G(\delta, \hat{\delta}) = 0$ also leads to the principle of minimum potential energy as we shall explain next:

If the spring is in its equilibrium position, δ, then the external force P is balanced by the reaction $k \, \delta$

$$P = k \, \delta$$

Hence, the principle of virtual displacements applies

$$G(\delta, \hat{\delta}) = P \, \hat{\delta} - \delta \, k \, \hat{\delta} = 0 \quad \forall \, \hat{\delta}$$

or, as well,

$$(-1) \, G(\delta, \hat{\delta}) = -P \, \hat{\delta} + \delta \, k \, \hat{\delta} = 0 \quad \forall \, \hat{\delta}$$

This expression is the first variation of the function

$$\Pi(\delta) = \frac{1}{2} \delta \, k \, \delta - P \, \delta$$

and we, therefore, conclude that the equilibrium position δ is a stationary point of the function $\Pi(\delta)$, i. e.

$$\delta \, \Pi(\delta, \hat{\delta}) = \delta \, k \, \hat{\delta} - P \, \hat{\delta} = 0 \quad \forall \, \hat{\delta} \tag{2.4}$$

Next let $\hat{\delta}$ be an arbitrary number. Then holds

$$\Pi(\delta + \hat{\delta}) - \Pi(\delta) = \delta\, k\, \hat{\delta} - P\, \hat{\delta} + \frac{1}{2}\hat{\delta}\, k\, \hat{\delta} = \hat{\delta}\, \Pi(\delta, \hat{\delta}) + \frac{1}{2}\hat{\delta}\, k\, \hat{\delta} \tag{2.5}$$

According to Eq. (2.4) the first term on the right-hand side is zero, hence, Eq. (2.5) simplifies to

$$\Pi(\delta + \hat{\delta}) - \Pi(\delta) = \frac{1}{2}\hat{\delta}\, k\, \hat{\delta} > 0 \qquad \forall\, \hat{\delta} \neq 0$$

But this expression is (k is greater than zero) evidently positive definite. Consequently $\Pi(\delta)$ must be the minimum.

This simple example of a spring contains already, as in a nutshell, all the principles of structural mechanics.

In what follows we shall replace the spring by bars, beams, plates and elastic bodies but the approach will be the same. If there is a difference then it is the difference between vector-statics and statics of continua.

A number, δ, has only one degree of freedom, a function, $u(x)$, as the displacement of a bar infinitely many degrees, hence, statics of continua requires the mathematics of continua, that is infinitesimal calculus.

2.1 Integral Identities

As an introduction to structural mechanics of continua consider the bar with constant cross section in Fig. 2.2 which translates the single force P into the wall.

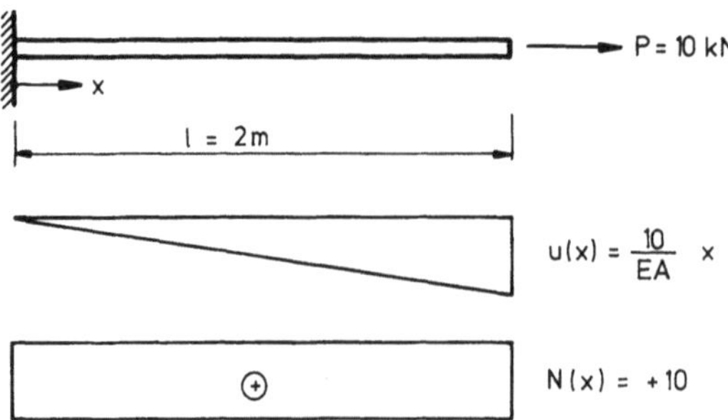

Figure 2.2

Let δu be an arbitrary admissible $(\delta u(0) = 0)$ virtual displacement of the bar. The principle of virtual displacements states that the external work done by the load P acting through the virtual end displacement $\delta u(2)$ at $x = 2$ is equal to the internal work of the stresses acting through the virtual strains

$$\int_0^2 \frac{N\, \delta N}{EA}\, dx = P\, \delta u(2) \tag{2.6}$$

A check of this statement with the function $\delta u = x$ renders

$$20.0 = 20.0$$

An additional test with $\delta u = \sin x$ renders

$$9.092974268 = 9.092974268$$

Sometimes it is proclaimed that the principle of virtual displacements only applies if the displacements are "small". But even if we restrict the principle to functions with small values, why is it that for all admissible virtual displacements the left-hand side of Eq. (2.6) has the same value as the right-hand side?

An equation as

$$\delta k \, \hat{\delta} - \delta k \, \hat{\delta} = 0$$

is valid for all numbers δ and $\hat{\delta}$, this is obvious. But Eq. (2.6) is not obvious. A function is a symbol, δu, with infinitely many degrees of freedom. Certainly there are infinitely many admissible δu. Why are we so sure that Eq. (2.6) is valid for every single δu?

The answer is: because we know that

(i) *the displacement $u(x)$ satisfies the differential equation*

$$-EA u'' = 0 \quad in \ 0 < x < l$$

(ii) *the displacement $u(x)$ belongs to C^2, and*

(iii) *all admissible δu belong to C^1*

To confirm that these conditions are sufficient let in the following u the displacement of the bar in Fig. 2.2 or any other arbitrary function in the class $C^2 [0, 2]$, e. g. $u = x$ or $u = 1$ or $u = \sin x$ or...

If we apply the operator $-EA \, d^2/dx^2$ to this function and multiply the result with a function $\hat{u}$ from the class $C^1 [0, 2]$, e. g. $\hat{u} = x$ or $\hat{u} = 10$ or $\hat{u} = \cos x$ or... and integrate the product over the interval $[0, 2]$ then we obtain a number

$$\int_0^2 -EA u'' \, \hat{u} \, dx$$

Because we presupposed that the functions u and $\hat{u}$ belong to C^2 and C^1 respectively, we may apply the integration by parts formula, see Eq. (1.10), to this definite integral. This yields

$$\int_0^2 -EA u'' \, \hat{u} \, dx = -[EA u' \, \hat{u}]_0^2 + \int_0^2 EA u' \, \hat{u}' \, dx \tag{2.7}$$

The functions u and $\hat{u}$ are in principle arbitrary elements of $C^2 \times C^1$. Hence, Eq. (2.7) holds for all pairs $\{u, \hat{u}\} \in C^2 \times C^1$; it is for all such pairs an identity. This is motivation for calling Eq. (2.7) the first identity of the operator $-EA d^2/dx^2$.

Let us apply, for a check, the identity to the pair of functions $u = x^2/EA$, $\hat{u} = x^3$. The result is

$$-8 = -32 + 24$$

One more try with the pair $u = (EA)^{-1} \sin x$, $\hat{u} = \cos x$, renders

$$0.4134109052 = 0.8268218104 - 0.4134109052$$

Next, we apply it to the bar in Fig. 2.2, the problem we started with, i. e. we identify the function u in the first identity with the displacement of the bar, and $\hat{u} = \delta u$ with a virtual displacement. On account of the equations

$$-EA\, u''(u) = 0, \quad EA\, u' = N = P, \quad EA\, \hat{u}' = \delta N, \quad 0 < x < 2$$

$$\delta u(0) = 0$$

the first identity simplifies to

$$0 = -P\,\delta u(2) + \int_0^2 \frac{N\,\delta N}{EA}\, dx$$

or

$$\int_0^2 \frac{N\,\delta N}{EA}\, dx = P\,\delta u(2)$$

which is just Eq. (2.6).

Obviously, the principle of virtual displacements is a verbal accord of the first identity of the operator $-EA\, d^2/dx^2$.

To set the stage for Betti's principle we, next, replace the single force at the end of the bar by a load, see Fig. 2.3, which is evenly distributed along the bar.

Betti's principle claims that the external work done by the single force P of system 1 in Fig. 2.2 acting through the displacements $\hat{u}$ of system 2 in Fig. 2.3 is equal to the work done by the distributed forces $\hat{p}$ of system 2 acting through the displacements u of system 1. In other words, we should have

$$P\,\hat{u}(2) = \int_0^2 p\,\hat{u}\, dx \tag{2.8}$$

A simple calculation confirms Betti's principle

$$\frac{100}{EA} = \frac{100}{EA}$$

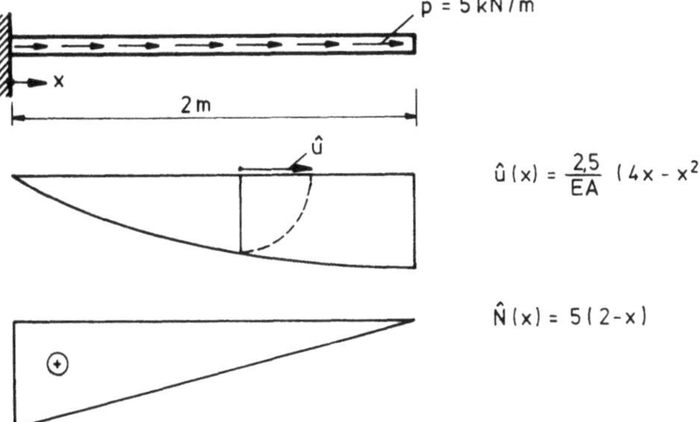

Figure 2.3

Is this pure chance? No, there is a mathematical law behind it, namely the law that every pair $\{u, \hat{u}\}$ of C^2-functions is a zero of the second identity of the operator $-EA\, d^2/dx^2$.

To find the expression of this second identity we interchange in the first identity the places of u and $\hat{u}$.

$$\int_0^2 -EA\,\hat{u}''\,u\,dx = -[EA\,\hat{u}'\,u]_0^2 + \int_0^2 EA\,\hat{u}'\,u'\,dx \tag{2.9}$$

If both functions belong to C^2 then this manipulation is according to the integration by parts formula, see Eq. (1.10), correct.

Subtracting Eq. (2.9) from Eq. (2.7) we obtain the second identity of the operator $-EA\,d^2/dx^2$.

$$\int_0^2 -EA\,u''\,\hat{u}\,dx - \int_0^2 u\,(-EA\,\hat{u}'')\,dx = -[EA\,u'\,\hat{u}]_0^2 + [u\,EA\,\hat{u}']_0^2$$

To verify that this identity is the mathematical basis of Betti's principle we identify the function u in this expression with the displacement of system 1 and $\hat{u}$ with the displacement of system 2 and consider the equations

$$-EA\,u'' = 0, \quad -EA\,\hat{u}'' = \hat{p} \quad 0 < x < l$$

$$u(0) = 0, \quad EA\,u'(2) = P, \quad \hat{u}(0) = 0, \quad EA\,\hat{u}'(2) = 0$$

The result is

$$-\int_0^2 u\,\hat{p}\,dx = -P\,\hat{u}(2)$$

which is (up to the factor (-1)) exactly Eq. (2.8).

Let us summarize: the principle of virtual displacements and Betti's principle are verbal accords of mathematical laws. Their range of applicability is wider than the engineer originally intended because displacements must not be small but their range is otherwise not unbounded because some real restrictions apply as, e. g., the restriction that the functions must be sufficiently smooth, i. e. belong to spaces as C^1 or C^2.

It is very important that the reader understands this basic feature:

The principles of structural mechanics of continua are statements about functions which satisfy certain conditions which are elements of two classes (every assemblage of properties constitutes a class). In a wider sense these principles are, therefore, statements about classes of functions.

And another important concept came to light in these first introductory examples, namely: duality.

We constantly operated with two mutually adjoint functions as $-EA\,u''$ and $\hat{u}$ (in domain integrals) or N and $\hat{u}$ (in boundary integrals). This duality of force and displacement permeats all of structural mechanics. Its source is the L_2-scalar product, i. e. the integral

$$\int -EA\,u''\,\hat{u}\,dx$$

and we think we do not exaggerate if we consider structural mechanics of continua an explication of this integral.

In the case of beams, plates and elastic bodies this integral has the form

$$\int EI\,w^{IV}\,\hat{w}\,dx, \quad \int_{\Omega} K\Delta\Delta\,w\hat{w}\,d\Omega, \quad \int_{\Omega} -L\boldsymbol{u}\cdot\hat{\boldsymbol{u}}\,d\Omega$$

and we continue in the following in a more systematic fashion what we started with the bar, that is we calculate the first and second identity of the operators $(EI(\ )'')''$ (beams), $K\Delta\Delta$ (Kirchhoff plates) and $-L$ (elastic plates and bodies).

2.1.1 The Beam

The differential equation which governs the deflection of a beam is

$$(EI\,w'')'' = p$$

Let $\{w, \hat{w}\}$ be a pair of functions from $C^4[a, b] \times C^2[a, b]$. If we apply integration by parts two times to the integral

$$\int_a^b (EI\,w'')''\,\hat{w}\,dx$$

then we obtain the first identity of the operator $(EI(\ )'')''$

$$\int_a^b (EI\,w'')''\,\hat{w}\,dx = [(EI\,w'')'\,\hat{w} - EI\,w''\,\hat{w}']_a^b + \int_a^b EI\,w''\,\hat{w}''\,dx \tag{2.10}$$

If the pair $\{w, \hat{w}\}$ is from $C^4 \times C^4$ then it is admissible to interchange the positions of w and $\hat{w}$ and we, thus, obtain

$$\int_a^b (EI\,\hat{w}'')''\, w\, dx = [(EI\,\hat{w}'')'\, w - EI\,\hat{w}''\, w']_a^b + \int_a^b EI\,\hat{w}''\, w''\, dx \qquad (2.11)$$

This Eq. (2.11) minus Eq. (2.10) is the second identity of the operator $(EI(\;)'')''$

$$\int_a^b (EI\,w'')''\, \hat{w}\, dx - \int_a^b w\, (EI\,\hat{w}'')''\, dx$$

$$= [(EI\,w'')'\, \hat{w} - EI\,w''\, \hat{w}']_a^b - [w\,(EI\,\hat{w}'')' - w'\,EI\,\hat{w}'']_a^b$$

With $V = -(EI\,w'')'$ and $M = -EI\,w''$ the first identity can also be formulated as

$$\int_a^b (EI\,w'')''\, \hat{w}\, dx = -[V\,\hat{w} - M\,\hat{w}']_a^b + \int_a^b \frac{M\,\hat{M}}{EI}\, dx$$

and the second identity as

$$\int_a^b (EI\,w'')''\, \hat{w}\, dx - \int_a^b w\,(EI\,\hat{w}'')''\, dx = -[V\,\hat{w} - M\,\hat{w}' + w'\,\hat{M} - w\,\hat{V}]_a^b$$

2.1.2 The Kirchhoff Plate

The differential equation which governs the deflection of a plate is

$$K\,\Delta\Delta\, w = p$$

Let $\{w, \hat{w}\}$ be a pair of functions from $C^4(\bar{\Omega}) \times C^2(\bar{\Omega})$. It is then permissible to apply the integration by parts formula to the integral

$$\int_\Omega K\,\Delta\Delta\, w\,\hat{w}\, d\Omega$$

and we, thus, obtain after some lengthy manipulations, see [H 6] p. 147, the first identity of the operator $K\,\Delta\Delta$.

$$\int_\Omega K\,\Delta\Delta\, w\,\hat{w}\, d\Omega = \int_\Gamma \left[M_n(w)\,\frac{\partial\,\hat{w}}{\partial n} - V_n(w)\,\hat{w} \right] ds - [[M_{nt}(w)\,\hat{w}]]$$
$$+ E(w, \hat{w})$$

where $M_n(w) = M_n$ is the bending moment, $V_n(w) = V_n$ the Kirchhoff-shear and $M_{nt}(w) = M_{nt}$ the twisting moment on the boundary Γ of the plate Ω, see section 1.9.2.

The bilinear form

$$E(w, \hat{w}) = K \int_\Omega [w,_{11} (\hat{w},_{11} + v\,\hat{w},_{22}) + 2(1 - v)\, w,_{12}\, \hat{w},_{12}$$

$$+ w,_{22} (\hat{w},_{22} + v\,\hat{w},_{11})]\, d\Omega \tag{2.12}$$

the strain energy of two deflections w and $\hat{w}$ is symmetric

$$E(w, \hat{w}) = E(\hat{w}, w)$$

and, therefore, the procedure applied so many times above to formulate the second identity can be repeated her as well. We formulate the first identity a second time, interchange the places of w and $\hat{w}$ and then subtract the two equations.

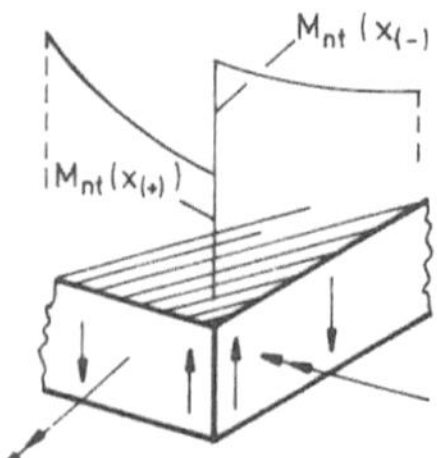

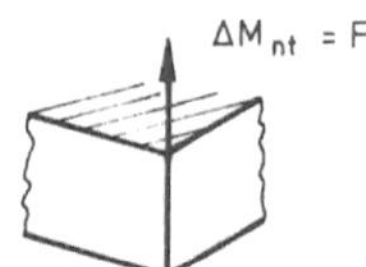

Figure 2.4

The result is the second identity (Rayleigh-Green identity) of the operator $K\Delta\Delta$.

$$\int_\Omega K\Delta\Delta\, w\,\hat{w}\, d\Omega - \int_\Omega w\, K\Delta\Delta\,\hat{w}\, d\Omega = \int_\Gamma \left[M_n(w)\, \frac{\partial \hat{w}}{\partial n} - V_n(w)\,\hat{w} \right] ds$$

$$- \int_\Gamma \left[\frac{\partial w}{\partial n}\, M_n(\hat{w}) - w\, V_n(\hat{w}) \right] ds - [[\, M_{nt}(w)\,\hat{w}\,]] + [[\, w\, M_{nt}(\hat{w})\,]]$$

where, see Eq. (1.9)

$$[[\, M_{nt}(w)\,\hat{w}\,]] = -\int_\Gamma \frac{d}{ds}\left(M_{nt}(w)\,\hat{w} \right) ds = \sum_{i=1}^{n} \Delta M_{nt}(w)\,(x^i)\,\hat{w}(x^i)$$

$$= \sum_{i=1}^{n} F(x^i)\,\hat{w}(x^i)$$

The points x^i, $i = 1, 2 \ldots n$ are the corner points of the plate and

$$\Delta M_{nt}(w)\,(x^i) = M_{nt}(w)\,(x^i_+) - M_{nt}(w)\,(x^i_-) = F(x^i)$$

is the jump in the twisting moment which is identified with corner forces $F(x^i)$.

Each of these forces acts through a virtual displacement $\hat{w}(x^i)$. The sum of all this virtual work is the number $[[\, M_{nt}(w)\,\hat{w}\,]]$.

2.1.3 The Elastic Plate and Body

The displacement field u of an elastic plate or body is governed by the system

$$-Lu = -\left[\mu\,\Delta\,u + \frac{\mu}{1-2v}\,\text{grad div } u\right] = p_{(3)}$$

Let $\{u, \hat{u}\} \in C^2(\bar{\Omega}) \times C^1(\bar{\Omega})$. Starting with the integral

$$\int_{\Omega} -Lu \cdot \hat{u}\, d\Omega$$

we obtain with integration by parts, s. [G 2] p. 95, the first identity of the operator $-L$

$$\int_{\Omega} -Lu \cdot \hat{u}\, d\Omega = -\int_{\Gamma} \tau(u) \cdot \hat{u}\, ds + E(u, \hat{u})$$

where $\tau(u) = Sn$ is the traction vector of the field u on the boundary Γ, see 1.9.3.
The bilinear form

$$E(u, \hat{u}) = \int_{\Omega} \varepsilon_{ij}(u)\, C^{ijkl}\, \varepsilon_{kl}(\hat{u})\, d\Omega = \int_{\Omega} E(u) \cdot C\,[E(\hat{u})]\, d\Omega \tag{2.13}$$

is the strain energy of two displacement fields u and $\hat{u}$. Because the elasticity tensor $C[\ ]$ is symmetric, see Eq. (1.37), also the form $E(u, \hat{u}) = E(\hat{u}, u)$ is and we obtain, therefore, as before the second identity (Betti's identity) by formulating the first identity twice and subtracting the two equations.

$$\int_{\Omega} -Lu \cdot \hat{u}\, d\Omega - \int_{\Omega} u \cdot (-L\hat{u})\, d\Omega = -\int_{\Gamma} \tau(u) \cdot \hat{u}\, ds + \int_{\Gamma} u \cdot \tau(\hat{u})\, ds$$

2.2 Summary

The identities state: two bilinear forms, $\mathscr{L}(u, \hat{u})$ and $\mathscr{R}(u, \hat{u})$, the two sides of the identities, have the same value for all admissible pairs $\{u, \hat{u}\}$

$$\mathscr{L}(u, \hat{u}) = \mathscr{R}(u, \hat{u})$$

This is equivalent with the statement

$$I(u, \hat{u}) = \mathscr{L}(u, \hat{u}) - \mathscr{R}(u, \hat{u}) = 0$$

In what follows we shall present the identities in this form. We denote the first identity by the letter G (as in Green) and the second identity by the letter B (as in Betti).

These identities represent "nil-forms",

$$G(u, \hat{u}) = 0 \quad \forall u, \hat{u} \in C^{2m} \times C^{m} \qquad\qquad B(u, \hat{u}) = 0 \quad \forall u, \hat{u} \in C^{2m} \times C^{2m}$$

All admissible pairs $\{u, \hat{u}\}$ are zeros of these bilinear forms.
The following theorem repeats our results in the new notation.

Theorem 2.1

Let $2m$ be the order of the operator and let $\{u, \hat{u}\}$ be a pair of functions from $C^{2m} \times C^{m}$ (first identity) or $C^{2m} \times C^{2m}$ (second identity) resp. It then holds

Bars

$$G(u, \hat{u}) = \int_a^b -(EA\, u')'\, \hat{u}\, dx + [N\, \hat{u}]_a^b - \int_a^b \frac{N\hat{N}}{EA}\, dx = 0 \tag{2.14}$$

$$B(u, \hat{u}) = \int_a^b -(EA\, u')'\, \hat{u}\, dx + [N\, \hat{u}]_a^b - [u\, \hat{N}]_a^b - \int_a^b u(-EA\, \hat{u}')'\, dx = 0 \tag{2.15}$$

Beams

$$G(w, \hat{w}) = \int_a^b (EI\, w'')''\, \hat{w}\, dx + [V\, \hat{w} - M\, \hat{w}']_a^b - \int_a^b \frac{M\hat{M}}{EI}\, dx = 0 \tag{2.16}$$

$$B(w, \hat{w}) = \int_a^b (EI\, w'')''\, \hat{w}\, dx + [V\, \hat{w} - M\, \hat{w}']_a^b - [w\, \hat{V} - w'\, \hat{M}]_a^b$$
$$- \int_a^b w(EI\, \hat{w}'')''\, dx = 0 \tag{2.17}$$

Kirchhoff plates

$$G(w, \hat{w}) = \int_\Omega K\Delta\Delta\, w\, \hat{w}\, d\Omega + \int_\Gamma \left[V_n(w)\, \hat{w} - M_n(w)\, \frac{\partial \hat{w}}{\partial n} \right] ds$$
$$+ \left[\left[M_{nt}(w)\, \hat{w} \right]\right] - E(w, \hat{w}) = 0 \tag{2.18}$$

$$B(w, \hat{w}) = \int_\Omega K\Delta\Delta\, w\, \hat{w}\, d\Omega + \int_\Gamma \left[V_n(w)\, \hat{w} - M_n(w)\, \frac{\partial \hat{w}}{\partial n} \right] ds$$
$$+ \left[\left[M_{nt}(w)\, \hat{w} \right]\right] - \left[\left[w\, M_{nt}(\hat{w}) \right]\right] - \int_\Gamma \left[w\, V_n(\hat{w}) - \frac{\partial w}{\partial n}\, M_n(\hat{w}) \right] ds$$
$$- \int_\Omega w\, K\Delta\Delta\, \hat{w}\, d\Omega = 0 \tag{2.19}$$

Elastic plates and bodies

$$G(u, \hat{u}) = \int_\Omega -L u \cdot \hat{u}\, d\Omega + \int_\Gamma \tau(u) \cdot \hat{u}\, ds - E(u, \hat{u}) = 0 \tag{2.20}$$

$$B(u, \hat{u}) = \int_\Omega - L\,u \cdot \hat{u}\, d\Omega + \int_\Gamma \tau(u) \cdot \hat{u}\, ds - \int_\Gamma u \cdot \tau(\hat{u})\, ds$$

$$- \int_\Omega u \cdot (-L\hat{u})\, d\Omega = 0 \qquad\qquad (2.21)$$

All this theorem states is: if $\{u, \hat{u}\}$ is a pair of admissible functions, admissible, e. g., with respect to a particular form $B(u, \hat{u})$, and if we calculate the integrals which belong to this form $B(u, \hat{u})$ then the sum of all these integrals is zero.

This simple theorem which can be formulated for other differential equations as well is the basis of structural mechanics of continua.

Before we discuss all this in more detail we must yet introduce the internal strain energy functionals $F(u)$.

The forms $E(w, \hat{w})$ and $E(u, \hat{u})$ were introduced to denote the strain energies of Kirchhoff plates and elastic plates and bodies. We shall often denote the strain energies of bars and beams by the same symbol

$$E(u, \hat{u}) = \int \frac{N\hat{N}}{EA}\, dx, \qquad E(w, \hat{w}) = \int \frac{M\hat{M}}{EI}\, dx$$

and instead of strain energy we shall more often simply speak of energy.

All energies are symmetric bilinear forms

(i) $\qquad\qquad E(u, \hat{u}) = E(\hat{u}, u)$

(ii) $\qquad\qquad E(a_1 u_1 + a_2 u_2, \hat{a}_1 \hat{u}_1 + \hat{a}_2 \hat{u}_2) = \sum_{i,j=1}^{2} a_i \hat{a}_j E(u_i, \hat{u}_j)$

and, therefore, each the first variation of a functional

$$F(u) = \frac{1}{2} E(u, u)$$

as the following equation confirms

$$\frac{d}{d\varepsilon} F(u + \varepsilon\,\hat{u})\Big|_{\varepsilon=0} = \frac{d}{d\varepsilon} \left\{ \frac{1}{2} E(u + \varepsilon\,\hat{u}, u + \varepsilon\,\hat{u}) \right\}\Big|_{\varepsilon=0}$$

$$= \frac{d}{d\varepsilon} \left\{ \frac{1}{2} E(u, u) + \varepsilon\, E(u, \hat{u}) + \frac{\varepsilon^2}{2} E(\hat{u}, \hat{u}) \right\}\Big|_{\varepsilon=0} = E(u, \hat{u})$$

These functionals $F(u)$ are the internal strain energy (or simply internal energy) functionals.

Bars

$$F(u) = \frac{1}{2} \int \frac{N^2}{EA}\, dx = \frac{1}{2} \int EA\, u'^2\, dx \geqslant 0 \qquad u \in C^1$$

Beams

$$F(w) = \frac{1}{2} \int \frac{M^2}{EI} \, dx = \frac{1}{2} \int EI \, w''^2 \, dx \geqslant 0 \quad w \in C^2$$

Kirchhoff plates

$$F(w) = \frac{1}{2} \int_\Omega \left(w,_{11}^2 + 2v \, w,_{11} \, w,_{22} + w,_{22}^2 + 2(1-v) \, w,_{12}^2 \right) d\Omega$$

$$\geqslant (1-v) \int_\Omega \left(w,_{11}^2 + 2 w,_{12}^2 + w_{22}^2 \right) d\Omega \geqslant 0 \quad w \in C^2 \tag{2.22}$$

Elastic plates and bodies

$$F(u) = \frac{1}{2} \int_\Omega \left(\varepsilon_{ij}(u)\,\varepsilon_{ij}(u)\,2\mu + \frac{2\mu v}{1-2v}\,(\varepsilon_{ii}(u))^2 \right) d\Omega \geqslant 0 \quad u \in C^1$$

For a proof of the inequality (2.22) see [R 1] p. 273.

With the internal energy functionals $F(u)$ defined we are now well prepared to substantiate our claim that Theorem 2.1 is the basis of structural mechanics. We only have to make the contents of Theorem 2.1 visible. We do this by formulating corollaries of Theorem 2.1.

2.3 Three Corollaries

Corollary 1, principle of virtual displacements

$$p: \ u \in C^{2m}$$

$$q: \ G(u, \hat{u}) = 0 \quad \forall \, \hat{u} \in C^m$$

Corollary 2, principle of virtual forces

$$p: \ u \in C^m$$

$$q: G(\hat{u}, u) = 0 \quad \forall \, \hat{u} \in C^{2m}$$

Corollary 3, Betti's principle

$$p: \ u, \hat{u} \in C^{2m}$$

$$q: \ B(u, \hat{u}) = 0$$

The bar is the simplest structural element and, therefore, appropriate in order to acquaint us with these corollaries, to make us see that these corollaries deserve these titles.

Consider the first identity $G(u, \hat{u})$ of the bar, Eq. (2.14), and assume we write it as follows

$$\int_0^l \frac{N\hat{N}}{EA}\,dx = \int_0^l (-EAu')'\,\hat{u}\,dx + [N\,\hat{u}]_0^l \qquad (2.23)$$

The left-hand side is the first variation of the internal energy

$$F(u) = \frac{1}{2}\int_0^l \frac{N^2}{EA}\,dx$$

and, hence, Eq. (2.23) can be interpreted as the statement:

The first variation of the internal energy is equal to the external work done by the distributed forces $-(EAu')'$ and end forces $N(0)$, $N(l)$ acting through the virtual displacements $\hat{u}$. This is the principle of virtual displacements.

Next we interchange u and $\hat{u}$ in Eq. (2.23), that is we formulate $G(\hat{u}, u)$

$$\int_0^l \frac{\hat{N}N}{EA}\,dx = \int_0^l (-EA\,\hat{u}')'\,u\,dx + [\hat{N}\,u]_0^l$$

The left-hand side is still (the strain energy is symmetric!) the first variation of the internal energy and with a view towards the right-hand side we may state:

The first variation of the internal energy is also equal to the work done by virtual forces $-(EA\,\hat{u}')'$ and $\hat{N}(0)$, $\hat{N}(l)$ acting through the real displacements u. This is the principle of virtual forces.

This duality, the fact that the first variation of the functional $F(u) = 1/2\,E(u, u)$ can be equated in two ways with expressions of external work, is a consequence of the symmetry of the strain energy, $E(u, \hat{u}) = E(\hat{u}, u)$.

In nonlinear mechanics the principle of virtual displacements still applies but the principle of virtual forces does not because in nonlinear mechanics the symmetry condition, $E(u, \hat{u}) = E(\hat{u}, u)$, is violated, see chapter 10.

Next, consider the second identity $B(u, \hat{u})$ of the bar, Eq. (2.15), and assume we write it as follows

$$\int_0^l -(EAu')'\,\hat{u}\,dx + [N\,\hat{u}]_0^l = \int_0^l u\,(-(EA\,\hat{u}')')\,dx + [u\,\hat{N}]_0^l$$

The left-hand side is the work done by the external forces $-(EA\,\hat{u}')'$ and $N(0)$, $N(l)$ acting through $\hat{u}$. The right-hand side is the work done by the external forces $-(EA\,\hat{u}')'$ and $\hat{N}(0)$, $\hat{N}(l)$ acting through u. Both are the same if $\{u, \hat{u}\} \in C^2$. This is Betti's principle.

What we did here with a bar can be done with other structural elements as well. The principle of virtual displacements, the principle of virtual forces and Betti's principle are simple corollaries of Theorem 2.1.

We exemplify this conclusion in the following also with a beam.

2.4 A Beam

2.4.1 Principle of Virtual Displacements

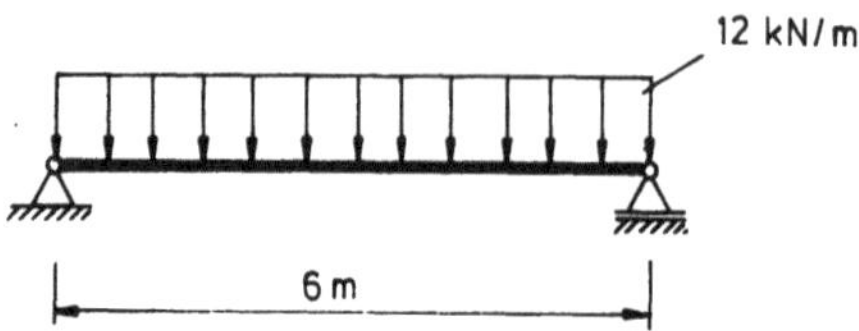

Figure 2.5

Consider the beam in Fig. 2.5 with constant stiffness EI. Its deflection satisfies the equations

$$EI\,w^{IV} = 12, \quad 0 < x < l, \quad w(0) = w(l) = M(0) = M(l) = 0 \tag{2.24}$$

The first identity of a beam is

$$G(w, \hat{w}) = \int_0^l EI\,w^{IV}\,\hat{w}\,dx + [V\hat{w} - M\hat{w}']_0^l - \int_0^l \frac{M\hat{M}}{EI}\,dx = 0 \tag{2.25}$$

If we identify the first function in this expression, w, with the deflection $w(x) = (EI)^{-1}(108x - 6x^3 + 0.5x^4)$ of the beam and $\hat{w}$ with an admissible virtual displacement, that is a function from C^2 with the properties

$$\hat{w}(0) = \hat{w}(l) = 0$$

—such a function is, e. g., $\hat{w}(x) = -6x + x^2$ —then Eq. (2.25) becomes

$$G(w, \hat{w}) = \int_0^6 p\,\hat{w}\,dx - \int_0^6 \frac{M\hat{M}}{EI}\,dx = -432 + 432 = 0$$

This result could be interpreted as the statement: "*The work done by the external forces acting through virtual displacements is equal to the strain energy of the pair* $\{w, \hat{w}\}$."

2.4.2 Principle of Virtual Forces

Corollary 2 is a simple modification of corollary 1. The functions w and $\hat{w}$ simply change places.

$$G(\hat{w}, w) = \int_0^l EI\,\hat{w}^{IV}\,w\,dx + [\hat{V}w - \hat{M}w']_0^l - \int_0^l \frac{\hat{M}M}{EI}\,dx = 0$$

If we identify w with the deflection of the beam and $\hat{w}$ with one of the (innumerable) solutions of the problem

$$EI\,\hat{w}^{IV} = \hat{p}, \quad 0 < x < l, \quad \hat{M}(0) = \hat{M}(l) = 0 \tag{2.26}$$

as, e. g.,

$$\hat{w} = 20 + 230\,x - 12\,x^3 + x^4, \qquad \hat{p} = 24\,EI$$

then we obtain

$$G(\hat{w}, w) = \int_0^6 \hat{p}\,w\,dx - \int_0^6 \frac{\hat{M}\,M}{EI}\,dx = 18\,662.4 - 18\,662.4 = 0$$

or in poor words: *"The external virtual work done by virtual forces $\hat{p}$ acting through the real displacements w of a beam is equal to the internal work done by the virtual stresses acting through the real strains."*

2.4.3 Betti's Principle

In the case of a beam corollary 3 becomes

$$B(w, \hat{w}) = \int_0^l EI\,w^{IV}\,\hat{w}\,dx + [V\,\hat{w} - M\,\hat{w}']_0^l - [w\,\hat{V} - w'\,\hat{M}]_0^l$$
$$- \int_0^l w\,EI\,\hat{w}^{IV}\,dx = 0$$

If we let w and $\hat{w}$ in this expression the functions

$$w(x) = \frac{1}{EI}\,(108\,x - 6\,x^3 + 0.5\,x^4), \qquad \hat{w}(x) = \frac{1}{EI}\,(25.2\,x - x^3 + 0.0083\,x^5)$$

—these functions satisfy in $(0, l)$ the differential equations

$$EI\,w^{IV} = 12, \qquad EI\,\hat{w}^{IV} = x$$

and comply with the support conditions of the beam in Fig. 2.5—then the second identity becomes

$$B(w, \hat{w}) = \int_0^l p\,\hat{w}\,dx - \int_0^l w\,\hat{p}\,dx = \frac{1}{EI}\,(2\,332.8 - 2\,332.8) = 0$$

This is Betti's principle: *"The work done by a first load p acting through the displacements $\hat{w}$ of a second load $\hat{p}$ is equal to the work done by the second load acting through the displacements w caused by the first load".*

2.5 Eigenwork = Internal Energy

A fourth corollary of Theorem 2.1 is obtained if we choose in the first identities $\hat{u} = u$ and multiply the identities with 1/2. The result is the statement: "eigenwork = int. energy", the external eigenwork is equal to the internal energy.

Corollary 4, "eigenwork = int. energy"

$$p: \ u \in C^{2m}(\bar{\Omega})$$

$$q: \ \frac{1}{2} G(u,u) = 0$$

In the case of a beam this corollary reads

$$\frac{1}{2} G(w,w) = \frac{1}{2} \int_0^l (EI w'')'' w\, dx + \frac{1}{2}[V w - M w']_0^l - \frac{1}{2} \int_0^l \frac{M^2}{EI}\, dx = 0$$

If we identify w with the deflection of the beam in Fig. 2.5 then we obtain

$$\frac{1}{2} G(w,w) = \frac{1}{2} \int_0^l p w\, dx - \frac{1}{2} \int_0^l \frac{M^2}{EI}\, dx = 0$$

which confirms the view *"that the external work is stored as internal energy"*.

To explain the factor 1/2 remember that we distinguish in structural mechanics between the work done when a force acts through its own displacements, $\hat{w} = w$, we call this work *"eigenwork"* und work done when a force acts through virtual displacements, $\hat{w} \neq w$. The latter we call *virtual work*. In structural mechanics eigenwork is multiplied with the factor 1/2 but virtual work is not.

This means, from a mathematical point of view, that, on the diagonal, $\hat{w} = w$, the first identity is multiplied with 1/2 but outside the diagonal it is not.

Naturally, there is a reason for it, the same as in the case of the spring.

Assume we load a bar with a force $P, 0 \leqslant P \leqslant P_e$ which increases continuously from 0 to a final value P_e, see Fig. 2.6.

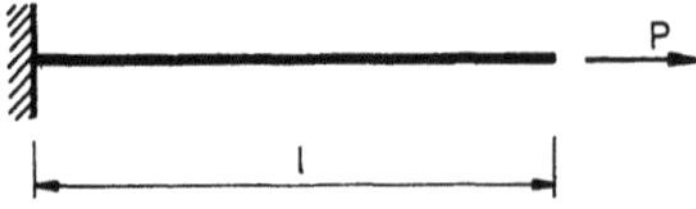

Figure 2.6

At every stage, P, the displacement u is determined by the equations

$$-EA u'' = 0, \quad 0 < x < l, \quad u(0) = 0, \quad N(l) = P \tag{2.27}$$

Let $u_e(x) = P_e x / A E$ be the solution of this *bvp* at the final stage, $P = P_e$, then we define as the *"external eigenwork done during the whole process"* the integral

$$W_e = \int_0^{u_e(l)} P\,du = \int_0^{u_e(l)} \frac{EA}{l}\,u(l)\,du = \frac{1}{2}\frac{EA}{l}\,u_e(l)^2 = \frac{1}{2}P_e\,u_e(l)$$

and as "*internal energy*" the integral

$$W_i = \int_V \left[\int_0^{\varepsilon_e} \sigma\,d\varepsilon\right] dV = \int_0^l \int_0^\varepsilon \sigma\,d\varepsilon\,A\,dx = \frac{1}{2}\int_0^l \frac{N_e^2}{EA}\,dx$$

where

$$\sigma = E\varepsilon, \quad \varepsilon = \frac{du}{dx}, \quad \varepsilon_e = \frac{du_e}{dx}, \quad N_e = EA\,u_e(x), \quad dV = A\,dx$$

and u and u_e respectively are the solutions of the bvp (2.27). Because u_e is a solution of the bvp the two integrals have the same value

$$W_e = W_i \quad \text{(first identity)}$$

that is the external eigenwork is equal to the internal energy.

This result is (if we place all terms on one side) just the expression $1/2$ $G(u,u) = 0$.

2.6 Equilibrium

If we translate a body or let it perform an infinitesimal rotation then the elastic body performs a rigid-body movement. The displacement vector r of such a rigid-body movement has the form

$$r = a + b \times x$$

where a is the translation vector and the vector b the axis of the infinitesimal rotation.

The components of r are

$$r_i = a_i + e_{ijk}\,b_j\,x_k$$

Consequently the rigid-body movements of a plate are

$$r = a_3 + b_1\,x_2 - b_2\,x_1$$

and of a beam

$$r = a_3 - b_2\,x_1$$

and of a bar

$$r = a_1$$

The rigid-body movements are homogeneous solutions of the governing differential equations

$$-(EA\,r')' = 0, \quad (EI\,r'')'' = 0, \quad K\Delta\Delta\,r = 0, \quad \boldsymbol{L}\boldsymbol{r} = \boldsymbol{0}$$

and their internal actions are zero:

Elastic plate or body $\tau(\boldsymbol{r}) = \boldsymbol{0}$

Kirchhoff plate $M_n(r) = V_n(r) = M_{nt}(r) = 0$

Beam $V(r) = M(r) = 0$

Bar $N(r) = 0$

In addition to this we have

$$E(r, r) = 0 \tag{2.28}$$

and

$$E(r, u) = 0 \quad \forall\, u \in C^m \tag{2.29}$$

which means that the internal energy of a rigid-body movement is zero, Eq. (2.28), and the strain energy ("*the energy inner product*") between a rigid-body movement and an arbitrary function $u \in C^m(\bar{\Omega})$ is zero as well, Eq. (2.29).

Besides, it is possible to define the rigid-body movements r by Eq. (2.28) because we have in the case of bars, beams and Kirchhoff plates

$$E(u, u) = 0 \;\leftrightarrow\; u \text{ is a rigid-body movement} \tag{2.30}$$

The rigid-body movements are exactly the functions at which the functional $F(u) = 1/2\, E(u, u)$, the internal energy, attains the value zero. In the case of elastic plates or bodies the same conclusion (from left to right) is not so easily reached. In addition to $E(\boldsymbol{u}, \boldsymbol{u}) = 0$ we must know that $\boldsymbol{u}$ is zero on at least a small straight part of the boundary. If this is guaranteed then we may draw, with the help of Korn's inequalities, see [V 1], the conclusion that $\boldsymbol{u}$ is a rigid-body movement.

Mathematically, the rigid-body movements are polynomials. Hence, they belong to $C^\infty(\bar{\Omega})$ and, a fortiori, to $C^m(\bar{\Omega})$. Substituting such a rigid-body movement r together with a function u from C^{2m} into the first identity renders the following important results.

Corollary 5, Equilibrium

p: $u,\, w \in C^{2m}, r$ *rigid-body movement*

q: *Bar*

$$G(u, r) = \int_0^l -(EA\,u')'\, r\, dx + [Nr]_0^l = 0 \tag{2.31}$$

Beam

$$G(w, r) = \int_0^l (EIw'')'' \, r \, dx + [Vr - Mr']_0^l = 0 \tag{2.32}$$

Kirchhoff plate

$$G(w, r) = \int_\Omega K \Delta\Delta w \, r \, d\Omega + \int_\Gamma \left(V_n(w) \, r - M_n(w) \frac{\partial r}{\partial n} \right) ds + [[M_{nt}(w) \, r]] = 0 \tag{2.33}$$

Elastic plate or body

$$G(\boldsymbol{u}, r) = \int_\Omega -\boldsymbol{L}\boldsymbol{u} \cdot \boldsymbol{r} \, d\Omega + \int_\Gamma \tau(\boldsymbol{u}) \cdot \boldsymbol{r} \, ds = 0 \tag{2.34}$$

These equations are the equilibrium conditions of structural mechanics:

The forces which act on a structural element are in equilibrium if and only if the work they perform when acting through all possible rigid-body movements is zero.

The important point is: every function $u \in C^{2m}(\bar{\Omega})$ satisfies this condition! In other words, every smooth function is "in equilibrium", e.g. every polynomial.

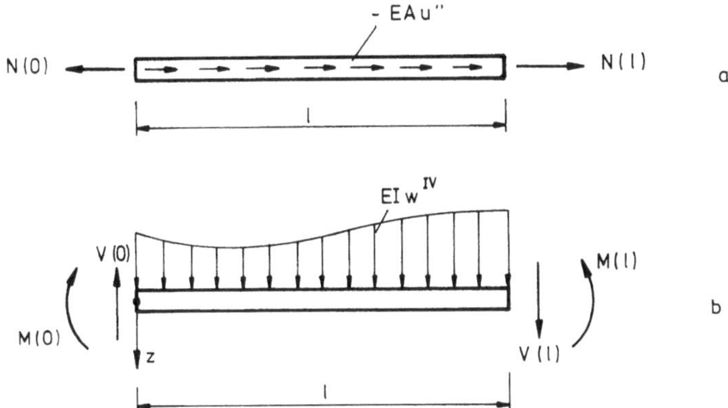

Figure 2.7

Examples:

a) Bar: Let $r = 1$ and $u(x) = 7x + x^2 + 0.1 x^4$ then Eq. (2.31) becomes ($EA = c.$)

$$\int_0^l -EA u'' r \, dx + N(l) - N(0) = \int_0^3 -EA (2 + 1.2 x^2) \, dx + 23.8 \, EA$$

$$-7 EA = (-16.8 - 23.8 - 7) \, EA = 0$$

that is there is equilibrium between the distributed forces, $-EA u'' = -EA (2 + 1.2 x^2)$ of the polynomial displacement $u(x)$ and its end forces $N(0) = EA u'(0)$, $N(l) = EA u'(l)$, see Fig. 2.7a.

b) Beam: Let $r = 1$ and $w(x) = -(36x^2 + 12x^3 + x^4)(72EI)^{-1}$ then Eq. (2.32) becomes

$$\int_0^l EIw^{IV} dx + V(l) - V(0) = \int_0^3 -\frac{1}{3} dx + 2 - 1 = -1 + 2 - 1 = 0$$

that is there is equilibrium between the distributed forces $EIw^{IV} = -0.33$ and the end forces $V(0) = -EIw'''(0)$, $V(l) = -EIw'''(l)$.

Finally, let $r = x$, a rotation of the beam, and w the same deflection then we obtain

$$\int_0^l EIw^{IV} x\, dx + V(l)l - M(l) + M(0) = \int_0^3 -\frac{1}{3} x\, dx + 2 \cdot 3 - 5.5 + 1$$
$$= -1.5 + 6 - 5.5 + 1 = 0$$

that is the moments of the polynomial deflection w are in equilibrium, see Fig. 2.7b.

2.7 Summation of Work and Energy

A well known trait of linear systems is the principle of superposition. If w_1 is the deflection under the action of the load p_1 and w_2 the deflection under the action of the load p_2 then $w_1 + w_2$ is the deflection if p_1 and p_2 act simultaneously.

But this principle does not apply when we add work. The reason is that the identities G and B are bilinear forms and their expansion—because that it is what we do when we add work or energy: we expand the identities—is not "linear", see Eq. (1.16).

If we evaluate, e. g., the first identity at two superimposed deflections $w = w_1 + w_2$ and $\hat{w} = \hat{w}_1 + \hat{w}_2$ then we obtain

$$G(w_1 + w_2, \hat{w}_1 + \hat{w}_2) = G(w_1, \hat{w}_1) + G(w_1, \hat{w}_2) + G(w_2, \hat{w}_1) + G(w_2, \hat{w}_2)$$

$$(2.35)$$

This rule applies to eigenwork as to virtual work though we modify it in structural mechanics (without changing the mathematics) if we apply it to eigenwork as we shall demonstrate next.

Assume a beam is loaded with a distributed load p_1 (deflection w_1) and then with a distributed load p_2 (deflection w_2), see Fig. 2.8. What is the amount of internal and external work done during the whole process?

In structural mechanics the argument runs as follows:
First the distributed load p_1 acts alone,

$$W_1^e = \frac{1}{2} \int p_1 w_1\, dx, \qquad W_1^i = \frac{1}{2} \int \frac{M_1^2}{EI}\, dx$$

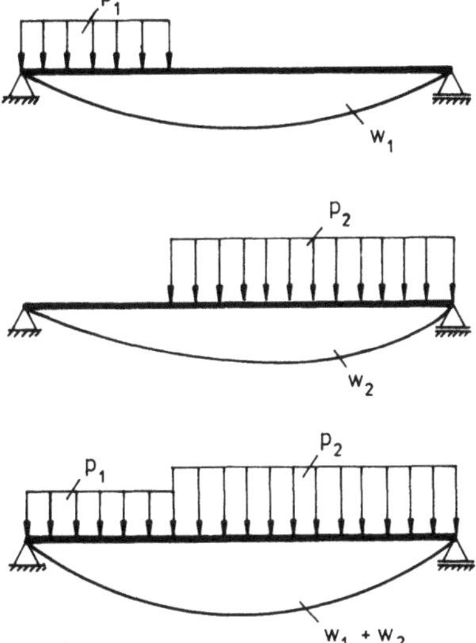

Figure 2.8

then we apply, in addition, the distributed load p_2. This load acting through its own deflection contributes eigenwork

$$W_2^e = \frac{1}{2} \int p_2\, w_2\, dx, \qquad W_2^i = \frac{1}{2} \int \frac{M_2^2}{EI}\, dx$$

but, in addition, the load p_1 which was present before we applied the load p_2 performs virtual work by acting through the deflection w_2,

$$W_{1,2}^e = \int p_1\, w_2\, dx, \qquad W_{1,2}^i = \int \frac{M_1 M_2}{EI}\, dx$$

The sum of all these single contributions is

$$W^e - W^i = \frac{1}{2} \int p_1\, w_1\, dx + \frac{1}{2} \int p_2\, w_2\, dx + \int p_1\, w_2\, dx$$

$$- \frac{1}{2} \int \frac{M_1^2}{EI}\, dx - \frac{1}{2} \int \frac{M_2^2}{EI}\, dx - \int \frac{M_1 M_2}{EI}\, dx = 0$$

which is equivalent with the expansion

$$\frac{1}{2} G(w_1 + w_2, w_1 + w_2) = \frac{1}{2} G(w_1, w_1) + \frac{1}{2} G(w_2, w_2) + G(w_1, w_2) \qquad (2.36)$$

If we compare this with Eq. (2.35) then, on first glance, something seems to be amiss. The answer lies in the equation

$$G(w_1, w_2) = \frac{1}{2}\,G(w_1, w_2) + \frac{1}{2}\,G(w_2, w_1) \tag{2.37}$$

Substituting the right-hand side of this equation into Eq. (2.36) we obtain

$$\frac{1}{2}\,G(w_1 + w_2, w_1 + w_2) = \frac{1}{2}\,G(w_1, w_1) + \frac{1}{2}\,G(w_2, w_2) + \frac{1}{2}\,G(w_1, w_2)$$
$$+ \frac{1}{2}\,G(w_2, w_1) = 0$$

which, if we multiply it with the factor 2, is just Eq. (2.35), that is the argument used in structural mechanics is correct.

It remains to prove our auxiliary lemma, Eq. (2.37).

To this end remember that both deflections, w_1 and w_2, satisfy the boundary conditions. Hence, the boundary integrals in the first identities $G(w_1, w_2)$ and $G(w_2, w_1)$ vanish.

$$G(w_1, w_2) = \int p_1 w_2 \, dx - \int \frac{M_1 M_2}{EI} \, dx = 0$$

$$G(w_2, w_1) = \int p_2 w_1 \, dx - \int \frac{M_2 M_1}{EI} \, dx = 0$$

And because the strain energy is symmetric

$$\int \frac{M_1 M_2}{EI} \, dx = \int \frac{M_2 M_1}{EI} \, dx$$

the external virtual work must be symmetric too.

$$\int p_1 w_2 \, dx = \int p_2 w_1 \, dx$$

Hence, the two expressions $G(w_1, w_2)$ and $G(w_2, w_1)$ coincide in each term, that is Eq. (2.37) is correct.

2.8 Rigid Supports and free Boundaries

The identities are sums of domain and boundary integrals. (We consider also $[Nu]$ or $[Vw - Mw']$ 'boundary integrals'). The integrands within the boundary integrals are pairs of conjugated quantities

Bar: u, N

Beam: w, V and w', M

Kirchhoff plate: $\qquad w, V_n$ and $\dfrac{\partial w}{\partial n}, M_n$

Elastic plate and body: $\tau(\boldsymbol{u}), \boldsymbol{u}$

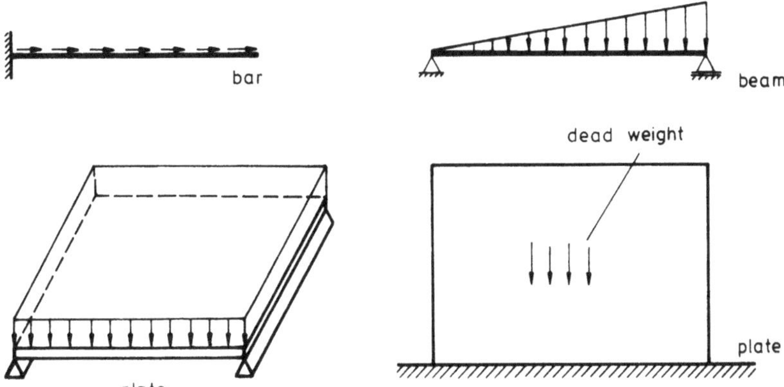

Figure 2.9

and it is obviously a rule in structural mechanics that on any part of the boundary of a plate, a beam etc. exactly one of two conjugated quantities is prescribed, never both at the same time. We call a *bvp* which satisfies this condition a *regular bvp*.

The *bvps* associated with the structures in Fig. 2.9, e. g., are regular. In addition these structures rest on rigid supports and no loads act at their free boundaries, the boundaries which are free to move. As a consequence all the boundary conditions are homogeneous (zero). Such *bvps* are called half-homogeneous *bvps*.

If the load in the domain would be zero too we would speak of a homogeneous *bvp*. Such *bvps* normally have, except for problems in stability theory, only the trivial solution $u = 0$.

2.9 Elastic Supports

The integrals in the first identity can be termed internal work (energy) or external work as, e. g., in the case of a beam

$$G(w, \hat{w}) = \underbrace{\int_0^l (EI\,w'')''\,\hat{w}\,dx + [V\,\hat{w} - M\,\hat{w}']_0^l}_{\text{external work}} - \int_0^l \underbrace{\frac{M\,\hat{M}}{EI}\,dx}_{\text{internal work}} = 0 \qquad (2.38)$$

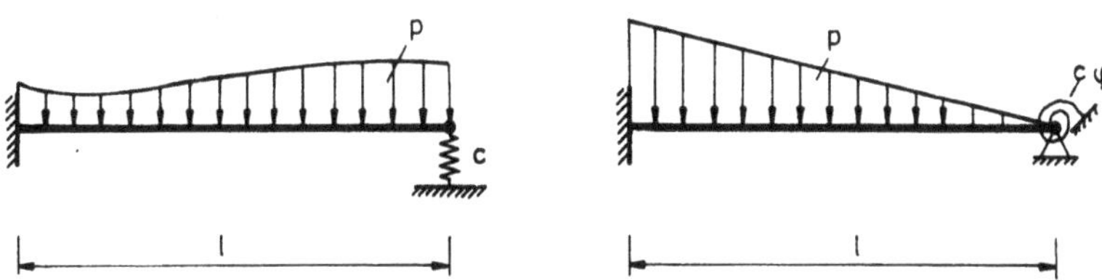

Figure 2.10

This classification is modified if the structure rests on one or more elastic supports.

Consider the two beams in Fig. 2.10. Their deflections are the solutions of the two *bvps*

$$(EI\,w'')'' = p, \; 0 < x < l, \qquad (EI\,w'')'' = p, \; 0 < x < l$$

$$w(0) = w'(0) = M(l) = 0, \qquad w(0) = w'(0) = w(l) = 0$$

$$V(l) = -c\,w(l), \qquad\qquad M(l) = c_\varphi\,w'(l)$$

Let w be the pertinent deflection and $\hat{w}$ an admissible virtual displacement (a displacement is admissible if it satisfies the geometric boundary conditions and if it is in $C^2[0,l]$) then Eq. (2.38) becomes

$$G(w, \hat{w}) = \int_0^l p\,\hat{w}\,dx - c\,w(l)\,\hat{w}(l) - \int_0^l \frac{M\,\hat{M}}{EI}\,dx = 0$$

$$G(w, \hat{w}) = \int_0^l p\,\hat{w}\,dx - c_\varphi\,w'(l)\,\hat{w}'(l) - \int_0^l \frac{M\,\hat{M}}{EI}\,dx = 0$$

The work done in the springs

$$-c\,w(l)\,\hat{w}(l) \quad \text{and} \quad -c_\varphi\,w'(l)\,\hat{w}'(l)$$

is classified, the negative sign suggests this already, as internal work. Thus the statement "eigenwork = int. energy" reads

$$\frac{1}{2}\int_0^l p\,w\,dx = \frac{1}{2}c\,w(l)\,w(l) + \frac{1}{2}\int_0^l \frac{M^2}{EI}\,dx$$

$$\frac{1}{2}\int_0^l p\,w\,dx = \frac{1}{2}c_\varphi\,w'(l)\,w'(l) + \frac{1}{2}\int_0^l \frac{M^2}{EI}\,dx$$

2.10 The Mathematical Basis of Structural Mechanics

It must have become clear by now that the mathematics of structural mechanics is of a uniform nature. To emphasize this uniform nature further we introduce a special notation for integrals and a series of symbols for the differential operators. The use of these symbols will be, mainly, restricted to this section, though we shall employ the notation later, on occasion, as a shorthand to sketch or illustrate something.

Integrals are denoted by

$$(f,g) = \int_\Omega f g\,d\Omega \qquad [f,g] = \int_\Gamma f g\,ds$$

Parantheses stand for domain integrals, brackets for boundary integrals. If Ω is an intervall, $\Omega = [a,b]$, then the bracket, naturally, retains its original meaning

$$[f,g] = f(b)g(b) - f(a)g(a)$$

If f and g are vector-valued functions or F and G matrix-valued functions then

$$(f,g) = \int_\Omega f \cdot g \, d\Omega = \int_\Omega f_i g_i \, d\Omega, \quad [f,g] = \int_\Gamma f \cdot g \, ds = \int_\Gamma f_i g_i \, ds$$

$$(F,G) = \int_\Omega F \cdot G \, d\Omega = \int_\Omega F_{ij} G_{ij} \, d\Omega, \quad [F,G] = \int_\Gamma F \cdot G \, ds = \int_\Gamma F_{ij} G_{ij} \, ds$$

The symbols

$$D; \partial^0, \partial^1 \ldots \partial^{2m-1}$$

denote differential operators. The small ∂^i are the boundary operators and the capital D is the operator in the domain.

Bars

$$D = -\left(EA(\ )'\right)'; \quad \partial^0 = \text{identity}, \quad \partial^1 = EA \frac{d}{dx} = N(\ )$$

Beams

$$D = \left(EI(\ )''\right)''; \quad \partial^0 = \text{id.}, \quad \partial^1 = \frac{d}{dx}, \quad \partial^2 = -EI(\ )'' = M(\ ),$$

$$\partial^3 = -\left(EI(\ )''\right)' = V(\ )$$

Kirchhoff plates

$$D = K\Delta\Delta; \quad \partial^0 = \text{id.}, \quad \partial^1 = \frac{\partial w}{\partial n}, \quad \partial^2 = M_n(\ ), \quad \partial^3 = V_n(\ )$$

Elastic plates and bodies

$$D = -L \ (\text{see Eq. } (1.38)); \quad \partial^0 = \text{id.}, \quad \partial^1 = \tau(\ )$$

The functions $\partial^i u, 0 \leqslant i < m$ are the *"displacements"* and the functions $\partial^i u, m \leqslant i \leqslant 2m-1$ the *"forces"*.

We call two boundary functions $\partial^i u, \partial^j u$ conjugated if their indices are conjugated, that is if $i + j = 2m - 1$.

The symbol

$$\varepsilon(u) = \{u; u_{,x_1}; u_{,x_2} \ldots\}$$

denotes the set of all the derivatives of u which appear in the internal energy $1/2$ $E(u,u)$. We call $\varepsilon(u)$ the (mathematical) strain of a displacement u.

With these new symbols the first identity of the different operators can be written uniformly as

$$G(u, \hat{u}) = (Du, \hat{u}) - \sum_{i=1}^{m} (-1)^i [\partial^{2m-i} u, \partial^{i-1} \hat{u}] - E(u, \hat{u}) = 0$$

and the second identity as

$$B(u, \hat{u}) = (Du, \hat{u}) - \sum_{i=1}^{2m} (-1)^i [\partial^{2m-i} u, \partial^{i-1} \hat{u}] - (u, D\hat{u}) = 0$$

These bilinear forms are sums of domain and boundary integrals. The integrands in these integrals, the functions

$$\{Du, \varepsilon(u); \partial^0 u, \partial^1 u \ldots \partial^{2m-1} u\}$$

constitute a set which we call *Green's data* of a function u, see Fig. 2.11.

The displacements of the structures are solutions of *bvps*. The single line

$$Du = p, \quad \partial^\lambda u = f_\lambda, \quad \lambda \in \Lambda$$

paraphrases the multitude of these *bvps*. The term p is the external force in the domain and f_λ is a prescribed function or number. The set Λ contains a selection of m indices from the full set $\{0, 1, 2 \ldots 2m-1\}$.

We assume throughout this book that the *bvp* in question is regular (for a definition see section 2.8).

The solution of a *bvp* is indicated by a label s as in u_s or w_s.

The boundary conditions are either geometric (*essential*) conditions or static (*natural*) conditions.

A function u is called a *geometrically admissible virtual displacement* if the function satisfies the homogeneous geometric boundary conditions of the problem.

A function u is called a *statically admissible virtual displacement* if it satisfies the homogeneous static boundary conditions and is a homogeneous solution of the governing differential equation, $-EA u'' = 0$, $EI w^{IV} = 0$ etc.

These definitions are independent of the status of the original boundary conditions, that is whether they are homogeneous or non-homogeneous. The test for a virtual displacement are the boundary conditions in their homogeneous form.

Though the mathematical analysis has not yet fully advanced we summarize already at this point the principal features of the mathematical structure of statics of continua.

The common basis

1) *The degree of all operators is even, $2m = 2$ or 4 (by degree we mean throughout the text the maximum order of the derivatives in an operator).*

2) *To every operator D belongs a set of $2m$ boundary operators $\partial^0, \partial^1 \ldots \partial^{2m-1}$ and a symmetric strain energy $E(u, u)$.*

3) *To every operator D belong two identities*

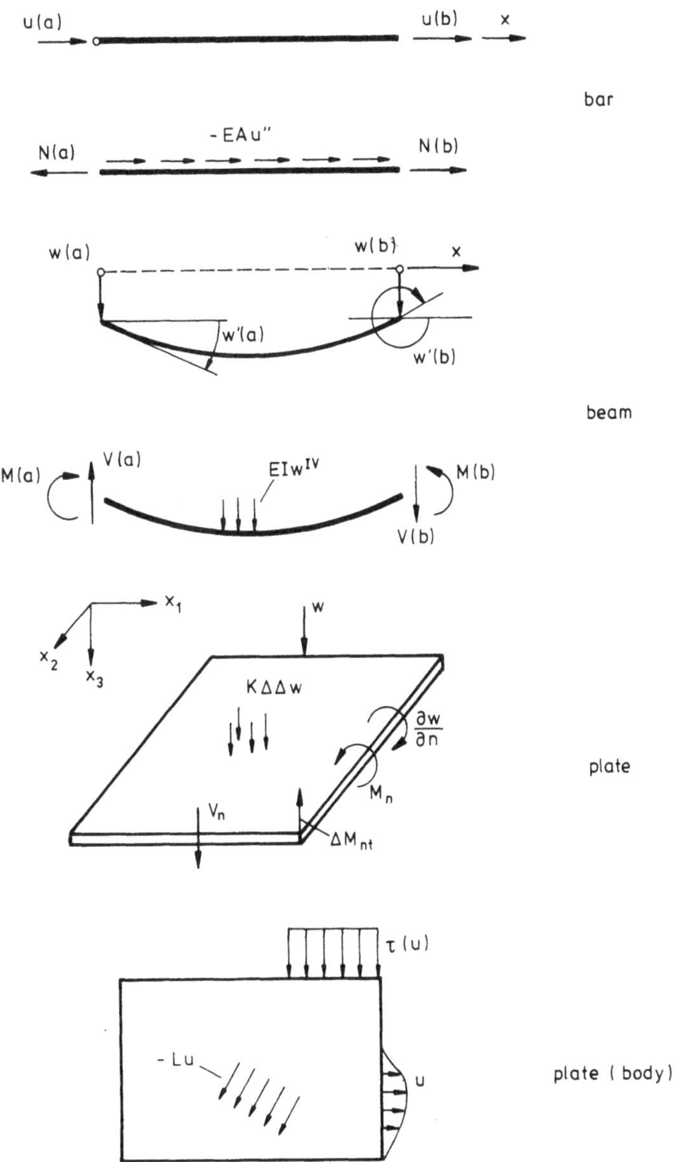

Figure 2.11

$$G(u, \hat{u}) = (Du, \hat{u}) - \sum_{i=1}^{m} (-1)^i \left[\partial^{2m-i}u, \partial^{i-1}\hat{u}\right] - E(u, \hat{u}) = 0$$

$$B(u, \hat{u}) = (Du, \hat{u}) - \sum_{i=1}^{2m} (-1)^i \left[\partial^{2m-i}u, \partial^{i-1}\hat{u}\right] - (u, D\hat{u}) = 0$$

(In the case of Kirchhoff plates the work of the concentrated forces at the corners, $[[\ldots]]$ *must be added).*

4) *The operators are self-adjoint,* $(Du, \hat{u}) = (u, D\hat{u})$, *on* $C_0^{2m}(\Omega)$.

5) *The functions in the boundary integrals*

$$[\partial^{2m-i}u, \partial^{i-1}\hat{u}]$$

are conjugated because

$$(2m-i)+(i-1)=2m-1$$

and all the integrals which appear in G and B can be interpreted as expressions of mechanical work

$$(Du, \hat{u}) = (force \times displacement)$$

$$E(u, \hat{u}) = \int \varepsilon_{ij}\,\hat{\sigma}_{ij}\,d\Omega$$

$$[\partial^{2m-i}u, \partial^{i-1}\hat{u}] = force \times displacement$$

6) *Every function u in $C^{2m}(\bar{\Omega})$ is "in equilibrium", that is its data Du, $\partial^i u$, $m \leqslant i \leqslant 2m-1$, the forces, satisfy the equilibrium conditions. (If the domain is unbounded then this must not be true).*

7) *The boundary operators which appear in the formulation of the bvps are the same operators as the operators in the boundary integrals of the first and second identity.*

8) *We know the fundamental solutions (a concentrated force acts on the structure) of all operators D, see chapter 5.*

9) *There are two ways to construct integral representations of a function if Green's data*

$$Du, \varepsilon(u); \partial^0 u, \ldots \partial^{2m-1}u$$

of this function are known, see chapter 6.

10) *The boundary data*

$$\partial^0 u, \partial^1 u \ldots \partial^{2m-1}u$$

of a function $u \in C^{2m}(\bar{\Omega})$ satisfy $2m$ compatibility conditions on the boundary, see chapter 6.

Everything said in these ten paragraphs applies, naturally, to all the structural elements whose equations are listed in section 1.9, not only to bars, beams, Kirchhoff plates and elastic plates and bodies. It is only that these elements are the best known and they, therefore, best illustrate the theory.

2.11 The Space $C^{2m}(\bar{\Omega})$ and its Limitations

Every structure displaces under the action of an external load. If the load p is continuous, $p \in C(\Omega)$, then the displacement u belongs to $C^{2m}(\Omega)$. This is a simple consequence of the equation $Du = p$.

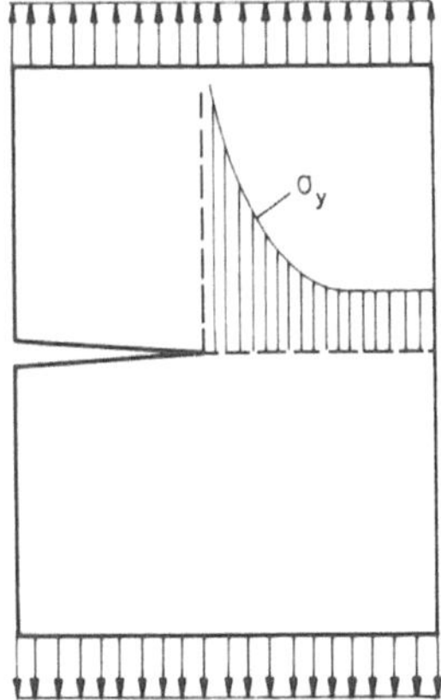

Figure 2.12

But we need to know more: is the displacement $2\,m$-times differentiable up to the boundary, that is does u belong to $C^{2m}(\bar{\Omega})$?

Only if this is guaranteed may we apply integration by parts, the basic tool for all our analysis.

What could cause trouble are irregular points on the boundary as in the case of the elastic plate in Fig. 2.12. The displacement field $\boldsymbol{u}$ satisfies in the interior the homogeneous equation $-\boldsymbol{L}\boldsymbol{u}=\boldsymbol{0}$, hence, belongs in the interior to $C^2(\Omega)$ but it does not belong to $C^2(\bar{\Omega})$ because the first derivatives, the stresses, become singular at the crack tip.

Problems such as these indicate that the class $C^{2m}(\bar{\Omega})$ which we most use in this book is too narrow because there are meaningful practical problems whose solutions belong to classes wider (i.e. not so restrictive) as $C^{2m}(\bar{\Omega})$.

It is certainly always possible to consider the crack tip rounded out and to replace, thus, the singular problem by a "nearly singular" problem.

But, if we are honest, we face a real problem: what is the expression of the principle of virtual work or of Betti's principle if the stresses become singular and consequently the displacement field no longer belongs to $C^{2m}(\bar{\Omega})$? Over what function spaces do we define Green's first and second identity?

The answer is: we replace $C^{2m}(\bar{\Omega})$ and $C^m(\bar{\Omega})$ by the Sobolev spaces $H^{2m}(\Omega)$ and $H^m(\Omega)$ introduced in section 1.8.

Recall that Q, the set of all rational numbers, $1/2$, $3/4$, $5/6$ etc., is not complete with respect to the arithmetic operations. The equation

$$x^2 = 2$$

has no solution in Q. Hence, the set Q is too narrow for practical purposes and must be extended by adding irrational numbers as, e. g., the square root of 2. In the same sense the space $C^{2m}(\bar{\Omega})$ must be extended to the Sobolev space $H^{2m}(\Omega)$.

The need for such an extension is also felt if we consider the problem of finding a number x which renders the function

$$\Pi(x) = \frac{1}{2} x^2 - x\sqrt{2}$$

a minimum.

If we search among all rational numbers then there is no such solution. Given any rational number x there is yet another rational number $\tilde{x}$ with the property that

$$\Pi(\tilde{x}) < \Pi(x)$$

The problem is only solved if we admit irrational numbers. Then there is a solution, namely $x = \sqrt{2}$.

In the same sense it would be impossible to find a displacement field $\boldsymbol{u}$ in $C^2(\bar{\Omega})$ which minimizes the potential energy

$$\Pi_1(\boldsymbol{u}) = \frac{1}{2} E(\boldsymbol{u}, \boldsymbol{u}) - \int_\Gamma \boldsymbol{t} \cdot \boldsymbol{u}\, ds$$

of the cracked elastic plate. But this problem should have a solution in $H^1(\Omega)$.

The rules of arithmetic do not change if we switch from Q to $\mathbb{R}$. Just as well Green's integral identities would remain valid if we would switch from $C^{2m}(\bar{\Omega}) \times C^m(\bar{\Omega})$ to $H^{2m}(\Omega) \times H^m(\Omega)$ (if the domain Ω is sufficiently regular, — what Ω is in practice).

Though Sobolev spaces occupy a prominent place nowadays in mathematics (they are an indispensable tool when it comes to proofs concerning the existence of solutions and minima of functionals, see [F 1]) we abstained from actually performing this extension (at least in the main part of the text) because this step, if done correctly, is burdened with technicalities which easily distract the engineer.

Most of what we want to say in this book can be said with C^{2m} and C^m functions as well and if a displacement field is not in C^{2m} we always may assume that a little smoothing of the corners and jumps is sufficient to make it well behaved.

3 Continuous Beams, Trusses and Frames

In the previous chapter we formulated the principles of virtual work and Betti's principle for rather simple structures.

A structure, usually, does not consist of just one bar, one beam, but of many single bars, beams, plates, slabs, columns etc. The structural elements considered in chapter 2 are only segments of the composite real-life structures.

Hence, we face the question: how do we extend the principles of virtual work and Betti's principle to composite structures?

This extension corrolates with the proof of Mohr's integral.

$$1\,\delta_j = \int \left[\frac{M\hat{M}}{EI} + \frac{M_x \hat{M}_x}{GI_p} + \frac{N\hat{N}}{EA} + \varkappa\, \frac{V\hat{V}}{GA} + \alpha_T \left(N\hat{T} + M\,\frac{\Delta\hat{t}}{h} \right) \right] dx$$
$$+ \, C_F \hat{c}_F - C_L \hat{c}_L$$

It is certainly possible to verify Mohr's integral for a simple beam by calculating $w(x)$ once analytically and then with Mohr's integral. But how do we check Mohr's integral if the structure is a 20-story building?

Such a building, certainly, consists of more than 3000 members. Each of these deforms and the statement is that the sum of all these influences at an arbitrary point and in an arbitrary direction is just the right-hand side of Mohr's integral.

How do we prove this?—By adding zeros! The equation

$$0 + 0 + 0 + \ldots + 0 = 0$$

is always correct, regardless of the number of zeros, 3000 or 10000, in it.

This equation is the key to the extension of the work and energy principles to composite structures.

The example of a simple two-span beam will best explain our approach.

3.1 Continuous Beams

Consider the continuous beam in Fig. 3.1

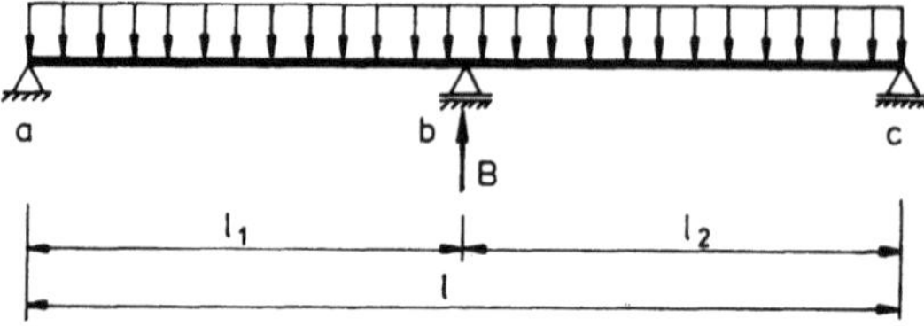

Figure 3.1

which consists of two single beams connected at the support b. The deflections, w_1 in the first span and w_2 in the second span, satisfy the differential equations

$$EI\, w_1^{IV} = p, \quad 0 < x < l_1, \qquad EI\, w_2^{IV} = p, \quad l_1 < x < l$$

and the boundary and interface conditions

support a $w_1 = 0, \; M_1 = 0$
$x = a$

support b $w_1 = 0, \; w_2 = 0, \; w_1' = w_2'$ (3.1)
$x = l_1$

$\qquad\qquad\qquad M_1 = M_2, \; V_1 + B = V_2$

support c $w_2 = 0, \; M_2 = 0$
$x = l$

Furthermore the deflection w_1 belongs to $C^4[0, l_1]$ and the deflection w_2 to $C^4[l_1, l]$. Hence, we obtain if we integrate over $[0, l_1]$ and $[l_1, l]$ respectively

$$G(w_1, w_1) = 0 \qquad G(w_2, w_2) = 0$$

Because both identities are zero their sum must be zero too

$$G(w_1, w_1) + G(w_2, w_2) = 0 + 0 = 0$$

or at full length

$$\int_0^{l_1} EI\, w_1^{IV}\, w_1\, dx + [V_1 w_1 - M_1 w_1']_0^{l_1} - \int_0^{l_1} \frac{M_1^2}{EI}\, dx$$

$$+ \int_{l_1}^{l} EI\, w_2^{IV}\, w_2\, dx + [V_2 w_2 - M_2 w_2']_{l_1}^{l} - \int_{l_1}^{l} \frac{M_2^2}{EI}\, dx = 0 \qquad (3.2)$$

The work done at the rigid supports at both ends of the beam is zero

$$\{-V_1\,w_1 + M_1\,w_1'\}|_{x=0} = 0, \qquad \{V_2\,w_2 - M_2\,w_2'\}|_{x=l} = 0$$

and also at the mid-support because the interface condition, $M_l = M_r$, is satisfied, see Eq. (3.1).

$$V_1\,w_1 - M_1\,w_1' - V_2\,w_2 + M_2\,w_2' = V_1\,0 - V_2\,0 - (M_1 - M_2)\,w_1' = 0$$

Hence, Eq. (3.2) reduces to

$$\int_0^{l_1} p\,w_1\,dx - \int_0^{l_1} \frac{M_1^2}{EI}\,dx + \int_{l_1}^{l} p\,w_2\,dx - \int_{l_1}^{l} \frac{M_2^2}{EI}\,dx = 0 \tag{3.3}$$

or if we no longer distinguish between w_1 and w_2 and multiply with $1/2$

$$\frac{1}{2}\int_0^{l} p\,w\,dx - \frac{1}{2}\int_0^{l} \frac{M^2}{EI}\,dx = 0$$

This is the energy balance, "eigenwork = int. energy", for the two-span beam.
As a second example consider the continuous beam in Fig. 3.2.

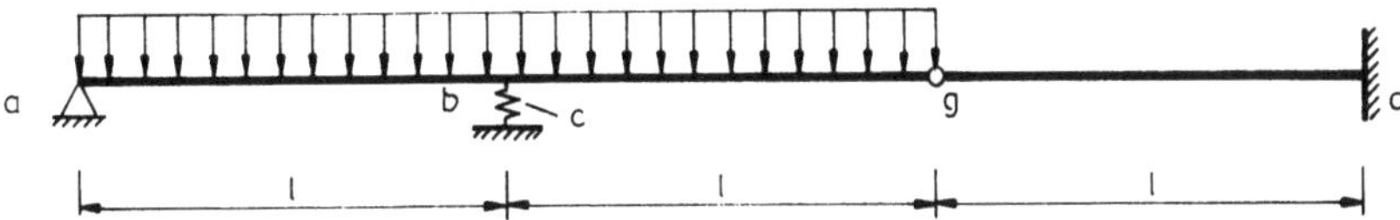

Figure 3.2

The single deflections w_i, $i = 1, 2, 3$ satisfy the differential equations

$$EI\,w_i^{IV} = p_i$$

and the boundary and interface conditions

support a $\qquad w_1 = M_1 = 0$ $\qquad\qquad\qquad\qquad \hat{w}_1 = 0$
$x = 0$

support b $\qquad w_1 = w_2 \quad M_1 = M_2$ $\qquad\qquad \hat{w}_1 = \hat{w}_2$
$x = l$ $\qquad\quad\; w_1' = w_2' \quad V_1 + c\,w_1 = V_2$ $\qquad \hat{w}_1' = \hat{w}_2'$

pin g $\qquad\quad\; w_2 = w_3 \quad V_2 = V_3$ $\qquad\qquad\quad \hat{w}_2 = \hat{w}_3$
$x = 2l$ $\qquad\;\; M_2 = M_3 = 0$

support c $\qquad w_3 = 0$ $\qquad\qquad\qquad\qquad\quad\;\; \hat{w}_3 = 0$
$x = 3l$ $\qquad\quad\; w_3' = 0$ $\qquad\qquad\qquad\qquad\quad \hat{w}_3' = 0$

Let $\hat{w} = \{\hat{w}_1, \hat{w}_2, \hat{w}_3\}$ be a set of admissible virtual displacements, that is $\hat{w}$ satisfies the geometric boundary and interface conditions and belongs in each span to C^2.

The deflection $w = \{w_1, w_2, w_3\}$ of the beam itself belongs in each span to C^4 (because the load p is smooth in each span) and, hence, we have if we couple w and $\hat{w}$

$$G(w_1, \hat{w}_1) + G(w_2, \hat{w}_2) + G(w_3, \hat{w}_3) = 0 + 0 + 0 = 0$$

or

$$\int_0^l p\,\hat{w}_1\,dx + \int_l^{2l} p\,\hat{w}_2\,dx + [V_1\,\hat{w}_1 - M_1\,\hat{w}_1']_0^l + [V_2\,\hat{w}_2 - M_2\,\hat{w}_2']_l^{2l}$$

$$+ [V_3\,\hat{w}_3 - M_3\,\hat{w}_3']_{2l}^{3l} - \int_0^l \frac{M_1\,\hat{M}_1}{EI}\,dx - \int_l^{2l} \frac{M_2\,\hat{M}_2}{EI}\,dx - \int_{2l}^{3l} \frac{M_3\,\hat{M}_3}{EI}\,dx = 0$$

Due to the rigid supports the virtual work done at the ends of the continuous beam is zero. The virtual work done at the support b, $x = l$, is

$$V_1\,\hat{w}_1 - M_1\,\hat{w}_1' - V_2\,\hat{w}_2 + M_2\,\hat{w}_2' = V_1\,\hat{w}_1 - (V_1 + cw_1)\,\hat{w}_1 = -cw_1\,\hat{w}_1$$

and the virtual work done at the pin is

$$V_2\,\hat{w}_2 - M_2\,\hat{w}_2' - V_3\,\hat{w}_3 + M_3\,\hat{w}_3' = 0$$

Hence, we obtain

$$\int_0^{2l} p\,\hat{w}\,dx - cw(l)\,\hat{w}(l) - \int_0^{3l} \frac{M\hat{M}}{EI}\,dx = 0 \tag{3.4}$$

where we replaced w_i and $\hat{w}_i$ by w and $\hat{w}$.

Eq. (3.4) expresses the fact that the virtual external work

$$\delta W = \int_0^{2l} p\,\hat{w}\,dx$$

done by the load p acting through the virtual displacement $\hat{w}$ is equal to the virtual internal energy

$$\delta U = cw(l)\,\hat{w}(l) + \int_0^{3l} \frac{M\hat{M}}{EI}\,dx$$

We trust that these two simple examples have acquainted the reader with the basic idea and we shall, in what follows, formulate identities for trusses, the simplest composite structures, in a more systematic fashion.

3.2 Trusses

A truss is a structure, see Fig. 3.3, which consists of single bars connected at their ends by (frictionless) pins.

To analyze the truss we employ the following notation:

n $=$ total number of bars,

K $=$ total number of joints,

$\boldsymbol{P}^k$ $= \{P^k_x, P^k_y, P^k_z\}$ load vector at joint k,

l_i $=$ length of bar i,

u_i $=$ displacement function of bar i,

$\boldsymbol{\delta}^k$ $= \{\delta^k_x, \delta^k_y, \delta^k_z\}$ displacement vector of joint k,

$\boldsymbol{\delta}^{i,k}$ $= \{\delta^{i,k}_x, \delta^{i,k}_y, \delta^{i,k}_z\}$ displacement vector of bar i at the end which connects with joint k.

$\boldsymbol{N}^{i,k}$ $= \{N^{i,k}_x, N^{i,k}_y, N^{i,k}_z\}$ vector of normal force components of bar i at the end which connects with joint k.

I_k $=$ set of all indices i which belong to joint k.

$i \in I_k \;\leftrightarrow\;$ the bar i connects with joint k.

All vector components refer to a global system of Cartesian coordinates.

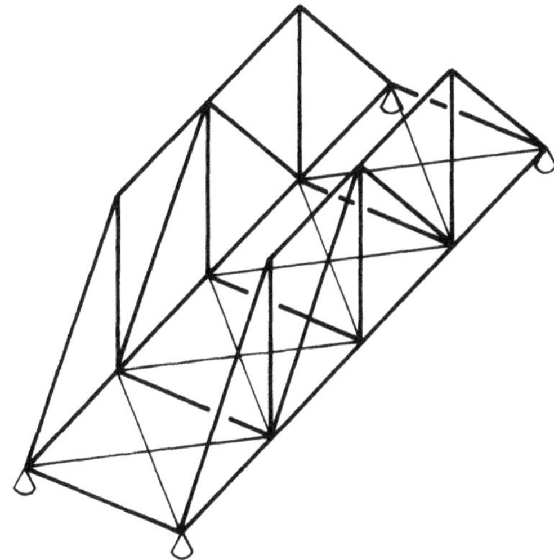

Figure 3.3

The set of all the displacements u_i of the single bars constitutes a vector

$$u = \{u_1, u_2 \ldots u_n\}^T$$

We say such a vector u belongs to C^q if each single component u_i belongs to $C^q[0, l_i]$ and we call the vector the displacement of the truss if:

a) the vector is *geometrically compatible*, i.e. the end displacements of the bars which connect with the same joint have the same value and u satisfies the construction constraints at the supports. In our notation the first condition reads

$$\delta_x^{i,k} = \delta_x^k, \qquad \delta_y^{i,k} = \delta_y^k, \qquad \delta_z^{i,k} = \delta_z^k \qquad i \in I_k$$

b) the vector u is *statically compatible*, i.e. the forces at each free joint are in equilibrium

$$P_x^k = R_x^k, \qquad P_y^k = R_y^k, \qquad P_z^k = R_z^k$$

(the forces R^k are defined below) and each single bar is in equilibrium, that is each displacement function u_i satisfies

$$-EA\, u_i'' = 0, \qquad 0 < x < l_i$$

Next, consider the element of the truss with label 1 and assume we formulate with two functions $\{u_1, \hat{u}_1\}$ from $C^2[0, l_1] \times C^1[0, l_1]$ defined along its axis the first identity

$$G(u_1, \hat{u}_1) = 0$$

Then we do the same with the element with label 2, and then with the element with label 3 and so on .... . All the single identities, the single zeros, can be added and the different functions $u_1, u_2, u_3, \ldots, u_n; \hat{u}_1, \hat{u}_2, \hat{u}_3, \ldots \hat{u}_n$ can be grouped as vectors u and $\hat{u}$. We obtain, thus, the result

$$p: \; u \in C^2, \; \hat{u} \in C^1$$

$$q: \; G(u, \hat{u}) = \sum_{i=1}^{n} G(u_i, \hat{u}_i)$$

$$= \sum_{i=1}^{n} \left\{ \int_0^{l_i} -EA\, u_i'' \,\hat{u}_i \, dx + [N_i \hat{u}_i]_0^{l_i} - \int_0^{l_i} \frac{N_i \hat{N}_i}{EA} \, dx \right\} = 0 \qquad (3.5)$$

(In what follows we shall neglect the subscript i on u_i and $\hat{u}_i$).

If the vector $\hat{u}$ is geometrically compatible (every admissible virtual displacement of the truss is, e.g., such a vector) the term

$$\sum_{i=1}^{n} [N\hat{u}]_0^{l_i} \qquad (3.6)$$

in the first identity $G(u, \hat{u}) = 0$ simplifies considerably.

For one i, for one single bar, the term

$$[N\hat{u}]_0^{l_i} = N(l_i)\,\hat{u}(l_i) - N(0)\,\hat{u}(0)$$

is the scalar product of the force and displacement vectors at both ends of the bar, see Fig. 3.4.

$$[N\hat{u}]_0^{l_i} = \boldsymbol{N}^{i,1} \cdot \hat{\boldsymbol{\delta}}^{i,1} + \boldsymbol{N}^{i,2} \cdot \hat{\boldsymbol{\delta}}^{i,2}$$

(let the nodes which lie opposite to the bar be numbered 1 and 2)

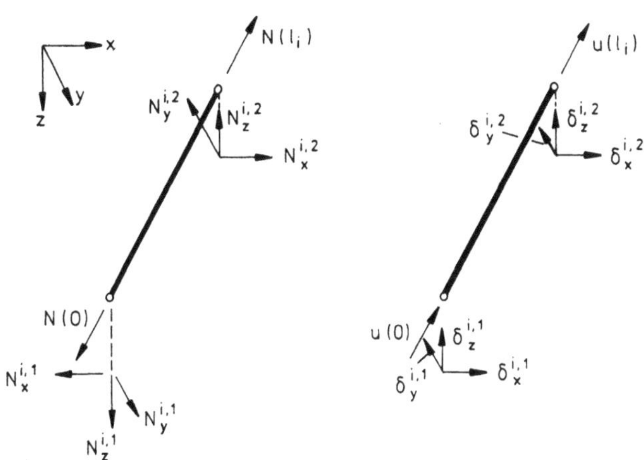

Figure 3.4

To form (3.6), that is to sum the work done at all the end cross sections of the single bars, we could start at joint 1 and count the work done at the faces which lie opposite to joint 1 then repeat this procedure at joint 2 etc. This would give

$$\sum_{i=1}^{n} [N\hat{u}]_0^{l_i} = \sum_{k=1}^{K} \sum_{i \in I_k} \boldsymbol{N}^{i,k} \cdot \hat{\boldsymbol{\delta}}^{i,k} \tag{3.7}$$

But if $\hat{\boldsymbol{u}} = \{\hat{u}_i\}$ is geometrically compatible then the bars at one joint all perform the same movement

$$\hat{\boldsymbol{\delta}}^{i,k} = \hat{\boldsymbol{\delta}}^{k} \qquad i \in I_k$$

and, therefore, Eq. (3.7) simplifies to

$$\sum_{i=1}^{n} [N\hat{u}]_0^{l_i} = \sum_{k=1}^{K} \left(\sum_{i \in I_k} \boldsymbol{N}^{i,k} \right) \cdot \hat{\boldsymbol{\delta}}^{k} = \sum_{k=1}^{K} \boldsymbol{R}^{k} \cdot \hat{\boldsymbol{\delta}}^{k} \tag{3.8}$$

where the resultant vector

$$R^k = \{R_x^k,\, R_y^k,\, R_z^k\}$$

is the sum of all the axial forces at the faces opposite to joint k. If an external force $P^k = \{P_x^k,\, P_y^k,\, P_z^k\}$ acts at joint k then the joint is in equilibrium if

$$P^k = R^k$$

Substituting now Eq. (3.8) into Eq. (3.5) we obtain the following result:

p: $u \in C^2$, $\hat{u} \in C^1$ and $\hat{u}$ is geometrically compatible

$$q:\ G(u,\hat{u}) = \sum_{i=1}^{n} \int_0^{l_i} - EA\, u''\, \hat{u}\, dx$$

$$+ \sum_{k=1}^{K} R^k \cdot \hat{\delta}^k - \sum_{i=1}^{n} \int_0^{l_i} \frac{N\hat{N}}{EA}\, dx = 0 \tag{3.9}$$

We call this bilinear form $G(u, \hat{u})$ the first identity for trusses. Repeating this equation with u and $\hat{u}$ changing places and subtracting the two equations, we obtain the second identity.

p: $u \in C^2$, $\hat{u} \in C^2$ and both geometrically compatible

$$q:\ B(u, \hat{u}) = G(u, \hat{u}) - G(\hat{u}, u) = 0$$

These identities are the mathematical basis of the work and energy principles for trusses.

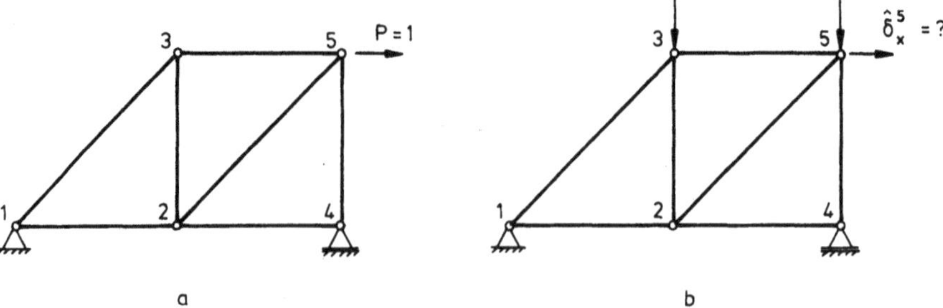

Figure 3.5

As an illustration of the application of the first identity $G(u, \hat{u}) = 0$ consider the plane truss in Fig. 3.5b loaded with nodal forces.

We want to know: how large is the horizontal displacement, δ_x^5, at joint 5 under the action of these forces?

To this end we load the same truss at joint 5 alone with a single horizontal force $P = 1$, see Fig. 3.5a, and we observe that the displacement $\hat{u}$ of the truss on the right is an admissible virtual displacement of the truss on the left.

The displacement $u = \{u_i\}$ of the truss on the left satisfies the equations

$$- E A_i u_i'' = 0 \qquad i = 1, 2 \ldots 7$$

and at joint 5 the horizontal component of the sum of all concurrent internal forces is equal to the external load

$$R_x^5 = 1$$

At all other joints either these components are zero, $R_x^k = 0$ or $R_z^k = 0$ or the conjugated virtual displacements are zero, $\delta_x^k = 0$ or $\delta_z^k = 0$. Hence, the work done at the joints, with the exception of joint 5, is zero and the first identity simplifies to

$$G(u, \hat{u}) = 1\,\delta_x^5 - \sum_{i=1}^{7} \int_0^{l_i} \frac{N \hat{N}}{E A}\, dx = 0$$

or

$$\delta_x^5 = \sum_{i=1}^{7} \int_0^{l_i} \frac{N \hat{N}}{E A}\, dx$$

This is the desired result.

Next, assume the nodal forces have disappeared but instead the temperature of the truss has increased by T degrees. How large is the horizontal displacement of joint 5?

The truss is statically determinate and, therefore, the temperature increase causes no internal actions, $E A\, \hat{u}_i' = 0$, that is, the single displacements are simply the integrals of the temperature strains $\hat{\varepsilon}_i = \alpha_T \hat{T}$ alone

$$\hat{u}_i = \alpha_T x \, \hat{T}$$

Evidently the displacement $\hat{u}$ caused by the change in the temperature in the truss on the right is an admissible virtual displacement of the truss on the left. With

$$\hat{N}^i = E A_i \frac{d\hat{u}_i}{dx} = E A_i\, \alpha_T \, \hat{T}$$

the first identity becomes

$$G(u, \hat{u}) = 1\,\delta_x^5 - \sum_{i=1}^{7} \int_0^{l_i} N\, \alpha_T \hat{T}\, dx = 0$$

or

$$\delta_x^5 = \alpha_T \hat{T} \sum_{i=1}^{7} \int_0^{l_i} N\, dx = \alpha_T\, \hat{T} \sum_{i=1}^{7} N_i l_i$$

Hence, the displacement of joint 5 is the integral of the normal forces in the truss on the left times $\alpha_T \hat{T}$.

3.3 Frames

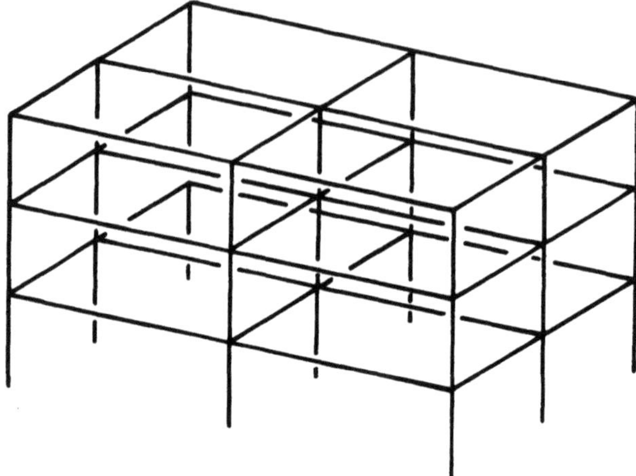

Figure 3.6

A frame consists of single frame elements. The displacements

$$u, v, v_\tau, w, w_\tau, \varphi$$

of a single element are governed by the differential equations

$$-(EAu')' = p_x, \quad (EI_z v'')'' = p_y, \quad (EI_y w'')'' = p_z$$

$$-(GI_p \varphi')' = m_x, \quad -\left(\frac{1}{\varkappa_y} G A v'_\tau\right)' = p_y, \quad -\left(\frac{1}{\varkappa_z} G A w'_\tau\right)' = p_z$$

The end displacements and end forces of the frame element form two vectors

$$S_l^{i,k} = \{N, V_y, V_z, M_x, M_y, M_z\}^T$$

$$\delta_l^{i,k} = \{u, v, w, \varphi_x, \varphi_y, \varphi_z\}^T$$

where the index i refers to the member and k to the particular joint which lies opposite to the end cross section at which the forces act and the displacements are observed.

The subscript l indicates that the components refer to a local system of coordinates. The transformation of the vectors S_l and δ_l from the local system into the global system of coordinates is done with transformation matrices T whose elements are the cosines of the Eulerian angles $\alpha^i, \beta^i, \gamma^i$.

$$S^{i,k} = T^i S_l^{i,k}, \quad \delta^{i,k} = T^i \delta_l^{i,k}$$

Each node has six degrees of freedom, three displacements and three rotations. These form the vector

$$\delta^k = \{\delta_x^k,\ \delta_y^k,\ \delta_z^k,\ \varphi_x^k,\ \varphi_y^k,\ \varphi_z^k\}$$

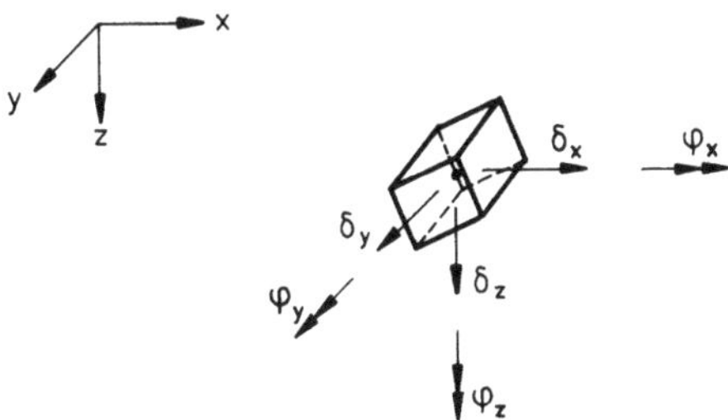

Figure 3.7

The sum of the end forces $S^{i,k}$ of all the elements which frame into one particular node k is the resultant force R^k

$$R^k = \sum_{i \in I_k} S^{i,k}$$

It has the components

$$R^k = \{R_x^k,\ R_y^k,\ R_z^k,\ R_{xx}^k,\ R_{yy}^k,\ R_{zz}^k\}$$

To describe the movements of one single member we need a set of six functions

$$u_i,\ v_i,\ v_{\tau_i},\ w_i,\ w_{\tau_i},\ \varphi_i$$

The sequence of all these $6 \times n$ functions forms the vector

$$u = \{u_1,\ v_1 \ldots \varphi_1,\ u_2,\ v_2 \ldots \varphi_n\}^T$$

and we write $u \in C^a$ if for each i holds

$$u_i,\ v_{\tau_i},\ w_{\tau_i},\ \varphi_i \in C^2,\quad v_i,\ w_i \in C^4$$

and we write $u \in C^b$ if for each i holds

$$u_i,\ v_{\tau_i},\ w_{\tau_i},\ \varphi_i \in C^1,\quad v_i,\ w_i \in C^2$$

Now assume a frame is loaded with distributed forces

$$p_{x_i},\ p_{y_i},\ p_{z_i},\ m_{x_i} \qquad i = 1,2 \ldots n$$

and nodal forces and couples

$$P^k = \{P^k_x, P^k_y, P^k_z\}, \qquad M^k = \{M^k_x, M^k_y, M^k_z\}$$

We call the vector

$$u = \{u_1, v_1 \ldots \varphi_1, u_2, v_2 \ldots \varphi_n\}^T$$

of $6 \times n$ functions the displacement of the frame if

a) u is geometrically compatible (see trusses) b) u is statically compatible, i.e. we have equilibrium at each free joint

$$P^k_x = R^k_x + c^k_x \delta^k_x, \qquad P^k_y = R^k_y + c^k_y \delta^k_y, \qquad P^k_z = R^k_z + c^k_z \delta^k_z$$

$$M^k_x = R^k_{xx} + c^k_{xx} \varphi^k_{xx}, \qquad M^k_y = R^k_{yy} + c^k_{yy} \varphi^k_{yy}, \qquad M^k_y = R^k_{zz} + c^k_{zz} \varphi^k_{zz}$$

(the terms c^k_x, c^k_{xx} are the stiffnesses of (possible) elastic supports),
and the single elements are in equilibrium, i.e. the components u_i, v_i etc. satisfy the equations

$$-(EAu'_i)' = p_{x_i}, \qquad (EI_z v''_i)'' = p_{y_i}, \qquad (EI_y w''_i)'' = p_{z_i}$$

$$-\left(\frac{1}{\varkappa_y} GA v'_{\tau_i}\right)' = p_{y_i}, \qquad -\left(\frac{1}{\varkappa_z} GA w'_{\tau_i}\right)' = p_{z_i}, \qquad -(GI_p \varphi'_i)' = m_{x_i}$$

To each of the six functions u, v, v_τ, w, w_τ, φ which express the movement of a single member belongs an identity $G(u, \hat{u})$, see chapter 1 and 2. By adding all these $6 \times n$ identities, these $6 \times n$ zeros, we obtain the following expression.

$$p: \quad u, \hat{u} \in C^a \times C^b$$

$$q: \quad G(u, \hat{u}) = \sum_{i=1}^{n} \{ G(u_i, \hat{u}_i) + G(v_{\tau_i}, \hat{v}_{\tau_i}) + G(v_i, \hat{v}_i) + G(w_i, \hat{w}_i)$$

$$+ G(w_{\tau_i}, \hat{w}_{\tau_i}) + G(\varphi_i, \hat{\varphi}_i) \} = \sum_{i=1}^{n} \left\{ \int_0^{l_i} [(-EAu')' \hat{u} \right.$$

$$+ (EI_z v'')'' \hat{v} - \left(\frac{1}{\varkappa_y} GA v'_\tau\right)' \hat{v}_\tau + (EI_y w'')'' \hat{w}$$

$$- \left(\frac{1}{\varkappa_z} GA w'_\tau\right)' \hat{w}_\tau - (GI_p \varphi')' \hat{\varphi}] \, dx$$

$$+ [N\hat{u} + V_y \hat{v} + M_z \hat{v}' + V_y \hat{v}_\tau + V_z \hat{w}$$

$$- M_y \hat{w}' + V_z \hat{w}_\tau + M_x \hat{\varphi}]_0^{l_i}$$

$$- \int_0^{l_i} \left(\frac{N\hat{N}}{EA} + \frac{M_z \hat{M}_z}{EI_z} + \varkappa_y \frac{V_y \hat{V}_y}{GA} + \varkappa_z \frac{V_z \hat{V}_z}{GA} + \frac{M_y \hat{M}_y}{EI_y}\right.$$

$$\left. + \frac{M_x \hat{M}_x}{GI_p}\right) dx \right\} = 0 \tag{3.10}$$

If the vector $\hat{u}$ is geometrically compatible then, as in the case of a truss, the sum of all the work done at the ends of the single members is

$$\sum_{i=1}^{n} [N\hat{u} + V_y\hat{v} + M_z\hat{v}' + V_y\hat{v}_\tau + V_z\hat{w} - M_y\hat{w}' + V_z\hat{w}_\tau + M_x\hat{\varphi}]_0^{l_i}$$

$$= \sum_{k=1}^{K} \boldsymbol{R}^k \cdot \hat{\boldsymbol{\delta}}^k$$

Substituting this result into Eq. (3.10) we obtain

$$p: \ \boldsymbol{u}, \hat{\boldsymbol{u}} \in C^a \times C^b \ \text{and} \ \hat{\boldsymbol{u}} \ \text{is geometrically compatible}$$

$$q: \ G(\boldsymbol{u}, \hat{\boldsymbol{u}}) = \sum_{i=1}^{n} \int_0^{l_i} \left[(-EAu')' \, \hat{u} + (EI_z v'')'' \, \hat{v} \right.$$

$$- \left(\frac{1}{\varkappa_y} GAv_\tau' \right)' \hat{v}_\tau + (EI_y w'')'' \, \hat{w}$$

$$\left. - \left(\frac{1}{\varkappa_z} GAw_\tau' \right)' \hat{w}_\tau - (GI\varphi')' \, \hat{\varphi} \right] dx$$

$$+ \sum_{k=1}^{K} \boldsymbol{R}^k \cdot \hat{\boldsymbol{\delta}}^k - \sum_{i=1}^{n} \int_0^{l_i} \left(\frac{N\hat{N}}{EA} + \frac{M_z\hat{M}_z}{EI_z} + \varkappa_y \frac{V_y\hat{V}_y}{GA} \right.$$

$$\left. + \frac{M_y\hat{M}_y}{EI_y} + \varkappa_z \frac{V_z\hat{V}_z}{GA} + \frac{M_x\hat{M}_x}{GI_p} \right) dx = 0 \tag{3.11}$$

We call this bilinear form the first identity of frames and the form

$$B(\boldsymbol{u}, \hat{\boldsymbol{u}}) = G(\boldsymbol{u}, \hat{\boldsymbol{u}}) - G(\hat{\boldsymbol{u}}, \boldsymbol{u}) = 0$$

the second identity of frames.

These two identities are the mathematical basis of the principle of virtual displacements, virtual forces, Betti's principle and they formulate the equilibrium conditions for framed structures.

To see how the first identity is applied to solve practical problems consider the plane frame, loaded with distributed forces, in Fig. 3.8 b. We want to calculate the displacement at the point x with the help of the first identity or, as textbooks would formulate it, with the principle of virtual displacements.

To this end we place a conjugated quantity, a concentrated force, at x and calculate the virtual work done by the external forces of the frame on the left, the frame in Fig. 3.8 a, when they act through the displacements $\hat{u}$ of the frame on the right; simply stated we evaluate $G(\boldsymbol{u}, \hat{\boldsymbol{u}})$.

The expression $G(\boldsymbol{u}, \hat{\boldsymbol{u}})$ simplifies considerably because no distributed forces act on the frame on the left and, consequently, all components u_i, v_i etc. of $\boldsymbol{u}$, the displacement of the frame on the left, are homogeneous solutions of the governing equations

$$-(EAu_i')' = 0, \quad (EI_z v_i'')'' = 0, \ldots -(GI_p \varphi_i')' = 0$$

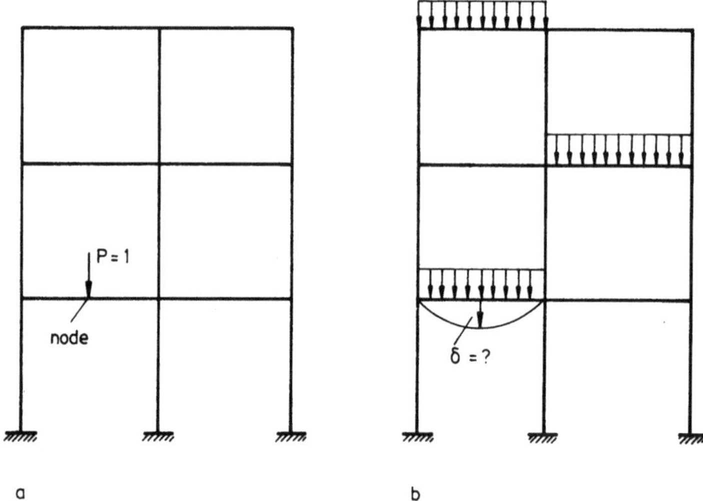

Figure 3.8

Furthermore, because only one joint is loaded the external work done at the joints is simply

$$\sum_{k=1}^{K} \boldsymbol{R}^k \cdot \hat{\boldsymbol{\delta}}^k = 1\hat{\delta}$$

where $\hat{\delta}$ is the unknown displacement.

If we place the unknown on the left-hand side Eq. (3.11) becomes

$$1\hat{\delta} = \sum_{i=1}^{n} \int_{0}^{l_i} \left(\frac{N\hat{N}}{EA} + \frac{M_z\hat{M}_z}{EI_z} + \varkappa_y \frac{V_y\hat{V}_y}{GA} + \frac{M_y\hat{M}_y}{EI_y} + \varkappa_z \frac{V_z\hat{V}_z}{GA} + \frac{M_x\hat{M}_x}{GI_p} \right) dx$$

$$(3.12)$$

Hence, $\hat{\delta}$ is equal to the strain energy, the energy inner product, of $\boldsymbol{u}$ and $\hat{\boldsymbol{u}}$.
If there were elastic supports then an additional term,

$$C_F \hat{c}_F = \sum_{k} \{ \delta_x^k c_x^k \hat{\delta}_x^k + \delta_y^k c_y^k \hat{\delta}_y^k + \ldots + \varphi_z^k c_{zz}^k \hat{\varphi}_z^k \}$$

the work done at all the elastic supports would appear on the right-hand side of Eq. (3.12).

Sometimes we have to calculate displacements caused by consolidations or by swelling of the soil, see Fig. 3.9. For $\hat{\boldsymbol{u}}$ to remain an admissible virtual displacement of the system on the left we remove the support which displaced at the system on the right and let the support reaction become an external force, see Fig. 3.9c.

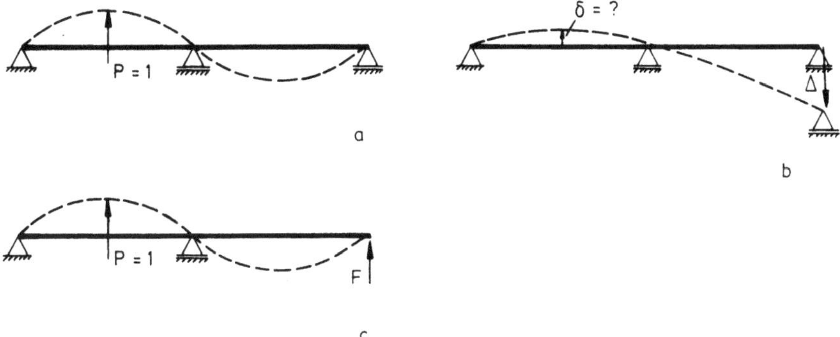

Figure 3.9

If we now evaluate $G(\boldsymbol{u}, \hat{\boldsymbol{u}})$ then we obtain the same result as in Eq. (3.12) plus, on the right-hand side, the additional term

$$- C_L \hat{c}_L = - \sum_k \{P_x^k \hat{\delta}_x^k + P_y^k \hat{\delta}_y^k + \ldots M_z^k \hat{\varphi}_z^k\}$$

where summation is done over all the degrees of freedom introduced by removing constraints. (In the case of Fig. 3.9 this term is simply $F\Delta$).

To calculate displacements caused by a change in temperature, poses no difficulty either.

Remember that the internal actions are defined as

$$N = E A u', \qquad M_y = - E I_y w'', \qquad M_z = E I_z v''$$

Let the virtual displacement have the form

$$\hat{u} = \hat{u}_{el} + \alpha_T \hat{T} x, \quad \hat{v} = \hat{v}_{el} + \alpha_T \frac{\Delta \hat{T}}{h_y} x^2, \quad \hat{w} = \hat{w}_{el} - \alpha_T \frac{\Delta \hat{T}}{h_z} x^2$$

where $\hat{u}_{el}$, $\hat{v}_{el}$ and $\hat{w}_{el}$ are the "elastic displacements", i.e. the part of $\hat{u}$, $\hat{v}$ and $\hat{w}$ which is not yet contained in the temperature terms. The corresponding internal actions become

$$\hat{N} = E A \hat{u}'_{el} + E A \alpha_T \hat{T} = \hat{N}_{el} + E A \alpha_T \hat{T}$$

$$\hat{M}_y = - E I_y \hat{w}''_{el} + E I_y \alpha_T \frac{\Delta \hat{T}_z}{h_z} = \hat{M}_{y,el} + E I_y \alpha_T \frac{\Delta \hat{T}_z}{h_z}$$

$$\hat{M}_z = E I_z \hat{v}''_{el} + E I_z \alpha_T \frac{\Delta \hat{T}_y}{h_y} = \hat{M}_{z,el} + E I_z \alpha_T \frac{\Delta \hat{T}_y}{h_y}$$

If we substitute these terms into Eq. (3.12) and drop the subscript "*el*" then we obtain

$$1\,\delta = \sum_{i=1}^{n} \int_0^{l_i} \left(\frac{N\hat{N}}{EA} + N\alpha_T \hat{T} + \frac{M_z \hat{M}_z}{EI_z} + M_z \alpha_T \frac{\Delta \hat{T}_y}{h_y} + \varkappa_y \frac{V_y \hat{V}_y}{GA} + \frac{M_y \hat{M}_y}{EI_y} \right.$$

$$\left. + M_z \alpha_T \frac{\Delta \hat{T}_z}{h_z} + \varkappa_z \frac{V_z \hat{V}_z}{GA} + \frac{M_x \hat{M}_x}{GI_p} \right) dx + C_F \hat{c}_F - C_L \hat{c}_L \qquad (3.13)$$

which is Mohr's integral in its most general form.

3.3.1 The Method of Reduction

This method exploits the fact that the system on the left must not be identical with the original system, the system on the right. It is sufficient if it is a subsystem of the original system.

According to this method a combination as in Fig. 3.10 would, e. g., be possible.

This method is easily understood if we remember that Mohr's integral is just another formulation of the first identity

$$G(\boldsymbol{u}, \hat{\boldsymbol{u}}) = \sum_{i=1}^{n} \{ G(u_i, \hat{u}_i) + G(v_i, \hat{v}_i) + G(v_{\tau_i}, \hat{v}_{\tau_i}) + G(w_i, \hat{w}_i)$$

$$+ G(w_{\tau_i}, \hat{w}_{\tau_i}) + G(\varphi_i, \hat{\varphi}_i) \} = 0$$

of the frame.

Because each curly bracket is zero the equation remains correct if we do not add all the curly brackets, i. e. if we do not integrate over the whole structure.

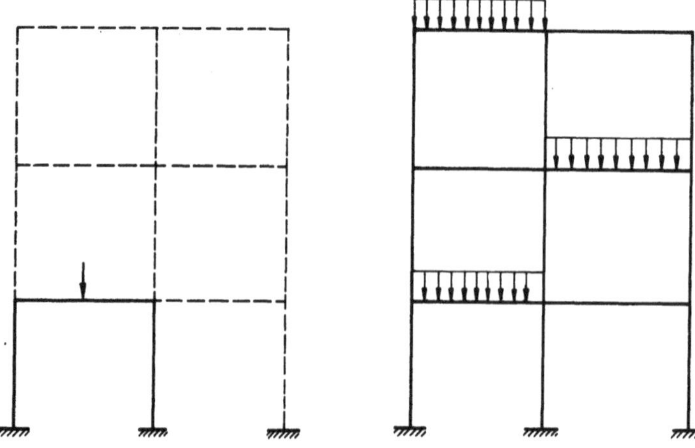

Figure 3.10

3.4 Stiffness Matrices

In this section we calculate the (12×12) stiffness matrix of the frame element. To this aim we first calculate the stiffness matrices of the single displacements, u, v, v_τ, w, w_τ, φ and then assemble the single entries to form the (12×12) matrix.

We start with the stiffness matrix of the axial displacement $u(x)$.

3.4.1 The Axial Displacement

The end forces and end displacements are now termed f_i, δ_i resp. and considered positive if they point in the direction of the positive x-axis, see Fig. 3.11.

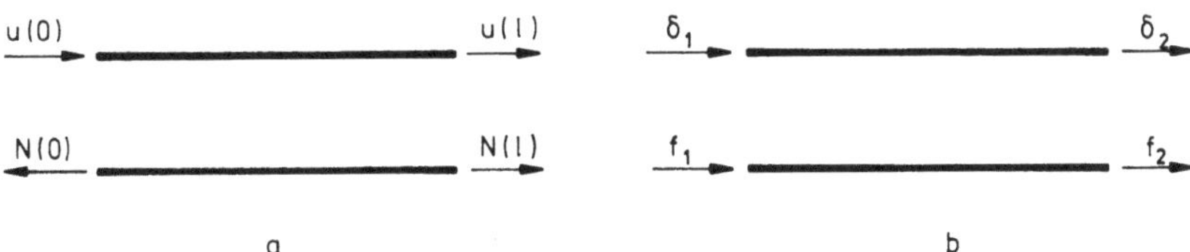

Figure 3.11

The general solution of the equation

$$-EAu'' = 0$$

is the function

$$u(x) = \varphi_1(x)\,\delta_1 + \varphi_2(x)\,\delta_2 \tag{3.14}$$

where

$$\varphi_1(x) = 1 - \frac{x}{l}, \qquad \varphi_2(x) = \frac{x}{l}$$

Note that, see Fig. 3.12,

$$u(0) = \delta_1, \qquad u(l) = \delta_2$$

Introducing the vectors

$$\boldsymbol{\Phi}(x) = \{\varphi_1, \varphi_2\}^T, \qquad \boldsymbol{\delta} = \{\delta_1, \delta_2\}^T$$

$u(x)$ can also be written as $u(x) = \boldsymbol{\Phi}^T \boldsymbol{\delta}$ and the first derivative as

$$u'(x) = \boldsymbol{\Phi}'^T \boldsymbol{\delta} = \varphi_1' \delta_1 + \varphi_2' \delta_2 = -\frac{1}{l}\delta_1 + \frac{1}{l}\delta_2$$

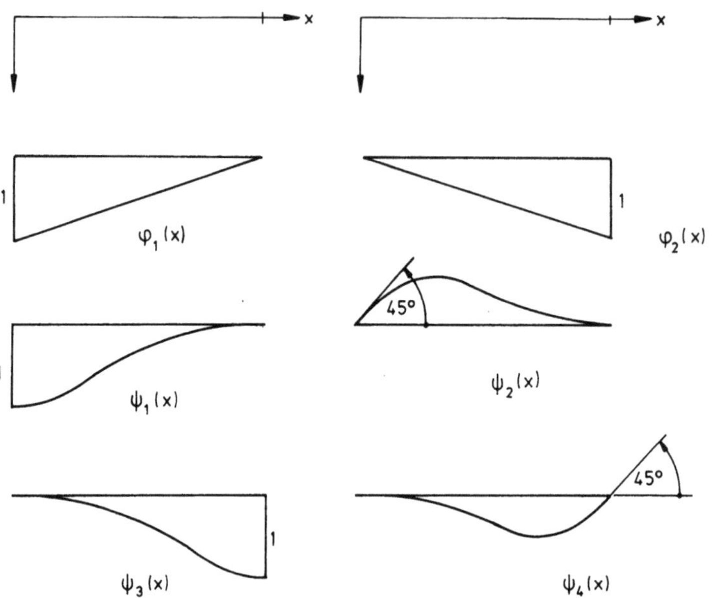

Figure 3.12

Now let $\hat{u}(x) = \boldsymbol{\Phi}^T \hat{\boldsymbol{\delta}}$ be a second homogeneous solution. The first identity of such a pair $\{u, \hat{u}\}$ of homogeneous solutions reads

$$G(u, \hat{u}) = [N\hat{u}]_0^l - \int_0^l \frac{N\hat{N}}{EA}\, dx = N(l)\,\hat{u}(l) - N(0)\,\hat{u}(0) - \int_0^l \frac{N\hat{N}}{EA}\, dx$$

$$= f_1\hat{\delta}_1 + f_2\hat{\delta}_2 - \boldsymbol{\delta}^T EA \int_0^l \boldsymbol{\Phi}'\boldsymbol{\Phi}'^T dx\,\hat{\boldsymbol{\delta}} = \boldsymbol{f}^T\hat{\boldsymbol{\delta}} - \boldsymbol{\delta}^T \boldsymbol{K}\hat{\boldsymbol{\delta}} = 0 \ \forall\ \boldsymbol{\delta}, \hat{\boldsymbol{\delta}} \in \mathbb{R}^2 \quad (3.15)$$

where $\boldsymbol{f} = \{f_1, f_2\}^T$ is the vector of the end forces and $\boldsymbol{K}$ the matrix

$$\boldsymbol{K} = EA \int_0^l \boldsymbol{\Phi}'\,\boldsymbol{\Phi}'^T dx = \frac{EA}{l}\begin{bmatrix} 1 & -1 \\ -1 & 1 \end{bmatrix} \quad (3.16)$$

If we choose in Eq. (3.15), consecutively, the vectors

$$\hat{\boldsymbol{\delta}} = \{1, 0\}^T \quad \hat{\boldsymbol{\delta}} = \{0, 1\}^T$$

then we obtain two equations which can be written as

$$\boldsymbol{f}^T - \boldsymbol{\delta}^T \boldsymbol{K} = \boldsymbol{0}^T$$

or, equivalently, as

$$\boldsymbol{K}\boldsymbol{\delta} = \boldsymbol{f}$$

The symmetric matrix K is the stiffness matrix of the bar. Its elements are end actions corresponding to unit values of displacements.

3.4.2 Shear Deformations

The general solution of the equation

$$-\frac{1}{\varkappa_z} G A w_\tau'' = 0$$

is

$$w_\tau = \varphi_1 \delta_1 + \varphi_2 \delta_2 = \boldsymbol{\Phi}^T \boldsymbol{\delta}$$

where the functions $\varphi_i(x)$ are the same as in Eq. (3.14).

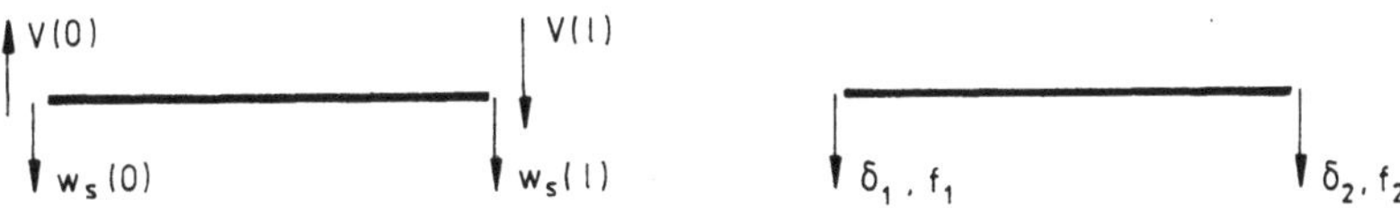

Figure 3.13

Let

$$w_\tau = \boldsymbol{\Phi}^T \boldsymbol{\delta} \qquad \hat{w}_\tau = \boldsymbol{\Phi}^T \hat{\boldsymbol{\delta}}$$

be two homogeneous solutions then the first identity, see Fig. 3.13, becomes

$$G(w_\tau, \hat{w}_\tau) = [V \hat{w}_\tau]_0^l - \varkappa_z \int_0^l \frac{V \hat{V}}{G A} dx = f^T \hat{\boldsymbol{\delta}} - \boldsymbol{\delta}^T K \hat{\boldsymbol{\delta}} = 0 \qquad \forall\, \boldsymbol{\delta}, \hat{\boldsymbol{\delta}} \in \mathbb{R}^2$$

where K is the matrix

$$K = \frac{1}{\varkappa_z} \frac{G A}{l} \begin{bmatrix} 1 & -1 \\ -1 & 1 \end{bmatrix} = \frac{1}{\varkappa_z} G A \int_0^l \boldsymbol{\Phi}' \boldsymbol{\Phi}'^T dx$$

3.4.3 Rotation

The general solution of the equation

$$- G I_p \varphi'' = 0$$

is

$$\varphi(x) = \varphi_1 \delta_1 + \varphi_2 \delta_2 = \boldsymbol{\Phi}^T \boldsymbol{\delta}$$

where, again, the functions $\varphi_i(x)$ are the same as in Eq. (3.14).

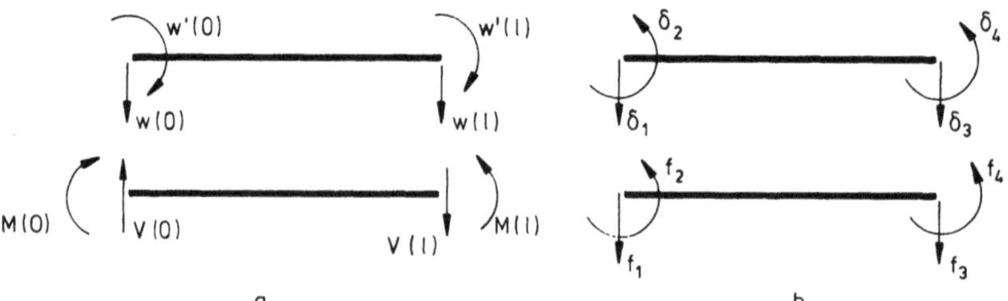

Figure 3.14

Let

$$\varphi = \boldsymbol{\Phi}^T \boldsymbol{\delta}, \qquad \hat{\varphi} = \boldsymbol{\Phi}^T \hat{\boldsymbol{\delta}}$$

be two homogeneous solutions. The first identity, see Fig. 3.14, then becomes

$$G(\varphi, \hat{\varphi}) = [M_x \hat{\varphi}]\,_0^l - \int_0^l \frac{M_x \hat{M}_x}{G I_p}\, dx = f^T \hat{\boldsymbol{\delta}} - \boldsymbol{\delta}^T \boldsymbol{K} \hat{\boldsymbol{\delta}} = 0 \qquad \forall\, \boldsymbol{\delta}, \hat{\boldsymbol{\delta}} \in \mathbb{R}^2$$

where $\boldsymbol{K}$ is the matrix

$$\boldsymbol{K} = \frac{G I_p}{l} \begin{bmatrix} 1 & -1 \\ -1 & 1 \end{bmatrix} = G I_p \int_0^l \boldsymbol{\Phi}' \boldsymbol{\Phi}'^T\, dx$$

3.4.4 Deflections v and w

Figure 3.15

We introduce new notations for end actions and end displacements, see Fig. 3.15 b. The general solution of the equation

$$E I w^{IV} = 0$$

is

$$w = \psi_1(x)\,\delta_1 + \psi_2(x)\,\delta_2 + \psi_3(x)\,\delta_3 + \psi_4(x)\,\delta_4 = \boldsymbol{\Psi}^T \boldsymbol{\delta}$$

where, see Fig. 3.12,

$$\psi_1(x) = 1 - \frac{3x^2}{l^2} + \frac{2x^3}{l^3}, \quad \psi_2(x) = -x + \frac{2x^2}{l} - \frac{x^3}{l^2}$$

$$\psi_3(x) = \frac{3x^2}{l^2} - \frac{2x^3}{l^3}, \qquad \psi_4(x) = \frac{x^2}{l} - \frac{x^3}{l^2} \tag{3.17}$$

and, consequently, we have

$$w(0) = \delta_1, \qquad w'(0) = -\delta_2, \qquad w(l) = \delta_3, \qquad w'(l) = -\delta_4$$

The second derivative is

$$w''(x) = \boldsymbol{\Psi}''(x)^T \boldsymbol{\delta} = \left(-\frac{6}{l^2} + \frac{12x}{l^3}\right)\delta_1 + \left(\frac{4}{l} - \frac{6x}{l^2}\right)\delta_2$$
$$+ \left(\frac{6}{l^2} - \frac{12x}{l^3}\right)\delta_3 + \left(\frac{2}{l} - \frac{6x}{l^2}\right)\delta_4$$

Let $\hat{w} = \boldsymbol{\Psi}^T \hat{\boldsymbol{\delta}}$ a second homogeneous solution, then the first identity becomes at $\{w, \hat{w}\}$

$$G(w, \hat{w}) = [V\hat{w} - M\hat{w}']\,_0^l - \int_0^l \frac{M\hat{M}}{EI}\, dx = V(l)\,\hat{w}(l) - V(0)\,\hat{w}(0)$$

$$- M(l)\,\hat{w}'(l) + M(0)\,\hat{w}'(0) - \int_0^l EIw''\,\hat{w}''\, dx = \boldsymbol{f}^T\hat{\boldsymbol{\delta}} - \boldsymbol{\delta}^T \boldsymbol{K}\hat{\boldsymbol{\delta}} = 0$$

$$\forall\, \boldsymbol{\delta}, \hat{\boldsymbol{\delta}} \in \mathbb{R}^4$$

where $\boldsymbol{f} = \{f_1, f_2, f_3, f_4\}^T$ is the vector of end actions and $\boldsymbol{K}$ the symmetric matrix

$$\boldsymbol{K} = \frac{EI}{l^3}\begin{bmatrix} 12 & -6l & -12 & -6l \\ & 4l^2 & 6l & 2l^2 \\ & & 12 & 6l \\ \text{sym.} & & & 4l^2 \end{bmatrix} = EI \int_0^l \boldsymbol{\Psi}''\,\boldsymbol{\Psi}''^T\, dx \qquad (3.18)$$

If we choose, consecutively, for $\hat{\boldsymbol{\delta}}$ the unit vectors $\hat{\boldsymbol{\delta}} = \{1, 0, 0, 0\}$ etc. then this results in four equations which can be written as

$$\boldsymbol{K}\boldsymbol{\delta} = \boldsymbol{f}$$

The stiffness matrices of the deflections v and v_τ are simply obtained by exchanging the constants in the foregoing equations.

Thus, the stiffness matrices of all the displacements u, v, w, v_τ, w_τ, φ are calculated and the entries can now be assembled to form a 12×12 matrix, the stiffness matrix of the frame element,

$$\boldsymbol{K}\boldsymbol{\delta} = \boldsymbol{f} \qquad (3.19)$$

where $\boldsymbol{K}$ is the matrix

$$K =$$

	1	2	3	4	5	6	7	8	9	10	11	12	
	$\frac{EA}{l}$												1
	0	$\frac{12EI_z}{l^3}+\alpha$											2
	0	0	$\frac{12EI_y}{l^3}+\beta$				sym.						3
	0	0	0	$\frac{GI_t}{l}$									4
	0	0	$-\frac{6EI_y}{l^2}$	0	$\frac{4EI_y}{l}$								5
	0	$\frac{6EI_z}{l^2}$	0	0	0	$\frac{4EI_z}{l}$							6
	$-\frac{EA}{l}$	0	0	0	0	0	$\frac{EA}{l}$						7
	0	$-\frac{12EI_z}{l^3}-\alpha$	0	0	0	$-\frac{6EI_z}{l^2}$	0	$\frac{12EI_z}{l^3}+\alpha$					8
	0	0	$-\frac{12EI_y}{l^3}-\beta$	0	$\frac{6EI_y}{l^2}$	0	0	0	$\frac{12EI_y}{l^3}+\beta$				9
	0	0	0	$-\frac{GI_t}{l}$	0	0	0	0	0	$\frac{GI_t}{l}$			10
	0	0	$-\frac{6EI_y}{l^2}$	0	$\frac{2EI_y}{l}$	0	0	0	$\frac{6EI_y}{l^2}$	0	$\frac{4EI_y}{l}$		11
	0	$\frac{6EI_z}{l^2}$	0	0	0	$\frac{2EI_z}{l}$	0	$-\frac{6EI_z}{l^2}$	0	0	0	$\frac{4EI_z}{l}$	12

$$\alpha = \frac{GA}{\varkappa_z\, l} \qquad\qquad \beta = \frac{GA}{\varkappa_y\, l}$$

The orientation of the displacements δ_i and forces f_i follows Fig. 3.16c.

While the composition of the frame element stiffness matrix was done by assembling the single stiffness matrices, the formulation of the first identity of the frame element is now done by adding the single identities

$$G(u, \hat{u}) + G(v, \hat{v}) + G(v_\tau, \hat{v}_\tau) + G(w, \hat{w}) + G(w_\tau, \hat{w}_\tau) + G(\varphi, \hat{\varphi})$$

$$= [N\hat{u} + V_y\hat{v} + M_z\hat{v}' + V_y\hat{v}_\tau + V_z\hat{w} - M_y\hat{w}' + V_z\hat{w}_\tau + M_x\hat{\varphi}]\,^l_0$$

$$- \int_0^l \left(\frac{N\hat{N}}{EA} + \frac{M_z\hat{M}_z}{EI_z} + \varkappa_y\frac{V_y\hat{V}_y}{GA} + \frac{M_y\hat{M}_y}{EI_y} + \varkappa_z\frac{V_z\hat{V}_z}{GA} + \frac{M_x\hat{M}_x}{GI_p} \right) dx$$

$$= f^T\hat{\delta} - \delta^T K\hat{\delta} = 0$$

Here, too, appears the stiffness matrix K. The scalar $\delta^T K\hat{\delta}$ is the strain energy of the frame element with respect to end displacements δ and $\hat{\delta}$ of homogeneous deformations.

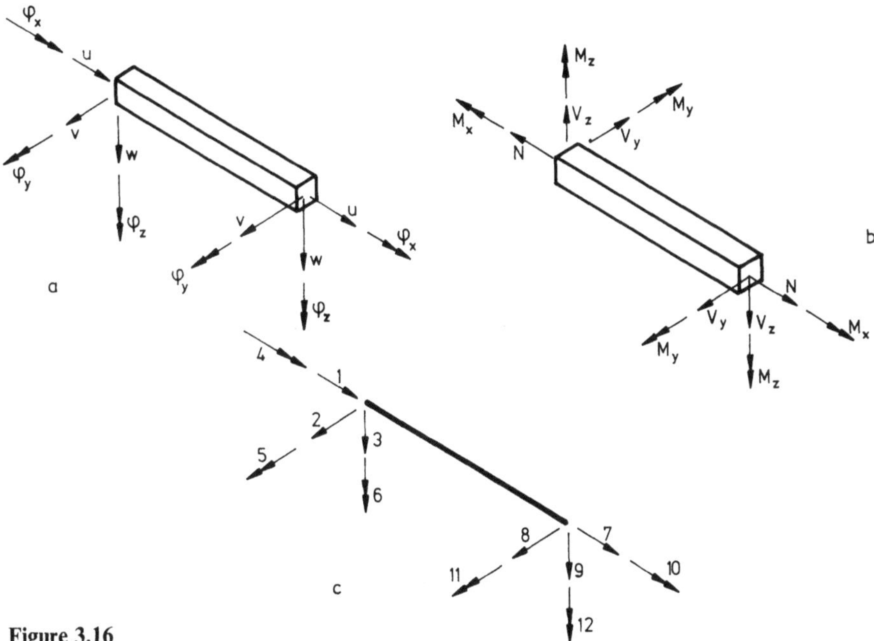

Figure 3.16

Let δ^0 denote the displacement vector of a rigid-body movement $r = a + b \times x$. The stiffness matrix K has then the following properties:

(P1) (Kernel) $K\delta^0 = 0$

(P2) (Equ.) $\delta^{0\,T} K\delta = 0$

(P3) (Sym.) $\delta^T K\hat{\delta} = \hat{\delta}^T K\delta$

(P4) (Pos. def.) $\delta^T K\delta > 0$ $\delta \neq \delta^0$

(P1) means that rigid-body movements do not result in end forces. (P2) states that the end forces are always in equilibrium. (P3) expresses the symmetry of K (Betti's principle). The property (P4) is a consequence of the fact that

$$\delta^T K\delta = \int_0^l \left(\frac{N^2}{EA} + \frac{M_z^2}{EI_z} + \varkappa_y \frac{V_y^2}{GA} + \frac{M_y^2}{EI_y} + \varkappa_z \frac{V_z^2}{GA} + \frac{M_x^2}{GI_p} \right) dx$$

and that this expression is only zero if each single term is zero. Because of

$$N = EA\,u', \quad M_z = EI_z v'', \quad V_y = \frac{1}{\varkappa_y} GA\,v'_\tau, \quad M_y = -EI_y w'',$$

$$V_z = \frac{1}{\varkappa_z} GA\,w'_\tau, \quad M_x = GI_p \varphi'$$

this can only happen if all functions u, v, w, v_τ, w_τ, φ are rigid-body movements but then the vector δ is the vector of a rigid-body movement, $\delta = \delta^0$.

4 Energy Principles

We formulate in this chapter the energy principles of structural mechanics.

How we proceed, which approach we use, is perhaps best illustrated with a small example.

Consider the equation

$$A x = b \tag{4.1}$$

where A is a symmetric $n \times n$ matrix. To A belongs the (trivial) identity

$$G(x, \hat{x}) = \hat{x}^T A x - \hat{x}^T A x = 0 \qquad \forall \, x, \hat{x} \in \mathbb{R}^n$$

If we let the first vector $x = x_s$, the solution of Eq. (4.1), then only the second vector is free to vary in $\mathbb{R}^n$ and we obtain the expression

$$G(x_s, \hat{x}) = \hat{x}^T A x_s - \hat{x}^T b = 0 \qquad \forall \, \hat{x} \in \mathbb{R}^n \tag{4.2}$$

which is the first variation of the function

$$\Pi(x) = \frac{1}{2} x^T A x - x^T b$$

at x_s

$$\delta \Pi(x_s, \hat{x}) = \frac{d\Pi}{d\varepsilon} (x_s + \varepsilon \, \hat{x}) \Big|_{\varepsilon = 0} = \hat{x}^T A \, x_s - \hat{x}^T b$$

Hence, Eq. (4.2) means that x_s, the solution of Eq. (4.1), is a stationary point of the function $\Pi(x)$.

If the matrix A is not only symmetric but also positive definite

$$x^T A \, x > 0 \qquad \forall \, x \neq 0$$

then x_s is even a minimum of the function $\Pi(x)$.

This is seen if we evaluate Π at a point $x_s + \hat{x}$

$$\Pi(x_s + \hat{x}) = \frac{1}{2} x_s^T A x_s + \hat{x}^T A x_s + \frac{1}{2} \hat{x}^T A \hat{x} - x_s^T b - \hat{x}^T b$$

$$= \Pi(x_s) + \delta \Pi(x_s, \hat{x}) + \frac{1}{2} \delta^2 \Pi(x_s, \hat{x}) = \Pi(x_s) + \frac{1}{2} \delta^2 \Pi(x_s, \hat{x})$$

and compare it with $\Pi(x_s)$.

(We dropped the term $\delta \Pi(x_s, \hat{x})$ because the first variation $\delta \Pi(x_s, \hat{x}) = G(x_s, \hat{x})$ of $\Pi(x)$ vanishes at x_s, see Eq. (4.2)).

The difference

$$\Pi(x_s + \hat{x}) - \Pi(x_s) = \frac{1}{2} \delta^2 \Pi(x_s, \hat{x}) = \frac{1}{2} \hat{x}^T A \hat{x} > 0 \quad \forall \hat{x} \neq 0$$

is positive because $\hat{x}^T A \hat{x}$ is a positive quadratic form. Hence, we conclude that $\Pi(x_s)$ must be the minimum.

In the following vectors are replaced by functions and matrices by differential operators but the approach itself is unchanged. The energy principles of structural mechanics are a simple consequence of the fact that the integral identities are nil-forms on function spaces as,

$$G(u, \hat{u}) = 0 \quad \forall u, \hat{u} \in C^{2m} \times C^m$$

Variational formulations are obtained if we substitute for u the solution of the bvp, $u = u_s$, and let the second argument, $\hat{u}$, vary among all virtual displacements.

If the functions $\hat{u}$ are geometrically admissible then we obtain the principle of minimum potential energy and if the functions $\hat{u}$ are statically admissible then we obtain the principle of complementary energy.

If the functions $\hat{u}$ satisfy no conditions at all (besides the smoothness condition $\hat{u} \in C^m$) then we obtain the basic principle ("Hu-Washizu principle").

We discuss all this, next, in more detail.

4.1 The Basic Principle

To every regular bvp belong three energy principles

a) *the basic principle*
b) *the principle of minimum potential energy*
c) *the complementary principle*

To become familiar with the first principle, the basic principle, consider the bar with constant cross section in Fig. 4.1.

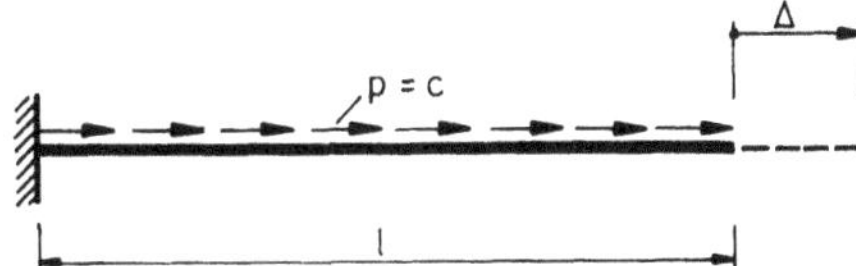

Figure 4.1

With regard to the displacement of this bar, the function u which satisfies the equations

$$-EAu'' = p, \quad 0 < x < l, \quad u(0) = 0, \quad u(l) = \Delta \tag{4.3}$$

the basic principle makes the following two statements:

The displacement u is a stationary point of the **basic functional** (also called **total energy functional**)

$$\Pi(u) = \frac{1}{2} \int_0^l \frac{N^2}{EA} \, dx - \int_0^l pu \, dx + N(l)\,(\Delta - u(l)) + N(0)\,u(0)$$

that is u satisfies the variational equation

$$\delta \Pi(u, \hat{u}) = 0 \quad \forall \, \hat{u} \in C^1[0, l] \tag{4.4}$$

Conversely, if u is a function from $C^2[0, l]$ which is a stationary point of the basic functional, that is a function which satisfies Eq. (4.4) with respect to all virtual displacements $\hat{u}$ from $C^1[0, l]$, then u is the displacement of the bar, that is it satisfies Eq. (4.3).

Before we prove these statements we formulate the basic principle in more general terms because this basic principle applies to all regular $bvps$.

Basic principle

If u is a solution of the regular bvp

$$Du = p \quad in \quad \Omega, \quad \partial^\lambda u = f_\lambda, \quad \lambda \in \Lambda, \quad on \quad \Gamma \tag{4.5}$$

and u from $C^{2m}(\bar{\Omega})$, then u is also a stationary point of the basic functional $\Pi(u)$ (or total energy functional), that is u satisfies the variational equation

$$\delta \Pi(u, \hat{u}) = \frac{d}{d\varepsilon} \Pi(u + \varepsilon \hat{u}) \bigg|_{\varepsilon = 0} = 0 \quad \forall \, \hat{u} \in C^m(\bar{\Omega}) \tag{4.6}$$

Conversely, if u is a stationary point of the functional $\Pi(u)$ in the sense of Eq. (4.6) and if u belongs to $C^{2m}(\bar{\Omega})$ then u is also a solution of the bvp, i.e. it satisfies the Eq. (4.4).

We call this energy principle the basic principle because the principle of minimum potential energy and the complementary principle are (simple) modifications of the basic principle. What distinguishes the basic principle from these two principles is the fact that no subsidiary conditions are imposed on the competing functions.

In the case of the operators A, see chapter 7, the basic principle is identical with the Hu-Washizu principle.

The importance of the basic principle rests on the fact that its energy functional is the basis of all other energy functionals. So as the other energy principles are modifications of the basic principle so their functionals are modifications of the basic functional $\Pi(u)$.

Strictly speaking there exists only one energy functional for every regular *bvp*, namely the basic functional. Everything hinges on this functional.

Hence, the first question we have to answer is: how do we find the basic functional?

We shall give the answer in terms of our model problem, the bar in Fig. 4.1.

The second identity of the bar was

$$p: u, \hat{u} \in C^2[0, l]$$

$$q: B(u, \hat{u}) = \int_0^l - EA u'' \hat{u}\, dx + [N\hat{u} - u\hat{N}]_0^l - \int_0^l u(-EA\hat{u}'')\, dx = 0 \quad (4.7)$$

We found this identity in chapter 2 by applying the integration by parts formula to the definite integral

$$\int_0^l - EA u'' \hat{u}\, dx \tag{4.8}$$

that is we shifted the operator $-EA\, d^2/dx^2$ from u onto $\hat{u}$.

Reversing the procedure we could now shift $-EA\, d^2/dx^2$ back onto u. This must be done in two steps. The first step yields

$$\int_0^l - EA u'' \hat{u}\, dx + [N\hat{u} - u\hat{N}]_0^l + [u\hat{N}]_0^l - \int_0^l \frac{N\hat{N}}{EA}\, dx = 0$$

(this is just $G(u, \hat{u})$) and the second step yields

$$\int_0^l - EA u'' \hat{u}\, dx + [N\hat{u} - u\hat{N}]_0^l + [u\hat{N} - N\hat{u}]_0^l - \int_0^l - EA u'' \hat{u}\, dx = 0$$

or

$$\int_0^l (-EA u'' + EA u'')\, \hat{u} + [(N - N)\hat{u} + (u - u)\hat{N}]_0^l = 0 \tag{4.9}$$

which is a trivial expression consisting of a series of zeros.

This shifting of the operator from u onto $\hat{u}$ and back again is a circular movement, see Fig. 4.2.

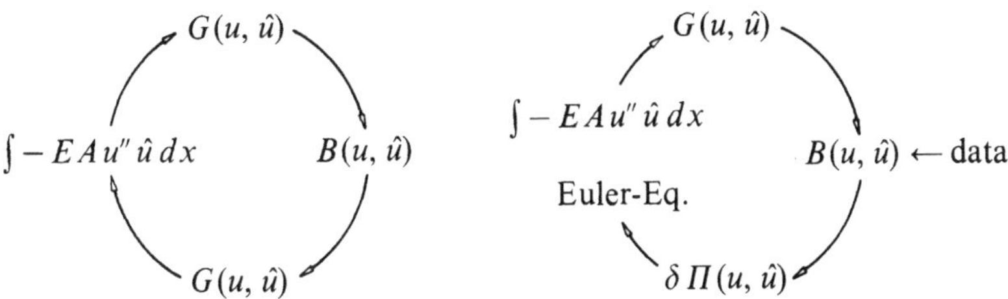

Figure 4.2

which makes a bit more sense if we substitute the data of the bvp into the form $B(u, u)$ before we move back to the starting point because then integration by parts applied to $B(u, \hat{u})$ results in the first variation, $\delta \Pi (u, \hat{u})$, of the basic functional.

Let us demonstrate this with the bar. The displacement u of the bar in Fig. 4.1 belongs to $C^2 [0, l]$ and hence we have

$$(-1) B(u, \hat{u}) = -\int_0^l - EA\, u''\, \hat{u}\, dx - [N\hat{u} - u\hat{N}]_0^l + \int_0^l u(-EA\, \hat{u}'')\, dx = 0$$

$$\forall\, \hat{u} \in C^2$$

or if we replace the terms of u by their actual values ($=$ data) as far as these appear in the formulation of the bvp.

$$(-1) B(u, \hat{u}) = -\int_0^l p\, \hat{u}\, dx - [N\hat{u}]_0^l + \Delta \hat{N}(l) + \int_0^l u(-EA\, \hat{u}'')\, \mathrm{d}x = 0$$

$$\forall\, \hat{u} \in C^2$$

Integration by parts of the last integral in this equation renders

$$\delta \Pi (u, \hat{u}) = -\int_0^l p\, \hat{u}\, dx - [N\hat{u}]_0^l + \Delta \hat{N}(l) - [u\hat{N}]_0^l + \int_0^l \frac{N\hat{N}}{EA}\, dx = 0$$

$$\forall\, \hat{u} \in C^2 \qquad (4.10)$$

which is identical with the expression

$$(-1) G(u, \hat{u}) + [u\hat{N}]_0^l - [u\hat{N}]_0^l = 0$$

and because $G(u, \hat{u})$ is zero for all pairs $\{u_s, \hat{u}\}$ as long as $\hat{u}$ is in $C^1 [0, l]$ we conclude that Eq. (4.10) must be even valid for all $\hat{u} \in C^1 [0, l]$.

The expression $\delta \Pi (u, \hat{u})$ in Eq. (4.10) is the first variation of the functional

$$\Pi (u) = \frac{1}{2} \int_0^l \frac{N^2}{EA}\, dx - \int_0^l p\, u\, dx + N(l)\, (\Delta - u(l)) + N(0)\, u(0) \qquad (4.11)$$

that is we have

$$\frac{d}{d\varepsilon} \Pi (u + \varepsilon\, \hat{u}) \bigg|_{\varepsilon = 0} = \delta \Pi (u, \hat{u}) = 0 \quad \forall\, \hat{u} \in C^1 \qquad (4.12)$$

and because, as we learnt above, all pairs $\{u_s, \hat{u}\}$, $\hat{u} \in C^1$, are zeros of $\delta \Pi (u, \hat{u})$ we conclude that the function u_s is a stationary point of $\Pi (u)$ with respect to all variations $\hat{u} \in C^1$.

The converse statement made in the basic principle is: if u renders $\Pi (u)$ stationary and if u belongs to $C^2 [0, l]$ then it is the displacement of the bar, that is u satisfies the Eq. (4.3).

This conclusion is based on the following lemma, the so-called fundamental lemma of the calculus of variation, see [G 2] p. 20.

Lemma 1
Let

a) *Ω a domain in $\mathbb{R}^n$, $n = 1, 2$ or 3*

b) *$f(x)$ continuous in Ω and assume*

$$\int_{\Omega} f(x)\,\hat{u}(x)\,d\Omega = 0 \qquad \forall\,\hat{u} \in C_0^{\infty}(\Omega)$$

then $f(x)$ is identical zero in $\bar{\Omega}$.

Now, let u a function from $C^2[0, l]$ which satisfies Eq. (4.10). If we apply integration by parts to the strain energy integral Eq. (4.10) becomes

$$-\int_0^l (p + EA\,u'')\,\hat{u}\,dx + \Delta\,\hat{N}(l) - [u\,\hat{N}]_0^l = 0 \quad \forall\,\hat{u} \in C^1 \tag{4.13}$$

As the class C_0^{∞} is a subset of C^1

$$C_0^{\infty} \subset C^1$$

it is admissible to choose in Eq. (4.13) functions $\hat{u}$ from C_0^{∞}. These test functions have zero end-forces, $\hat{N}(0) = EA\,\hat{u}'(0) = \hat{N}(l) = EA\,\hat{u}'(l) = 0$, consequently, the virtual work done on the boundary is zero in such a case and, hence, Eq. (4.13) simplifies to

$$-\int_0^l (p + EA\,u'')\,\hat{u}\,dx = 0 \quad \forall\,\hat{u} \in C_0^{\infty}$$

But by virtue of the fundamental lemma this variational statement is equivalent with the statement

$$-EA\,u'' = p \quad \text{in } [0, l]$$

that is the domain integral in Eq. (4.13) can be dropped and Eq. (4.13) becomes the statement that the expression

$$\Delta\,\hat{N}(l) - [u\,\hat{N}]_0^l = (\Delta - u(l))\,\hat{N}(l) + u(0)\,\hat{N}(0) = 0 \quad \forall\,\hat{u} \in C^1$$

is zero for all $\hat{u}$ in C^1.

It only remains to choose now a function $\hat{u} \in C^1$ with the properties $\hat{N}(0) = 1$ and $\hat{N}(l) = 0$ to see that

$$u(0) = 0$$

and to choose a function $\hat{u} \in C^1$ with the properties $\hat{N}(0) = 0$, $\hat{N}(l) = 1$ to learn that

$$u(l) = \Delta$$

We are, thus, convinced that every stationary point $u \in C^2$ of the functional $\Pi(u)$ is also a solution of the *bvp*, that it satisfies Eq. (4.3).

This result in connection with the converse statement, the one we proved first, confirms that the functional $\Pi(u)$, constructed above, is the basic functional.

In the formulation of $\Pi(u)$ we followed a step by step procedure which can be applied to every regular *bvp*.

a) We substitute the data into the second identity $B(u, \hat{u})$.

b) We apply integration by parts to the integral $(u, D\hat{u})$, the last integral in $B(u, \hat{u})$, and, thus, obtain the form $\delta\Pi(u, \hat{u})$, the first variation of the basic functional.

c) We formulate the basic functional $\Pi(u)$ (this is simple once its first variation is known).

In order not to have to do integration by parts every time anew whenever we consider a new *bvp* we perform the integration by parts once and for all in advance. To this end we separate the main part of the second identity, the part that stays, with two curly brackets $\{\ldots \ldots\}$ from the last integral, the integral that changes,

$$(-1)\, B(u, \hat{u}) = \left\{ -\int_0^l -EA u'' \, \hat{u} \, dx - [N\hat{u} - u\hat{N}]_0^l \right\}$$

$$+ \int_0^l u(-EA\hat{u}'') \, dx = 0$$

and apply integration by parts to the last integral in advance. We, thus, obtain an expression we call

$$V(u, \hat{u}) = \{\ldots \ldots\} - [u\hat{N}]_0^l + \int_0^l \frac{N\hat{N}}{EA} \, dx = 0$$

and which becomes, if we substitute into the curly brackets the data of the *bvp*, the first identity of the basic functional

$$V(u, \hat{u}) + \text{data} = \delta\Pi(u, \hat{u})$$

The forms $V(u, \hat{u})$ of the single structural elements are listed in the following.

Bars

$$V(u, \hat{u}) = \left\{ -\int_0^l (-EAu')' \, \hat{u} \, dx - [N\hat{u} - u\hat{N}]_0^l \right\} - [u\hat{N}]_0^l + \int_0^l \frac{N\hat{N}}{EA} \, dx = 0$$

$$(4.14)$$

Beams

$$V(w, \hat{w}) = \left\{ -\int_0^l (EIw'')'' \, \hat{w} \, dx - [V\hat{w} - M\hat{w}']_0^l + [w\hat{V} - w'\hat{M}]_0^l \right\}$$

$$- [w\hat{V} - w'\hat{M}]_0^l + \int_0^l \frac{M\hat{M}}{EI} \, dx = 0 \qquad (4.15)$$

Kirchhoff plates

$$V(w, \hat{w}) = \left\{ -\int_\Omega K \Delta \Delta w\, \hat{w}\, d\Omega - \int_\Gamma \left[V_n\, \hat{w} - M_n \frac{\partial \hat{w}}{\partial n} + \frac{\partial w}{\partial n} \hat{M}_n - w \hat{V}_n \right] ds \right.$$

$$\left. - [[M_{nt}\,\hat{w} - w\,\hat{M}_{nt}]] \right\} - \int_\Gamma \left[w \hat{V}_n - \frac{\partial w}{\partial n} \hat{M}_n \right] ds - [[w \hat{M}_{nt}]] + E(w, \hat{w}) = 0$$

$$(4.16)$$

Elastic plates and bodies

$$V(\boldsymbol{u}, \hat{\boldsymbol{u}}) = \left\{ -\int_\Omega \boldsymbol{L}\boldsymbol{u} \cdot \hat{\boldsymbol{u}}\, d\Omega - \int_\Gamma [\tau(\boldsymbol{u}) \cdot \hat{\boldsymbol{u}} - \boldsymbol{u} \cdot \tau(\hat{\boldsymbol{u}})]\, ds \right\} - \int_\Gamma \boldsymbol{u} \cdot \tau(\hat{\boldsymbol{u}})\, ds$$

$$+ E(\boldsymbol{u}, \hat{\boldsymbol{u}}) = 0 \qquad\qquad (4.17)$$

Every pair $\{\boldsymbol{u}, \hat{\boldsymbol{u}}\}$ of functions from $C^{2m} \times C^m$ is a zero of $V(\boldsymbol{u}, \hat{\boldsymbol{u}})$.

Note: we multiply the second identity of all operators D (and later also A) with (-1) to obtain the principle of minimum potential energy and not, as we would otherwise, the principle of maximum potential energy.

With these forms $V(u, \hat{u})$ it is now a simple matter to formulate the first variation of the basic principle for a regular bvp as

$$Du = p \quad \text{in } \Omega, \quad \partial^\lambda u = f_\lambda, \quad \lambda \in \Lambda, \quad \text{on } \Gamma$$

In the integrals within the curly brackets appear the terms Du, $\partial^i u$ etc. If these terms appear also in the formulation of the bvp then we replace these terms by the data on the right-hand side of the bvp. After that we delete the curly brackets and strike all the terms which cancel.

The result is the first variation of the basic functional.

4.2 Examples

We illustrate the formulation of the basic functional with some *bvps*.

4.2.1 An Elastic Plate

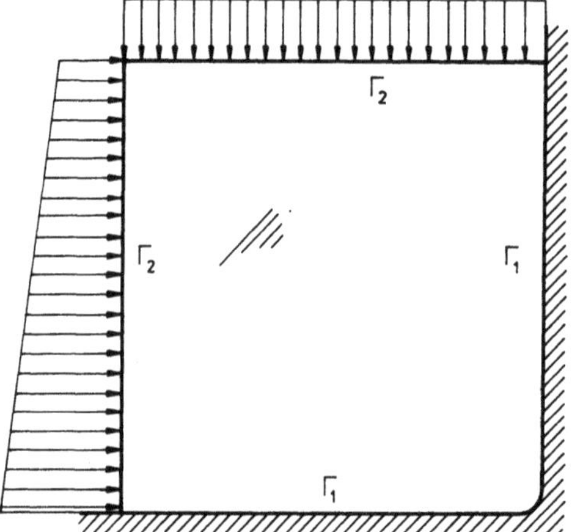

Figure 4.3

Consider the elastic plate in Fig. 4.3 whose displacement field satisfies the equations

$$-Lu = p \quad \text{in } \Omega, \qquad u = 0 \quad \text{on } \Gamma_1, \qquad \tau(u) = \bar{t} \quad \text{on } \Gamma_2$$

where the vector $\bar{t}$ represents the boundary forces.

Substituting the data of this *bvp* into the form $V(u, u)$, see Eq. (4.17), gives

$$\begin{aligned}
\delta\,\Pi(u, \hat{u}) = {} & \left\{ -\int_\Omega p \cdot \hat{u}\, d\Omega - \int_{\Gamma_1} \tau(u) \cdot \hat{u}\, ds - \int_{\Gamma_2} \bar{t} \cdot \hat{u}\, ds \right. \\
& \left. + \int_{\Gamma_1} 0 \cdot \tau(\hat{u})\, ds + \int_{\Gamma_2} u \cdot \tau(\hat{u})\, ds \right\} - \int_{\Gamma_1} u \cdot \tau(\hat{u})\, ds \\
& - \int_{\Gamma_2} u \cdot \tau(\hat{u})\, ds + E(u, \hat{u}) = -\int_\Omega p \cdot \hat{u}\, d\Omega - \int_{\Gamma_1} [\tau(u) \cdot \hat{u} \\
& + u \cdot \tau(\hat{u})]\, ds - \int_{\Gamma_2} \bar{t} \cdot \hat{u}\, ds + E(u, \hat{u}) = 0 \qquad \forall\, \hat{u} \in C^1(\bar{\Omega})
\end{aligned}$$

$$\tag{4.18}$$

which is the first variation of the basic functional

$$\Pi(u) = \frac{1}{2} E(u, u) - \int_\Omega p \cdot u\, d\Omega - \int_{\Gamma_1} \tau(u) \cdot u\, ds - \int_{\Gamma_2} \bar{t} \cdot u\, ds \tag{4.19}$$

that is it holds that

$$\frac{d}{d\varepsilon}\,\Pi\,(u+\varepsilon\,\hat{u})\bigg|_{\varepsilon=0}=\delta\,\Pi\,(u,\,\hat{u})=0\qquad\forall\,\hat{u}\in C^{1}\,(\bar{\Omega})$$

Hence, the displacement field of the plate makes the basic functional stationary.
Conversely, let $u\in C^{2}\,(\bar{\Omega})$ be a stationary point of the functional, that is u satisfies the variational equation

$$\delta\,\Pi\,(u,\,\hat{u})=0\qquad\forall\,\hat{u}\in C^{1}\,(\bar{\Omega})$$

If we apply integration by parts to the strain energy integral in Eq. (4.18) then the expression $\delta\,\Pi\,(u,\,\hat{u})$ becomes

$$\int\limits_{\Omega}(-L\,u-p)\cdot\hat{u}\,d\Omega-\int\limits_{\Gamma_{2}}(\bar{t}-\tau(u))\cdot\hat{u}\,ds-\int\limits_{\Gamma_{1}}u\cdot\tau(\hat{u})\,ds=0\qquad(4.20)$$

and in case $\hat{u}$ belongs to $C_{0}^{\infty}\,(\Omega)$ $\big($which is a subset of $C^{1}\,(\bar{\Omega})\big)$ this equation reduces to

$$\int\limits_{\Omega}(-L\,u-p)\cdot\hat{u}\,d\Omega=0$$

Consequently, see lemma 1, we must have,

$$-L\,u=p\quad\text{in }\Omega$$

To conclude that u satisfies also the boundary conditions we need an additional lemma, see [G 2] p. 115.

Lemma 2

a) *Let f be at least piecewise continuous on Γ_{2} and assume that for all $\hat{u}\in C^{\infty}\,(\bar{\Omega})$ which vanish on Γ_{1}*

$$\int\limits_{\Gamma_{2}}f\,\hat{u}\,ds=0$$

then f is zero on Γ_{2}.

b) *Let f be at least piecewise continuous on Γ_{1} and assume that for all tensors $\hat{S}\in C^{\infty}\,(\bar{\Omega})$ or displacement fields $\hat{u}\in C^{\infty}\,(\bar{\Omega})$ which vanish on Γ_{2}*

$$\int\limits_{\Gamma_{1}}f\cdot\hat{S}n\,ds=0,\qquad\int\limits_{\Gamma_{1}}f\cdot\tau(\hat{u})\,ds=0$$

then f is zero on Γ_{1}.

Before we apply this lemma note that Eq. (4.20) due to $-L\,u=p$ simplifies to

$$-\int\limits_{\Gamma_{2}}(\bar{t}-\tau(u))\cdot\hat{u}\,ds-\int\limits_{\Gamma_{1}}u\cdot\tau(\hat{u})\,ds=0\qquad\forall\,\hat{u}\in C^{1}\,(\bar{\Omega})$$

Choosing now the appropriate virtual displacements $\hat{u}$ we may with the help of lemma 2 conclude that u satisfies the equations

$$u = 0 \quad \text{on } \Gamma_1, \quad \tau(u) = \bar{t} \quad \text{on } \Gamma_2$$

that is all the boundary conditions.

Four additional examples will demonstrate how we construct the basic functional.

4.2.2 A Bar

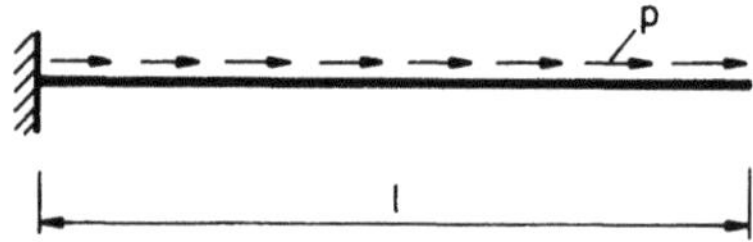

Figure 4.4

The displacement of the bar in Fig. 4.4 satisfies the equations

$$-(EA u')' = p, \quad u(0) = N(l) = 0$$

Substituting the data into the form $V(u, \hat{u})$, see Eq. (4.14), we obtain

$$\delta \Pi(u, \hat{u}) = \left\{ -\int_0^l p\hat{u}\, dx + N(0)\, \hat{u}(0) + u(l)\, \hat{N}(l) \right\} - u(l)\, \hat{N}(l)$$

$$+ u(0)\, \hat{N}(0) + \int_0^l \frac{N\hat{N}}{EA}\, dx = -\int_0^l p\hat{u}\, dx + N(0)\, \hat{u}(0)$$

$$+ u(0)\, \hat{N}(0) + \int_0^l \frac{N\hat{N}}{EA}\, dx = 0 \qquad \forall\, \hat{u} \in C^1$$

This is the first variation of the basic functional

$$\Pi(u) = \frac{1}{2} \int_0^l \frac{N^2}{EA}\, dx - \int_0^l pu\, dx + N(0)\, u(0)$$

4.2.3 A Cantilever Beam

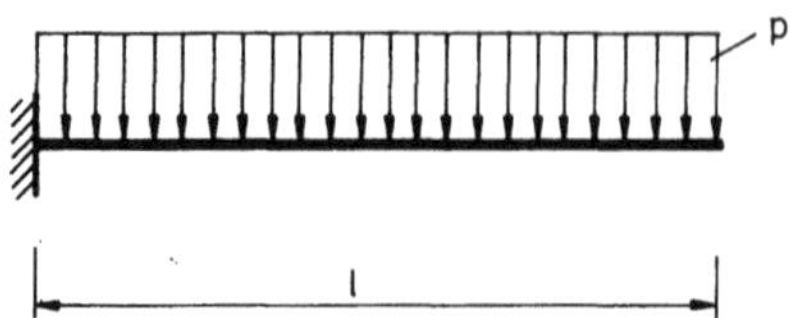

Figure 4.5

The deflection w of the cantilever beam satisfies the equations

$$(EIw'')'' = p, \quad 0 < x < l, \quad w(0) = w'(0) = M(l) = V(l) = 0$$

If we substitute the data into the form $V(w, \hat{w})$, see Eq. (4.15), this results in

$$\delta \Pi (w, \hat{w}) = - \int_0^l p\hat{w}\, dx + V(0)\, \hat{w}(0) - M(0)\, \hat{w}'(0) - w'(0)\, \hat{M}(0)$$

$$+ w(0)\, \hat{V}(0) + \int_0^l \frac{M\hat{M}}{EI}\, dx = 0 \quad \forall\, \hat{w} \in C^2$$

which is the first variation of the basic functional

$$\Pi(w) = \frac{1}{2} \int_0^l \frac{M^2}{EI}\, dx + V(0)\, w(0) - M(0)\, w'(0) - \int_0^l pw\, dx$$

4.2.4 A Beam on a Spring

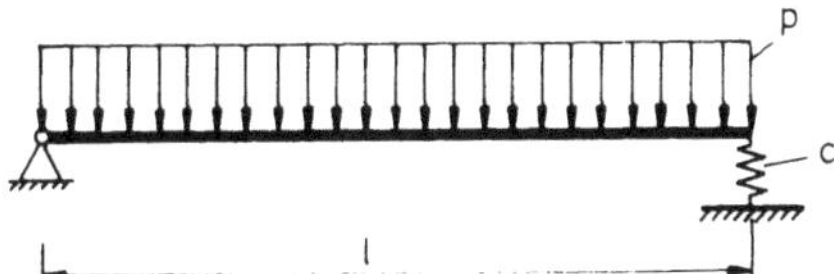

Figure 4.6

The deflection w of the beam satisfies the equations

$$(EIw'')'' = p, \quad 0 < x < l, \quad w(0) = M(0) = M(l) = 0,\ V(l) + cw(l) = 0$$

Substituting these data into the form $V(w, \hat{w})$, see Eq. (4.15), we obtain

$$\delta \Pi (w, \hat{w}) = - \int_0^l p\hat{w}\, dx + cw(l)\, \hat{w}(l) + V(0)\, \hat{w}(0) + w(0)\, \hat{V}(0)$$

$$+ \int_0^l \frac{M\hat{M}}{EI}\, dx = 0 \quad \forall\, \hat{w} \in C^2$$

which is the first variation of the basic functional

$$\Pi(w) = \frac{1}{2} \int_0^l \frac{M^2}{EI}\, dx + \frac{1}{2} cw(l)^2 + V(0)\, w(0) - \int_0^l pw\, dx$$

4.2.5 A Kirchhoff Plate

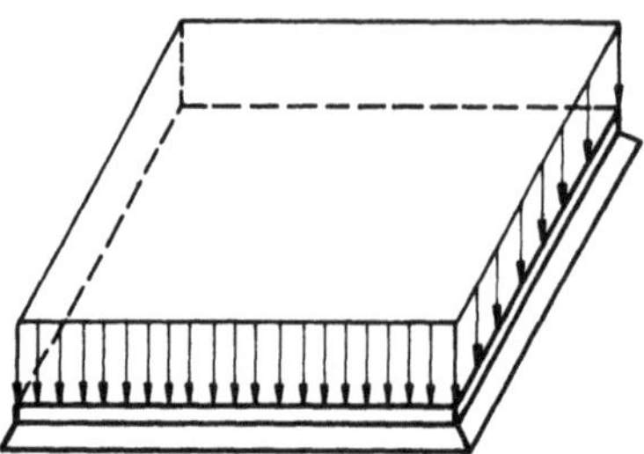

Figure 4.7

The deflection w of the rectangular simply supported plate, see Fig. 4.7, satisfies the equations

$$K \Delta\Delta w = p \ \text{ in } \Omega, \qquad w = M_n = 0 \ \text{ on } \Gamma$$

Substituting these data into the form $V(w, \hat{w})$, see Eq. (4.16), renders

$$\delta \Pi(w, \hat{w}) = - \int_\Omega p\hat{w} \, d\Omega - \int_\Gamma [V_n\hat{w} + w\hat{V}_n] \, ds - [[M_{nt}\hat{w}]]$$

$$- [[w\hat{M}_{nt}]] + E(w, \hat{w}) = 0 \qquad \forall \ \hat{w} \in C^2$$

which is the first variation of the basic functional.

$$\Pi(w) = \frac{1}{2} E(w, w) - \int_\Omega pw \, d\Omega - \int_\Gamma V_n w \, ds - [[M_{nt}w]]$$

Note: To formulate the basic principle in both directions we need, as in the case of the elastic plate in Fig. 4.3, an additional lemma.

4.3 The Principle of Minimum Potential Energy

This principle, a modification of the basic principle, is obtained if we restrict the competition to those functions which satisfy the geometric boundary conditions of the problem.

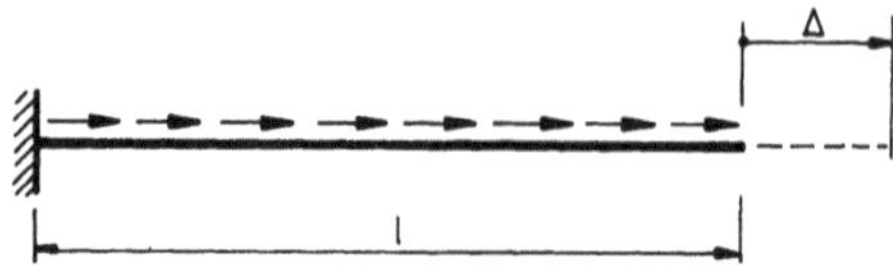

Figure 4.8

Consider the bar in Fig. 4.8. The functions in C^1 which satisfy its geometric boundary conditions constitute the class or, more precisely, the linear manifold

$$R_1 = \{u \in C^1 | u(0) = 0,\ u(l) = \Delta\}$$

Linear manifold, because the sum of two elements, $u + \hat{u}$, is no longer an element of R_1 and the difference of two elements, $u - \hat{u}$, belongs to the space

$$R_{1,0} = \{u \in C^1 | u(0) = u(l) = 0\}$$

Mathematically this is expressed as

$$R_1 = u^{(1)} \oplus R_{1,0}$$

Which means if $u^{(1)}$ is an arbitrary fixed element in R_1, then every u in R_1 can be expressed as the sum of $u^{(1)}$ plus a function from $R_{1,0}$. In other words, we obtain R_1 if we add, consecutively, all the functions in $R_{1,0}$ to $u^{(1)}$. This is equivalent with the statement that for all $u \in R_1$

$$(u - u^{(1)}) \in R_{1,0}$$

The functions in the subspace $R_{1,0}$ (subspace of C^1) are exactly the admissible virtual displacements of the bar.

Before we study the basic principle on this class R_1, on this manifold, it is perhaps helpful to repeat some simple results of calculus.

Consider the function $f(x) = 1/2\,ax^2 - bx$ which has a minimum at $x_0 = b/a$, see Fig. 4.9.

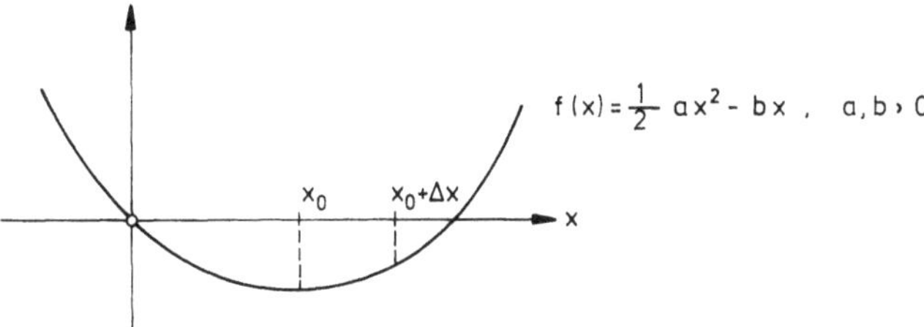

Figure 4.9

The Taylor series of $f(x)$ at x_0 is

$$f(x_0 + \Delta x) = f(x_0) + f'(x_0)\,\Delta x + \frac{1}{2} f''(x_0)\,\Delta x^2$$

and, hence, if we subtract $f(x_0)$ on both sides then this results in

$$f(x_0 + \Delta x) - f(x_0) = f'(x_0)\,\Delta x + \frac{1}{2} f''(x_0)\,\Delta x^2$$

where the right-hand side is a polynomial in Δx which attains its minimum at $\Delta x = 0$ and whose first derivative, $f'(x_0)$, therefore, must be zero. Hence, we are left with

$$f(x_0 + \Delta x) - f(x_0) = \frac{1}{2} f''(x_0)\, \Delta x^2$$

and this expression is only greater zero if $f''(x_0)$ is positive.

We are, thus, reminded that the properties $f'(x_0) = 0$ and $f''(x_0) > 0$ are necessary conditions for $f(x)$ to attain its minimum at $x = x_0$.

Enough about functions. Now back to functionals.

The restriction of the basic functional $\Pi(u)$, Eq. (4.11), to the class R_1 is the expression

$$\Pi(u)|_{R_1} =: \Pi_1(u) = \frac{1}{2} \int_0^l \frac{N^2}{EA}\, dx - \int_0^l pu\, dx + N(l)\, (\Delta \overset{\circ}{\not{u}})$$

$$+ N(0)\, \overset{\circ}{\not{u}}(0) = \frac{1}{2} \int_0^l \frac{N^2}{EA}\, dx - \int_0^l pu\, dx$$

We term this functional, $\Pi_1(u)$, **the potential energy functional.**

Simple algebra shows that the potential energy of the sum of two functions, $u + \hat{u}$, expresses as

$$\Pi_1(u + \hat{u}) = \underbrace{\frac{1}{2} \int \frac{N^2}{EA}\, dx}_{0} + \underbrace{\int \frac{N\hat{N}}{EA}\, dx}_{1} + \underbrace{\frac{1}{2} \int \frac{\hat{N}^2}{EA}\, dx}_{2}$$

$$\underbrace{- \int pu\, dx}_{0} \underbrace{- \int p\,\hat{u}\, dx}_{1} \tag{4.21}$$

where the underlined terms are the zero-th, first and second variation of $\Pi_1(u)$. Hence, Eq. (4.21) is equivalent with

$$\Pi_1(u + \hat{u}) = \Pi_1(u) + \delta \Pi_1(u, \hat{u}) + \frac{1}{2} \delta^2 \Pi_1(u, \hat{u})$$

Subtracting $\Pi_1(u)$ on both sides we obtain

$$\Pi_1(u + \hat{u}) - \Pi_1(u) = \delta \Pi_1(u, \hat{u}) + \frac{1}{2} \delta^2 \Pi_1(u, \hat{u}) \tag{4.22}$$

Up to now u and $\hat{u}$ were considered arbitrary functions. Now we let $u = u_s$ and $\hat{u}$ in $R_{1,0}$.

This causes the term $\delta \Pi_1(u_s, \hat{u})$ in Eq. (4.22) to drop out because, as we shall demonstrate, the first variation of the potential energy $\Pi_1(u)$ vanishes in the direction of all functions $\hat{u} \in R_{1,0}$, i.e. we have

$$\delta \Pi_1(u_s, \hat{u}) = 0 \qquad \forall\, \hat{u} \in R_{1,0}$$

To see this let $\hat{u} \in R_{1,0}$ and ε an arbitrary number then, according to the definition of $\Pi_1(u)$,

$$\Pi(u_s + \varepsilon\,\hat{u}) = \Pi_1(u_s + \varepsilon\,\hat{u})$$

Both sides are, if we keep $u = u_s$ and $\hat{u}$ fixed, polynomials in ε. Consequently their derivatives at $\varepsilon = 0$ must be the same, that is we must have

$$\delta\,\Pi(u_s, \hat{u}) = \delta\,\Pi_1(u_s, \hat{u})$$

The left-hand side is the first variation of the basic functional in the direction of a virtual displacement from $R_{1,0}$. But according to the basic principle the variations with respect to the class C^1 are zero. $R_{1,0}$ is a subset of C^1 and, hence, this variation must be zero, too. What applies to the left-hand side applies to the right-hand side as well, hence, also the right-hand side is zero.

Thus, Eq. (4.22) simplifies at $u = u_s$ to

$$\Pi_1(u_s + \hat{u}) - \Pi_1(u_s) = \frac{1}{2}\,\delta^2\,\Pi_1(u_s, \hat{u}) = \frac{1}{2}\int_0^l \frac{\hat{N}^2}{EA}\,dx = F(\hat{u})$$

which means that the difference in potential energy between $u_s + \hat{u}$ and u_s is just the internal energy $F(\hat{u})$ of the increase in displacement, $\hat{u}$. But as $F(u)$ is positive definite on $R_{1,0} - \{0\}$ because $R_{1,0}$ contains no rigid-body movements we conclude that

$$\Pi_1(u_s + \hat{u}) - \Pi_1(u_s) > 0 \qquad \forall\,\hat{u} \in R_{1,0} - \{0\}$$

that is the only point where $\Pi_1(u)$ attains its minimum is the point $u = u_s$.

This result which we derived here for a bar, applies to all our structural elements. We formulate this as a theorem.

Theorem 4.1 *Principle of minimum potential energy*

If u is the solution of the regular bvp

$$Du = p \quad in\ \Omega, \quad \partial^\lambda u = f_\lambda, \quad \lambda \in \Lambda, \quad on\ \Gamma \tag{4.23}$$

and if u belongs to C^{2m} then the potential energy $\Pi_1(u)$, the restriction of the basic functional to the class R_1, satisfies at u the inequality

$$\Pi_1(u + \hat{u}) - \Pi_1(u) = \frac{1}{2}\,E(\hat{u}, \hat{u}) \geqslant 0 \quad \forall\,\hat{u} \in R_{1,0} \tag{4.24}$$

Conversely, let u a function from R_1 which satisfies the inequality with respect to all virtual displacements $\hat{u}$ from $R_{1,0}$ and assume u belongs to C^{2m} then u solves the bvp in Eq. (4.23).

The internal energy $F(\hat{u}) = 1/2\,E(\hat{u}, \hat{u})$ is only zero if the virtual displacement $\hat{u}$ is a rigid-body movement, see Eq. (2.30). If $R_{1,0}$ contains no such $\hat{u}$ (besides the trivial

$\hat{u}=0$), that is if the structure rests on rigid supports so that rigid-body movements are excluded, then the displacement u renders the functional $\Pi_1(u)$ a minimum in the class R_1.

This principle of minimum potential energy rests on the algebraic properties of the functional $\Pi_1(u)$ as the next theorem will explain.

Theorem 4.2

If the bvp

$$Du = p \text{ in } \Omega, \quad \partial^\lambda u = f_\lambda, \quad \lambda \in \Lambda, \quad on \ \Gamma$$

is regular then the restriction, Π_1, of the basic functional to the class R_1 is of the form

$$\Pi(u)|_{R_1} = \Pi_1(u) = \frac{1}{2} E(u, u) - (p, u) + \sum_{\lambda \in \Lambda_f} (-1)^i \, [f_\lambda, \partial^{i-1} u] \qquad (4.25)$$

Here, Λ_f is the set of all indices, $\lambda \in \Lambda$, which denote forces

$$\lambda \in \Lambda_f \ \rightarrow \ \partial^\lambda u = \text{force}$$

and i is the conjugated index, $i + \lambda = 2m - 1$.

In other words, the functional $\Pi_1(u)$ consists of a quadratic form, the internal energy $F(u) = 1/2\, E(u, u)$, and a series of linear forms, the (negative) work of the external forces acting through u. In simpler terms, $\Pi_1(u)$ has the form $'f(x) = 1/2\, ax^2 - bx'$.

Note: if the structure rests on elastic supports then additional internal energy terms appear as $1/2\, c\, w(x)^2$, $1/2\, c_\varphi\, w'(x)^2$ etc. see section 4.10.

With the help of this Theorem 4.2 the proof of Theorem 4.1 is now readily reached:

On account of Eq. (4.25) the potential energy Π_1 of the sum of two displacements, $u + \hat{u}$, expresses as

$$\Pi_1(u + \hat{u}) = \Pi_1(u) + \delta\, \Pi_1(u, \hat{u}) + \frac{1}{2}\, \delta^2\, \Pi(u, \hat{u}) \qquad (4.26)$$

where

$$\frac{1}{2}\, \delta^2\, \Pi_1(u, \hat{u}) = \frac{1}{2}\, E(\hat{u}, \hat{u})$$

If $u = u_s$ and $\hat{u} \in R_{1,0}$ then we may drop the term $\delta\, \Pi_1(u_s, \hat{u})$ in this expansion because the first variation at u_s with respect to all virtual displacements from C^1 and, hence, from $R_{1,0}$ is zero (see the basic principle). It only remains to subtract $\Pi_1(u)$ on both sides to see that Eq. (4.26) is identical with Eq. (4.24).

To prove the converse statement in Theorem 4.1 let u a function from R_1 which satisfies Eq. (4.24), that is which, given a function $\hat{u}$ from $R_{1,0}$ satisfies the inequality

$$\Pi_1(u + \varepsilon\, \hat{u}) - \Pi_1(u) \geqslant 0 \qquad (4.27)$$

for all $\varepsilon \in (-\infty, +\infty)$. Due to Eq. (4.26) the first term can be written as

$$\Pi_1(u+\varepsilon\hat{u}) = \Pi_1(u) + \delta\Pi_1(u,\hat{u})\,\varepsilon + \frac{1}{2}\,E(\hat{u},\hat{u})\,\varepsilon^2$$

and, hence, the inequality (4.27) is equivalent with the statement

$$\delta\Pi_1(u,\hat{u})\,\varepsilon + \frac{1}{2}\,E(\hat{u},\hat{u})\,\varepsilon^2 \geqslant 0$$

The left-hand side is a polynomial in ε which attains its minimum at $\varepsilon = 0$, consequently, the derivative at $\varepsilon = 0$ must vanish, that is we must have

$$\delta\Pi_1(u,\hat{u}) = 0$$

and this must be true for all $\hat{u}$ in $R_{1,0}$.

If u now, in addition, belongs to C^{2m} then we may apply integration by parts to the first variation $\delta\Pi_1(u,\hat{u})$ and we obtain thus the result

$$(Du - p, \hat{u}) + \sum_{\lambda \in \Lambda_f} (-1)^i\,[f_\lambda - \partial^\lambda u, \partial^{i-1}\hat{u}] = 0 \qquad \forall\,\hat{u} \in R_{1,0}$$

from which, with the help of lemma 1, we conclude that u satisfies the equation $Du = p$ and the static boundary conditions.

4.4 The Complementary Principle

The principle of minimum potential energy was obtained by restricting the total energy Π, the basic functional, to the class R_1, the class of all geometrically admissible functions.

The principle of complementary energy is now obtained by restricting the total energy Π to the class R_2, the class of all those functions in C^{2m} which are statically admissible.

$$R_2 = \{u \in C^{2m} \mid Du = p,\ \partial^\lambda u = f_\lambda,\ m \leqslant \lambda \leqslant 2m-1\}$$

The functions in R_2 satisfy, with the exception of the geometric boundary conditions, all conditions of the bvp.

R_2 is, if the conditions are inhomogeneous, a linear manifold.

The functions which satisfy the homogeneous conditions constitute the space

$$R_{2,0} = \{u \in C^{2m} \mid Du = 0,\ \partial^\lambda u = 0,\ m \leqslant \lambda \leqslant 2m-1\}$$

and in the same sense as above we have

$$R_2 = u^{(2)} \oplus R_{2,0}$$

The functions in $R_{2,0}$ are exactly the statically admissible virtual displacements. In the case of the bar in Fig. 4.1 these two classes are

$$R_2 = \{u \in C^2 \,|\, -EAu'' = p\}, \quad R_{2,0} = \{u \in C^2 \,|\, -EAu'' = 0\}$$

The restriction of the basic functional in Eq. (4.11) to the class R_2 is now the **complementary energy functional**

$$\Pi(u)|_{R_2} =: \Pi_2(u) = \frac{1}{2} \int_0^l \frac{N^2}{EA} \, dx - \int_0^l pu \, dx + N(l)(\Delta - u) + N(0)\, u(0)$$

$$= \frac{1}{2} \int_0^l \frac{N^2}{EA} \, dx - \int_0^l -EAu''\, u \, dx + N(l)\Delta - [Nu]_0^l$$

$$= \frac{1}{2} \int_0^l \frac{N^2}{EA} \, dx + [Nu]_0^l - \int_0^l \frac{N^2}{EA} \, dx + N(l)\Delta - [Nu]_0^l$$

$$= -\frac{1}{2} \int_0^l \frac{N^2}{EA} \, dx + N(l)\Delta$$

Given a function u the increase in complementary energy due to an increase in displacement, $\hat{u}$, is

$$\Pi_2(u + \hat{u}) - \Pi_2(u) = \delta\, \Pi_2(u, \hat{u}) - \frac{1}{2} \int_0^l \frac{\hat{N}^2}{EA} \, dx$$

At $u = u_s$ the first variation of $\Pi_2(u)$ with respect to all virtual displacements $\hat{u} \in R_{2,0}$ is zero

$$\delta\, \Pi_2(u_s, \hat{u}) = 0 \quad \forall\, \hat{u} \in R_{2,0}$$

(this is a consequence of the basic principle where this was established for the larger class $C^1 \supset R_2$).

Hence, at $u = u_s$ the increase is negative (or zero, depending on $\hat{u}$)

$$\Pi_2(u_s + \hat{u}) - \Pi_2(u_s) = -\frac{1}{2} \int_0^l \frac{\hat{N}^2}{EA} \, dx \;\leqslant 0 \quad \forall\, \hat{u} \in R_{2,0}$$

In other words:

1) If we add to $u = u_s$ a rigid-body movement, $\hat{u} = c$, then the value of $\Pi_2(u)$ does not change because the strain energy of such a movement is zero.
2) If we add to $u = u_s$ a function $\hat{u} \neq c$ then $\Pi_2(u)$ decreases.

Hence, if $R_2 - "r"$ is the class R_2 without the rigid-body movements then, because of 2), the functional Π_2 attains its maximum at $u = u_s$. This is the only maximum on $R_2 - "r"$.

What we demonstrated here with a bar applies to all structural elements as the following theorem confirms.

Theorem 4.3 The complementary energy principle

If u belongs to C^{2m} and solves the regular bvp

$$Du = p \quad in\ \Omega, \quad \partial^\lambda u = f_\lambda, \quad \lambda \in \Lambda, \quad on\ \Gamma$$

then the complementary energy $\Pi_2(u)$, the restriction of the basic functional to the class R_2, satisfies at u the inequality

$$\Pi_2(u + \hat{u}) - \Pi_2(u) = -\frac{1}{2}E(\hat{u}, \hat{u}) \leqslant 0 \qquad \forall\ \hat{u} \in R_{2,0} \tag{4.28}$$

Conversely, if u belongs to R_2 and satisfies the inequality with respect to all virtual displacements from $R_{2,0}$ then u satisfies the geometric boundary conditions, hence, is a solution of the bvp.

This theorem rests on the algebraic properties of the functional $\Pi_2(u)$ as the following theorem will explain.

Theorem 4.4

If the bvp

$$Du = p \quad in\ \Omega, \quad \partial^\lambda u = f_\lambda, \quad \lambda \in \Lambda, \quad on\ \Gamma$$

is regular then the restriction of the basic functional Π to the class R_2 has the form

$$\Pi(u)|_{R_2} = \Pi_2(u) = -\frac{1}{2}E(u, u) + \sum_{\lambda \in \Lambda_d}(-1)^i\,[f_\lambda, \partial^{i-1}u] \tag{4.29}$$

Here, Λ_d is the set of all indices $\lambda \in \Lambda$ which denote displacements and i is the conjugated index, $i + \lambda = 2m - 1$.

In other words, the functional $\Pi_2(u)$ consists of a quadratic form, the negative internal energy, $-1/2\,E(u, u)$, and a series of linear forms, the positive external work done by the forces of u acting through the prescribed displacements, simply stated it has the form $'f(x) = -1/2\,ax^2 + cx'$.

Note: if the structure rests on elastic supports then additional internal energy terms as $-1/2\,c\,w(x)^2$, $-1/2\,c_\varphi\,w'(x)^2$ etc. appear.

The complementary principle is not one of the popular principles of structural mechanics but it is, as the name already indicates, complementary to the principle of minimum potential energy and, therefore, deserves our attention as well.

In the last two sections we have learnt that the displacement of the bar makes the functional Π_1 a minimum and the functional Π_2 a maximum. That is the problems:

find two functions u which render $\Pi_1(u)$ or $\Pi_2(u)$ a minimum or a maximum on R_1 or $R_2 - "r"$, resp.

$$\Pi_1(u) \to \text{Min on } R_1, \qquad \Pi_2(u) \to \text{Max on } R_2 - "r"$$

have the same solution $u = u_s$, the solution of the *bvp*. In addition we have

$$\Pi_2(\ddot{u}) \leqslant \Pi_2(u_s) = \Pi(u_s) = \Pi_1(u_s) \leqslant \Pi_1(\dot{u}) \tag{4.30}$$

where $\dot{u}$ and $\ddot{u}$ are arbitrary elements from R_1 and R_2 respectively.

This means that the complementary energy, $\Pi_2(\ddot{u})$, and the potential energy, $\Pi_1(\dot{u})$, of arbitrary functions $\ddot{u}$ and $\dot{u}$ from R_2 and R_1 respectively are lower and upper bounds of the total energy of the displacement u_s of the structure.

To visualize these results we identify the points $x = (x_1, x_2)$ of the plane, see Fig. 4.10, with the functions u. In particular the x_1-axis with the set $R_{1,0}$ and the x_2-axis with the set $R_{2,0} - "r"$. The linear manifolds R_1 and $R_2 - "r"$ are obtained by translating the subspaces $R_{1,0}$ and $R_{2,0} - "r"$ i.e. we add to all vectors $x = u$ a fixed vector $x^{(1)} = u^{(1)}$ or $x^{(2)} = u^{(2)}$.

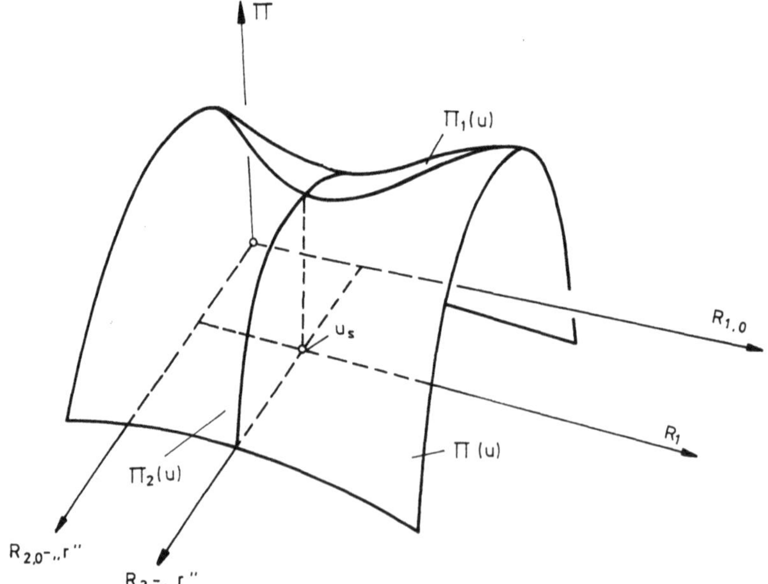

Figure 4.10

The two manifolds, the two lines, intersect at u_s, the solution of the *bvp*. At this point the function $\Pi(u)$ has a saddle point. The restrictions of $\Pi(u)$ to the manifolds R_1 and $R_2 - "r"$ are all those points on the surface $\Pi(u)$ which lie directly above the lines R_1 and $R_2 - "r"$. These points form the curves $\Pi_1(u)$ and $\Pi_2(u)$ respectively.

It should have become clear by now why we have chosen the names complementary principle and complementary energy for the second principle and its functional. Unfortunately, in structural mechanics it is $(-1)\Pi_2(u)$ and not $\Pi_2(u)$ which is termed the complementary energy of a function and, hence, all the symmetry inherent in the formulation of the energy principles is lost.

Furthermore, the principle of minimum complementary energy, as formulated in the literature, is formulated with stresses and not with displacements. Our principle is formulated with displacements.

Strictly speaking, there are two principles of minimum (or maximum) complementary energy, one formulated with displacements (it rarely appears in the literature) and one formulated with stresses. This second, very popular, principle belongs to the operators A, not to the operators D. We shall learn more about it in chapter 7.

Naturally, the value of the complementary energy of a problem is the same, whether we calculate it with stresses or displacements, that is the two principles coincide numerically.

4.5 The Formulation of $\Pi_1(u)$ and $\Pi_2(u)$

The potential energy functional $\Pi_1(u)$ (not the basic functional $\Pi(u)$) is the most important functional in structural mechanics. We think it is, therefore, appropriate when we introduce in this section a method which sidesteps the formulation of the basic functional and allows to formulate the potential energy functional and also the complementary energy functional directly and, furthermore, very quickly.

The method consists of just one step:

We substitute the solution of the bvp, $u = u_s$, and a virtual displacement $\hat{u}$ from $R_{1,0}$ or $R_{2,0}$ into the first identity

$$(-1)\, G(u, \hat{u}) = 0 \quad \text{or} \quad G(\hat{u}, u) = 0 \tag{4.31}$$

(please note the inversion of u and $\hat{u}$ in the second equation)
that is we replace u by its data (as far as these appear in the formulation of the bvp) and let $\hat{u}$ be an arbitrary function from $R_{1,0}$ or $R_{2,0}$ respectively.

The expression (4.31a) thus obtained is the first variation, $\delta \Pi_1(u, \hat{u})$, of the functional $\Pi_1(u)$ and the expression (4.31b) is the first variation, $\delta \Pi_2(u, \hat{u})$, of the functional $\Pi_2(u)$.

As an example consider the bvp of the elastic plate in Fig. 4.3. The basic functional was found to be

$$\Pi(u) = \frac{1}{2} E(u, u) - \int_\Omega p \cdot u \, d\Omega - \int_{\Gamma_1} \tau(u) \cdot u \, ds - \int_{\Gamma_2} \bar{t} \cdot u \, ds$$

and R_1 was the class

$$R_1 = R_{1,0} = \{ u \in C^1(\bar{\Omega}) \,|\, u = 0 \text{ on } \Gamma_1 \}$$

Hence, $\Pi_1(u)$, the restriction of $\Pi(u)$ to R_1 is

$$\Pi(u)|_{R_1} = \Pi_1(u) = \frac{1}{2} E(u, u) - \int_\Omega p \cdot u \, d\Omega - \int_{\Gamma_2} \bar{t} \cdot u \, ds \tag{4.32}$$

This was the old way of doing it.

The alternative proposed in this section is now to substitute u_s into the form $(-1)\,G(u, \hat{u}) = 0$ and to assume that $\hat{u}$ is from $R_{1,0}$. This furnishes

$$(-1)\,G(u, \hat{u}) = -\int_\Omega p \cdot \hat{u}\,d\Omega - \int_{\Gamma_2} \bar{t} \cdot \hat{u}\,ds + E(u, \hat{u}) = 0$$

which, indeed, is just $\delta\Pi_1(u, \hat{u})$.

Analogously $\Pi_2(u)$ can be obtained in two ways.

The old approach is to restrict $\Pi(u)$ to the class

$$R_2 = \{u \in C^2(\bar{\Omega}) \mid -Lu = p,\ \tau(u) = \bar{t}\text{ on }\Gamma_2\}$$

This restriction reads, if we use Eq. (2.20),

$$\Pi(u)|_{R_2} = \Pi_2(u) = \frac{1}{2}E(u, u) + \int_\Omega -Lu \cdot u\,d\Omega - \int_{\Gamma_1} \tau(u) \cdot u\,ds$$

$$- \int_{\Gamma_2} \bar{t} \cdot u\,ds = \frac{1}{2}E(u, u) + \int_\Gamma \tau(u) \cdot u\,ds - E(u, u)$$

$$- \int_{\Gamma_1} \tau(u) \cdot u\,ds - \int_{\Gamma_2} \bar{t} \cdot u\,ds = -\frac{1}{2}E(u, u)$$

But it is far simpler to substitute the pair $\{u_s, \hat{u}\}$ where $\hat{u}$ belongs to $R_{2,0}$ into the first identity and to, thus, obtain

$$G(\hat{u}, u) = \int_\Omega -\overset{\circ}{L}\hat{u} \cdot u\,d\Omega + \int_{\Gamma_2} \tau(\overset{\circ}{\hat{u}}) \cdot u\,ds - E(\hat{u}, u) = -E(\hat{u}, u)$$

which is, exactly, the first variation of $\Pi_2(u)$.

4.6 The Sign of the Total Energy

We know that at $w = w_s$ the three energies coincide, see Eq. (4.30)

$$\Pi_2(w_s) = \Pi(w_s) = \Pi_1(w_s) \tag{4.33}$$

The total energy, Π, of the displacement of a structure is equal to its complementary energy, Π_2, and also equal to its potential energy, Π_1.

This result implies, as we shall demonstrate in this section, that there are two groups of *bvps* where the sign of the total energy is either always positive or always negative.

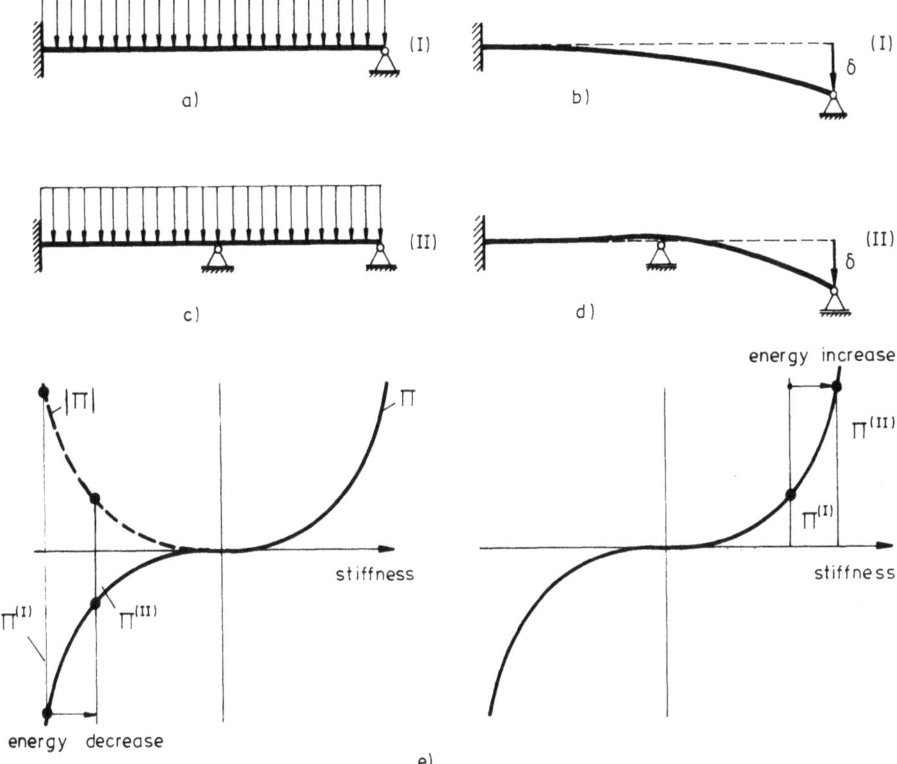

Figure 4.11

Consider the two beams in Fig. 4.11 a and b whose deflections satisfy the equations

$$(EIw'')'' = p, \quad 0 < x < l, \quad w(0) = w'(0) = w(l) = M(l) = 0$$

$$(EIw'')'' = 0, \quad 0 < x < l, \quad w(0) = w'(0) = M(l) = 0, \ w(l) = \delta$$

We call the first bvp an $f - bvp$ because all inhomogeneous terms are forces and the second one a $d - bvp$ because all inhomogeneous terms are displacements.

Let us start with the $f - bvp$. The energy functionals Π_1 and Π_2 are

$$\Pi_1(w) = \frac{1}{2} \int_0^l \frac{M^2}{EI}\, dx - \int_0^l pw\, dx, \quad \Pi_2(w) = -\frac{1}{2} \int_0^l \frac{M^2}{EI}\, dx$$

At $w = w_s$ we have

$$-\frac{1}{2} \int_0^l \frac{M^2}{EI}\, dx = \Pi_2(w) = \Pi(w) = \Pi_1(w)$$

and we, thus, learn that the total energy Π is negative.

Consider now the $d-bvp$. Its energy functionals Π_1 and Π_2 are

$$\Pi_1(w) = \frac{1}{2}\int_0^l \frac{M^2}{EI}\,dx, \quad \Pi_2(w) = -\frac{1}{2}\int_0^l \frac{M^2}{EI}\,dx + V(l)\,\delta$$

At $w = w_s$ we have

$$\Pi_1(w) = \Pi(w) = \Pi_2(w) = \frac{1}{2}\int_0^l \frac{M^2}{EI}\,dx$$

i.e. the total energy Π is positive.

These results can immediately be generalized:

The total energy of the solution of an $f-bvp$ is negative, $\Pi < 0$, and the total energy of the solution of a $d-bvp$ is positive, $\Pi > 0$.

forces	displacements	
inhomogeneous	homogeneous	$\Pi_2 = \Pi = \Pi_1 < 0$
homogeneous	inhomogeneous	$\Pi_2 = \Pi = \Pi_1 > 0$

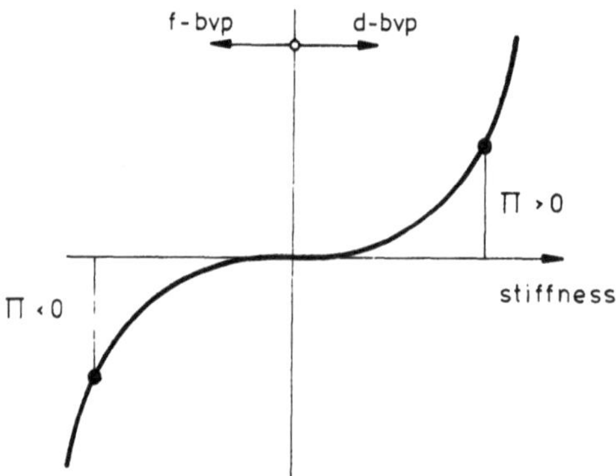

Figure 4.12

The proof of this statement rests on the algebraic properties of $\Pi_1(u)$ and $\Pi_2(u)$ as manifested in Eqs. (4.25) and (4.29).

If the inhomogeneous data are only forces, i.e. $f_\lambda = 0$, $\lambda \in \Lambda_d$, then

$$\Pi_1(u) = \Pi(u) = \Pi_2(u) = -\frac{1}{2}E(u, u) < 0 \qquad (u \neq r)$$

and if the inhomogeneous data are only displacements, i.e. $p = 0, f_\lambda = 0$, $\lambda \in \Lambda_f$, then

$$\Pi_2(u) = \Pi(u) = \Pi_1(u) = \frac{1}{2}E(u, u) > 0 \qquad (u \neq r)$$

But before we imagine now to have found an important law of structural mechanics we better study the cantilever beam in Fig. 4.13.

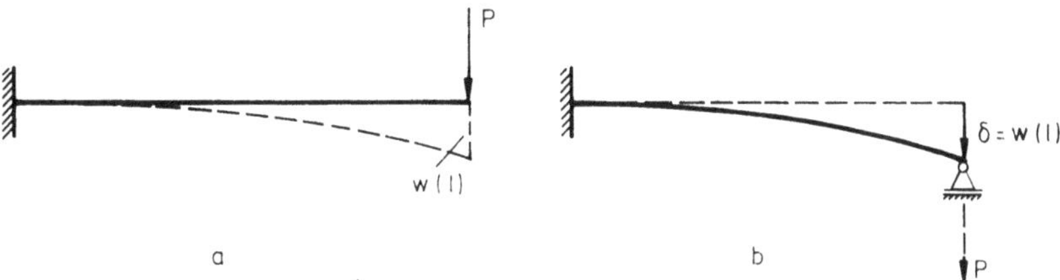

Figure 4.13

Its deflection w can be considered (at the same time!) the solution of an $f - bvp$

$$(EIw'')'' = 0, \ 0 < x < l, \ w(0) = w'(0) = M(l) = 0, \ V(l) = P$$

and a $d - bvp$

$$(EIw'')'' = 0, \ 0 < x < l, \ w(0) = w'(0) = M(l) = 0, \ w(l) = \delta$$

In the first case the total energy of the solution is negative, $\Pi < 0$, and in the second case positive, $\Pi > 0$, though the solution is the same.

This duality, the possibility to consider the displacement u the solution of an $f - bvp$ or a $d - bvp$ always exists if no forces act on the structure in the domain, i. e. if u is a homogeneous solution, $Du = 0$, of the governing equation.

This is equivalent with the fact that all inhomogeneous data are boundary data. It seems then that the distinction between positive and negative energy is artificial and that the sign of the total energy cannot have any essential meaning.

But this is only partially true. The sign of the energy determines, e. g., whether the strain in the material increases or decreases when the stiffness of a structure increases. We shall learn more about this in the next section.

4.7 The Point Π (w) and the Classes R_1 and R_2

If we support the beam in Fig. 4.11a at $x = l/2$ (see Fig. 4.11 c) then the geometrically admissible functions must satisfy, additionally, the condition

$$w\left(\frac{l}{2}\right) = 0$$

Because not all functions in the old class R_1 satisfy this additional condition the new class R_1 is smaller, R_1 decreases.

Analogously does the size of the class R_2 increase because in the new problem the shear force V is allowed to jump at $x = l/2$. Before that the shear force had to be continuous at $x = l/2$, $V_l = V_r$.

Since R_1 decreases and R_2 increases, and since the original deflection no longer is in R_1 and because Eq. (4.30) holds, the point $\Pi_2\,(w_s) = \Pi\,(w_s) = \Pi_1\,(w_s)$ must move to the right, see Fig. 4.11 e.

This means that the absolute value, $|\Pi\,(w)|$, of the total energy decreases; the support lessens the strain in the beam.

In the case of the second bvp an additional support, see Fig. 4.11 d, (R_1 decreases, R_2 increases), effects that

$$\Pi\,(w_s) = \frac{1}{2}\int \frac{M^2}{EI}\,dx > 0\cdot$$

moves to the right; the strain in the beam increases.

The same considerations apply in the case of the elastic plate in Fig. 4.14a. The two classes R_1 and R_2 associated with this bvp are

$$R_1 = \{u \in C^1\,(\bar{\Omega})\}$$
$$R_2 = \{u \in C^2\,(\bar{\Omega})| -L\,u = 0,\, \tau\,(u) = \bar{t}\}$$

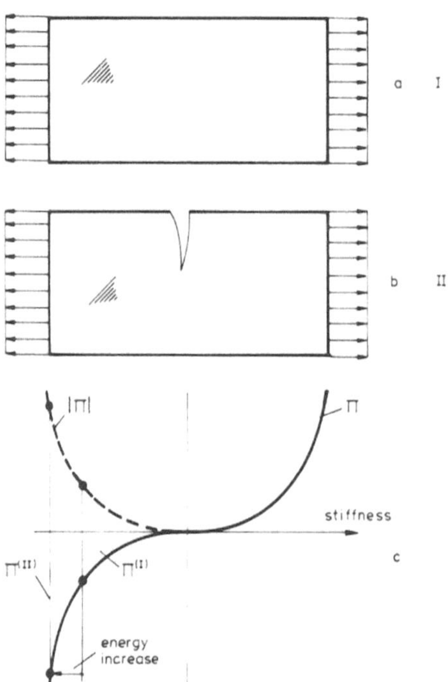

Figure 4.14

If the stresses cause the plate to tear apart, see Fig. 4.14b, then the size of the class R_1 increases because then those functions u are also admissible which are discontinuous on the flanks of the crack and the class R_2 decreases because additionally the condition $\tau(u) = 0$ must be satisfied on the flanks of the crack. Consequently the point

$$\Pi(u_s) = -\frac{1}{2} E(u_s, u_s) < 0$$

moves to the left, the strain increases.

We, thus, learn that the point $\Pi(u_s)$, the total energy, is an "equilibrium point" between the two classes R_1 and R_2 and that the "size" of the two classes (depending on the sign of Π) is proportional or inverse proportional to the strain in the material.

If R_1 decreases (the stiffness of the system increases) then R_2 must increase and consequently $\Pi(u_s)$ must move to the right; if R_1 increases (the stiffness of the system lessens) then R_2 must decrease and consequently $\Pi(u_s)$ must move to the left, see Fig. 4.15.

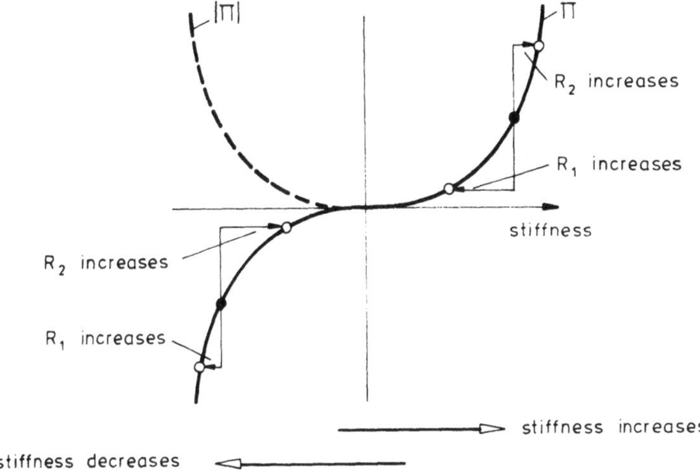

Figure 4.15

Whether such a movement effects an increase or a decrease in the strain, the absolute value of the total energy, $|\Pi(u_s)|$, depends on the position of $\Pi(u_s)$, that is whether it lies to the left or to the right of $\Pi = 0$.

If $\Pi(u_s)$ lies to the left then a movement to the left lets the strain increase while a movement to the right lessens the strain. If $\Pi(u_s)$ lies to the right of $\Pi = 0$ then the opposite is true. (Note: the stiffness—naturally—has no zero at the point where its axis intersects the vertical line, the energy axis; the stiffness is a strictly positive "quantity").

4.8 Displacement Method and Force Method

The classes R_1 and R_2 which we associate with every regular bvp and which have become so important in this chapter are also responsible for the distinction between the displacement method and the force method of structural mechanics as we want to show in this section.

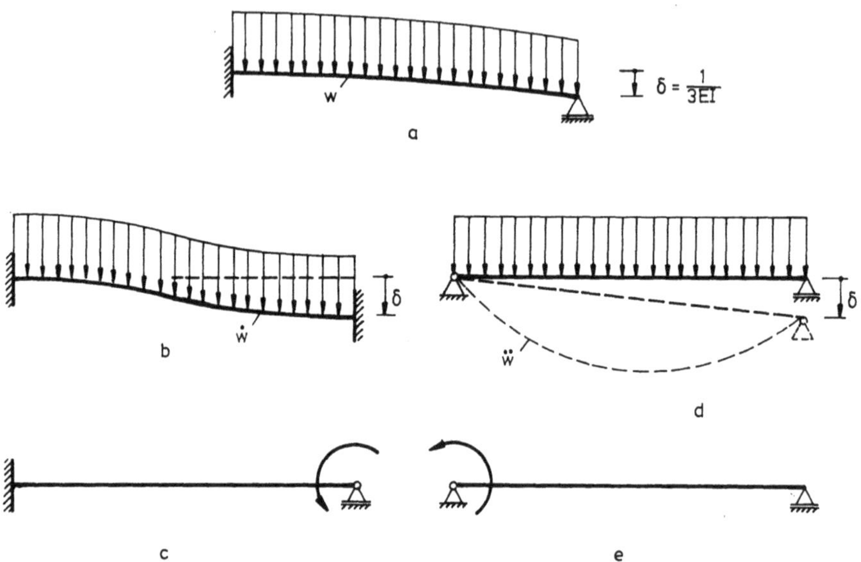

Figure 4.16

Consider the beam in Fig. 4.16a whose deflection satisfies the equations

$$(EIw'')'' = p, \quad 0 < x < l, \quad w(0) = w'(0) = M(l) = 0, \quad w(l) = \delta$$

The classes associated with this bvp are

$$R_1 = \{w \in C^2 | w(0) = w'(0) = 0, \, w(l) = \delta\}$$
$$R_{1,0} = \{w \in C^2 | w(0) = w'(0) = w(l) = 0\}$$
$$R_2 = \{w \in C^4 | (EIw'')'' = p, \, M(l) = 0\}$$
$$R_{2,0} = \{w \in C^4 | (EIw'')'' = 0, \, M(l) = 0\}$$

The deflection of the beam is the only function which belongs to R_1 and to R_2

$$w = R_1 \cap R_2$$

Hence, it is possible to compute w either with the force method $(f - m)$ or the displacement method $(d - m)$. The $d - m$ uses only functions which lie in R_1 while the $f - m$ uses only functions which lie in R_2.

The $d - m$ starts with the kinematically determinate system in Fig. 4.16b and adds to it the deflection of the system in Fig. 4.16c.

The $f - m$ begins, e.g., with the statically determinate system in Fig. 4.16d and adds to it the deflection of the system in Fig. 4.16e.

This solution procedure, the expansion of the solution into a series of $R_{i,0}$-displacements plus an R_i-displacement reflects the algebraic properties of the classes R_1 and R_2; they are linear manifolds.

$$R_1 = \{\text{set of all geometrically admissible states}\} = w^{(1)} \oplus R_{1,0}$$

$$R_2 = \{\text{set of all statically admissible states}\} \qquad = w^{(2)} \oplus R_{2,0}$$

In our case the function $w^{(1)}$ can be identified with the deflection of the kinematically determinate system in Fig. 4.16b and the function $w^{(2)}$ with the deflection of the statically determinate system in Fig. 4.16d.

Before we close this section let us check for once the inequalities (4.30)

$$\Pi_2(\ddot{w}) \leqslant \Pi_2(w) = \Pi(w) = \Pi_1(w) \leqslant \Pi_1(\dot{w})$$

with the geometrically admissible deflection $\dot{w}$ in Fig. 4.16b and the statically admissible deflection $\ddot{w}$ in Fig. 4.16d.

According to these inequalities the complementary energy of a statically admissible deflection, $\ddot{w} \in R_2$, is a lower bound and the potential energy of a geometrically admissible deflection, $\dot{w} \in R_1$, is an upper bound of the total energy, $\Pi(w)$, of the true deflection w in Fig. 4.16a.

In the case of our beam in Fig. 4.16a these single energy functionals are

$$\Pi(w) = \frac{1}{2} \int_0^l \frac{M^2}{EI}\, dx - \int_0^l pw\, dx + V(0)\, w(0) - M(0)\, w'(0)$$
$$+ V(l)\, (\delta - w(l))$$

$$\Pi_1(w) = \frac{1}{2} \int_0^l \frac{M^2}{EI}\, dx - \int_0^l pw\, dx$$

$$\Pi_2(w) = -\frac{1}{2} \int_0^l \frac{M^2}{EI}\, dx + V(l)\, \delta$$

The complementary energy of the statically admissible deflection $\ddot{w}$, the deflection of the statically determinate beam in Fig. 4.16d, $(p = 8, l = 1)$,

$$\ddot{w}(x) = \frac{1}{3EI}(2x - 2x^3 + x^4) \tag{4.34}$$

is $\Pi_2(\ddot{w}) = -1.6/EI$. The potential energy of the geometrically admissible deflection $\dot{w}$, the deflection of the kinematically determinate beam in Fig. 4.16b,

$$\dot{w}(x) = \frac{1}{3EI}(4x^2 - 4x^3 + x^4)$$

is $\Pi_1(\dot{w}) = -0.711/EI$.

The total energy of the deflection of the beam itself

$$w(x) = \frac{1}{6EI}(6x^2 - 6x^3 + 2x^4)$$

is $\Pi(w) = -0.933/EI$ and a comparison of these three numbers, these three energies.

$$-1.6 < -0.933 < -0.711 \qquad\qquad \times (EI)^{-1}$$

shows that, indeed, the inequalities (4.30) are satisfied.

4.9 Energy Principles for Continuous Beams, Trusses and Frames

Continuous beams and trusses are special frames, hence, it is sufficient to discuss the formulation of energy principles for frames alone.

The procedure is virtually the same as for a single structural element.

We only have to substitute the data of the frame into the second identity $B(u, \hat{u})$ (derived for frames in section 3.3) and we then have to apply integration by parts to the integrals $(u, D\hat{u})$. The result is the first variation of the basic functional.

To do this integration by parts not again and again (whenever a new frame has to be analyzed) we perform this integration by parts in advance once and for all, that is we introduce again the form $V(u, \hat{u})$.

$p: u, \hat{u} \in C^a$ *and both are geometrically compatible*

$$q: V(u, \hat{u}) = \left\{ -\sum_{i=1}^{n} \int_{0}^{l_i} \left(-(EAu')' \hat{u} + (EI_z v'')'' \hat{v} - \left(\frac{1}{\varkappa_y} GAv_\tau'\right)' \hat{v}_\tau \right. \right.$$

$$\left. + (EI_y w'')'' \hat{w} - \left(\frac{1}{\varkappa_z} GAw_\tau'\right)' \hat{w}_\tau - (GI_p \varphi')' \hat{\varphi} \right) dx - \sum_{k=1}^{K} R^k \cdot \hat{\delta}^k$$

$$+ \sum_{k=1}^{K} \delta^k \cdot \hat{R}^k \right\} - \sum_{k=1}^{K} \delta^k \cdot \hat{R}^k + \sum_{i=1}^{n} \int_{0}^{l_i} \left(\frac{N\hat{N}}{EA} + \frac{M_z \hat{M}_z}{EI_z} + \varkappa_y \frac{V_y \hat{V}_y}{GA} \right.$$

$$\left. + \frac{M_y \hat{M}_y}{EI_y} + \varkappa_z \frac{V_z \hat{V}_z}{GA} + \frac{M_x \hat{M}_x}{GI_p} \right) dx = 0$$

As a demonstration of the application of this form consider the truss in Fig. 3.5.

The displacement u of the truss (remember that u is a vector which contains all the displacement components of the single frame elements) consists only of the axial deformations of the single members), hence, the form $V(u, \hat{u})$ simplifies considerably,

$$V(u, \hat{u}) = \left\{ -\sum_{i=1}^{n} \int_{0}^{l_i} -EAu'' \hat{u}\, dx - \sum_{k=1}^{K} R^k \cdot \hat{\delta}^k + \sum_{k=1}^{K} \delta^k \cdot \hat{R}^k \right\}$$

$$- \sum_{k=1}^{K} \delta^k \cdot \hat{R}^k + \sum_{i=1}^{n} \int_{0}^{l_i} \frac{N\hat{N}}{EA}\, dx = 0$$

where

$$\boldsymbol{R}^k \cdot \boldsymbol{\delta}^k = R_x^k \, \delta_x^k + R_y^k \, \delta_y^k + R_z^k \, \delta_z^k$$

If we substitute into this form the data of the truss in Fig. 3.5a then we obtain the expression

$$\delta \Pi (\boldsymbol{u}, \hat{\boldsymbol{u}}) = - R_x^1 \, \delta_x^1 - R_z^1 \, \delta_z^1 - R_z^4 \, \delta_z^4 - P \, \delta_x^5 - \delta_x^1 \, \hat{R}_x^1 - \delta_z^1 \, \hat{R}_z^1 - \delta_z^4 \, \hat{R}_z^4$$

$$+ \sum_{i=1}^{7} \int_0^{l_i} \frac{N \hat{N}}{EA} \, dx$$

which is the first variation of the basic functional

$$\Pi (\boldsymbol{u}) = \frac{1}{2} \sum_{i=1}^{7} \int_0^{l_i} \frac{N^2}{EA} \, dx - \{R_x^1 \, \delta_x^1 + R_z^1 \, \delta_z^1 + R_z^4 \, \delta_z^4\} - P \, \delta_x^5$$

In case $\boldsymbol{u} = \{u_i\}$ satisfies the geometric boundary conditions,

$$\delta_x^1 = \delta_z^1 = \delta_z^4 = 0$$

that is if $\boldsymbol{u}$ belongs to R_1 the basic functional becomes

$$\Pi (\boldsymbol{u}) = \frac{1}{2} \sum_{i=1}^{7} \int_0^{l_i} \frac{N^2}{EA} \, dx - P \, \delta_x^5$$

and because of

$$\Pi_1 (\boldsymbol{u} + \hat{\boldsymbol{u}}) - \Pi_1 (\boldsymbol{u}) = \frac{1}{2} \sum_{i=1}^{7} \int_0^{l_i} \frac{\hat{N}^2}{EA} \, dx > 0 \qquad \forall \, \hat{\boldsymbol{u}} \in R_{1,0} - \{\boldsymbol{0}\}$$

the displacement $\boldsymbol{u}$ of the truss makes the functional $\Pi_1 (\boldsymbol{u})$ a minimum in the class of all geometrically admissible displacements $\boldsymbol{u}$.

The two classes, R_1 and R_2, which we associate with every frame have the same meaning as in the case of a single structural element

$$R_1 = \{\boldsymbol{u} \in C^b | \boldsymbol{u} \text{ is geometrically compatible}\}$$

$$R_2 = \{\boldsymbol{u} \in C^a | \boldsymbol{u} \text{ is statically compatible}\}$$

and as before are the displacements $\boldsymbol{u}$ in $R_{1,0}$ or $R_{2,0}$ the displacements which satisfy the homogeneous conditions.

Thus, a virtual displacement $\hat{\boldsymbol{u}} \in C^a$ is statically admissible if it satisfies the homogeneous field equations ($EI \, \hat{w}_i^{IV} = 0$ etc.) and if its resultant nodal forces at the free joints are zero, $\hat{\boldsymbol{R}}^k = \boldsymbol{0}$.

This is a long list of constraints and we, rightly, could ask: if no distributed forces are allowed to act on the frame elements and no nodal forces at the joints what else could be the cause of a statically admissible virtual displacement?

The answer is: the displacements $\hat{u}$ in $R_{2,0}$ are the displacements we observe when the structure performs a rigid-body movement (the constraints at the supports are neglected) and the displacements of the statically determinate released structure.

In other words, the displacements $\hat{u}$ in $R_{2,0}$ are the displacements caused by redundants.

Complementary to this are the displacements $\hat{u}$ in $R_{1,0}$ all the displacements caused by external forces.

$$\hat{u} \in R_{1,0} \quad cause:\ forces$$

$$\hat{u} \in R_{2,0} \quad cause:\ displacements$$

The geometrically admissible virtual displacements of a structure are the response of the structure to virtual forces and the statically admissible virtual displacements of a structure are the response of the structure to virtual displacements.

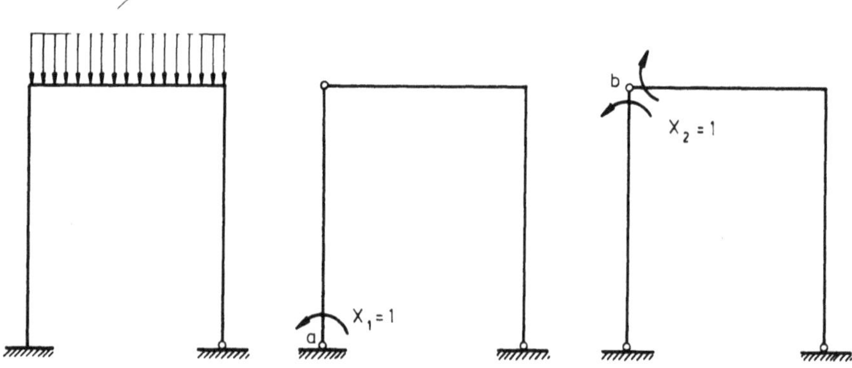

Figure 4.17

Consider Fig. 4.17. The couple X_1 at pin-joint a is statically admissible, the corresponding displacement $\hat{u}$ belongs to $R_{2,0}$. The couple simulates a rotation of the clamped end of the frame member. The two couples at node b, too, are statically admissible, because the resultant nodal moment

$$\hat{R}_{yy} = X_2 - X_2 = 1 - 1 = 0$$

is zero.

4.10 The Formulation of the Functionals "by hand"

In the previous sections we learnt how to construct the single functionals by a step by step procedure. As this method becomes tedious if the structure (and, therewith, the form $V(u, \hat{u})$) consists of too many single elements, we propose here an alternative method which is simple and direct.

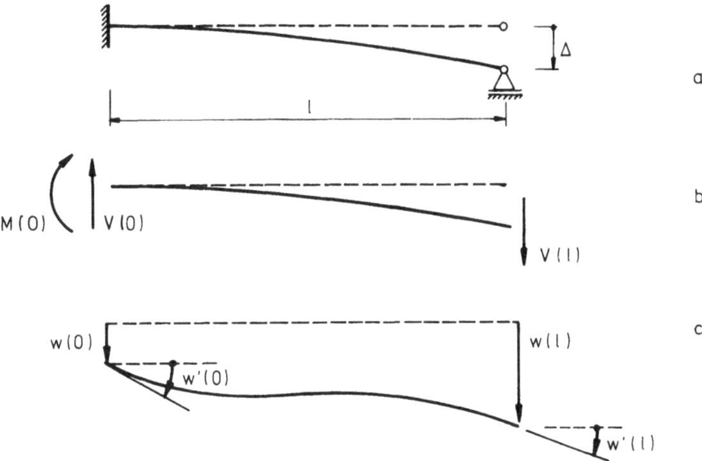

Figure 4.18

a) basic functional

We separate the structure from its support, i. e. we study the so-called free-body diagram of the structure and we calculate the work, W_f, done by all external forces (these include now also the support reactions) acting through an arbitrary displacement u or w. (Note that we write throughout the following text u or w though we might term the function a virtual displacement).

Next, we formulate the work, W_d, done by the support reactions acting through the prescribed displacements, δ, or rotations, φ.

The basic functional is then the expression

$$\Pi(u) = \frac{1}{2} E(u, u) - W_f + W_d$$

As an application consider the beam in Fig. 4.18a whose deflection w satisfies the equations

$$EI\,w^{IV} = 0, \quad 0 < x < l, \quad w(0) = w'(0) = M(l) = 0, \quad w(l) = \delta$$

If we separate the beam from its supports then the internal actions become external forces, see Fig. 4.18b. The work done by these forces acting through an arbitrary deflection w, see Fig. 4.18c, is

$$W_f = M(0)\,w'(0) - V(0)\,w(0) + V(l)\,w(l) \tag{4.35}$$

The work done by the shear force $V(l)$ acting through the prescribed displacement δ is

$$W_d = V(l)\,\delta$$

Thus, the basic functional is the expression

$$\Pi(w) = \frac{1}{2}\int \frac{M^2}{EI}\,dx - M(0)\,w'(0) + V(0)\,w(0) + V(l)\,(\delta - w(l))$$

b) the functional Π_1

The construction is done as in the case of the basic functional only that the functions u or w are restricted to $R_{1,0}$, i.e. they are considered to be geometrically admissible virtual displacements, and that we neglect the work W_d.

Hence, the functional Π_1 is the expression

$$\Pi_1(u) = \frac{1}{2}E(u, u) - W_f$$

We try this with the beam in Fig. 4.18a. A function w from $R_{1,0}$ satisfies

$$w(0) = w'(0) = w(l) = 0$$

and, therefore, the work W_f is zero, see Eq. (4.35), and the functional Π_1, in this case, simply

$$\Pi_1(w) = \frac{1}{2}\int \frac{M^2}{EI}\,dx$$

c) the functional Π_2

We, again, separate the structure from its supports but this time the displacements of the structure observed at the cuts are the basic quantities, not the forces.

We then calculate the work done by the forces of a statically admissible virtual displacement, i.e. a function from $R_{2,0}$, acting through these displacements. It is not too difficult to see that this work is just the work W_d defined in paragraph a). The functional Π_2 is then the expression

$$\Pi_2(u) = -\frac{1}{2}E(u, u) + W_d$$

Fig. 4.19a shows the beam with its end displacements and Fig. 4.19b the end forces of a statically admissible virtual displacement.

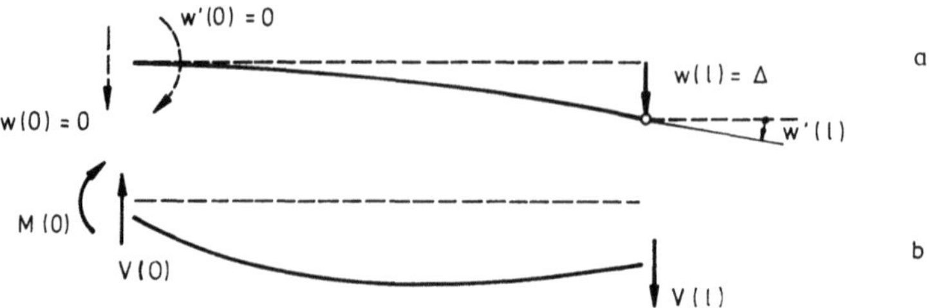

Figure 4.19

The work of these forces acting through the real displacements is just, as we claimed, $V(l)\,\delta = W_d$ and, hence, Π_2 has the form

$$\Pi_2(w) = -\frac{1}{2}\int \frac{M^2}{EI}\,dx + V(l)\,\delta$$

If there are elastic supports, i.e. if we encounter conditions such as

$$N(x) = c\,u(x), \quad V(x) = c\,w(x), \quad M(x) = c_\varphi\,w'(x), \quad \text{etc.}$$

then these rules are modified as follows: the forces set free by cutting through elastic supports are neglected when we calculate W_f. Instead the internal energy is extended by adding the terms.

$$\frac{1}{2}\,cu(x)^2, \quad \frac{1}{2}\,cw(x)^2, \quad \frac{1}{2}\,c_\varphi\,w'(x)^2, \quad \text{etc.}$$

One such modified functional would, e.g., be

$$\Pi(u) = \frac{1}{2}\,E(u,\,u) + \frac{1}{2}\,cu(x)^2 - W_f + W_d$$

Note that in a regular bvp no displacement is prescribed at an elastic support; hence, this modification does not regard the work W_d.

4.11 Lagrange Multipliers

In section 4.3 we have learnt that the displacement u of the bar in Fig. 4.1 makes the functional

$$\Pi_1(u) = \frac{1}{2}\int_0^l \frac{N^2}{EA}\,dx - \int_0^l pu\,dx$$

a minimum in the class

$$R_1 = \{u \in C^1 \,|\, u(0) = 0,\, u(l) = \Delta\}$$

Instead of the functional $\Pi_1(u)$ it would be possible to consider the functional

$$\Pi(u,\,\lambda_1,\,\lambda_2) = \frac{1}{2}\int_0^l \frac{N^2}{EA}\,dx - \int_0^l pu\,dx + \lambda_1\,u(0) + \lambda_2\,(\Delta - u(l)) \tag{4.36}$$

where the numbers λ_i are the so-called Lagrange multipliers.

Now the competing functions must no longer satisfy the geometric boundary conditions but instead the variational problem consists of three unknowns, the displacement u and the multipliers λ_1 and λ_2.

Let us assume we have found a function u and two numbers λ_1, λ_2 which render this modified functional stationary then a simple analysis would show that the function u satisfies the equations

$$- EA\, u'' = p, \quad u(0) = 0, \quad u(l) = \Delta$$

and that the multipliers are just the end forces of u

$$\lambda_1 = N(0), \quad \lambda_2 = N(l)$$

To see how this classical method of Lagrange multipliers incorporates into our scheme, consider the basic functional of the bar which we formulated in Eq. (4.11).

$$\Pi(u) = \frac{1}{2} \int_0^l \frac{N^2}{EA}\, dx + N(0)\, u(0) + N(l)\, (\Delta - u(l)) - \int_0^l pu\, dx$$

If we replace in this expression the forces $N(0)$ and $N(l)$ by parameters λ_1, λ_2, then we obtain just the functional (4.36).

This can be formulated as a rule:

It is admissible to replace in the basic functional all the force-terms of u or w by parameters λ_i (numbers, functions).

If $u, \lambda_1, \lambda_2, \ldots \lambda_n$ is a stationary point of the thus modified functional

$$\Pi(u, \lambda_1, \lambda_2 \ldots \lambda_n)$$

that is if

$$\delta\, \Pi(u, \lambda_1, \lambda_2 \ldots \lambda_n; \hat{u}, \hat{\lambda}_1, \hat{\lambda}_2 \ldots \hat{\lambda}_n)$$

$$= \left(\frac{d}{d\varepsilon} \Pi + \frac{d}{d\eta} \Pi + \ldots \frac{d}{d\rho} \Pi \right)\Bigg|_{\varepsilon = \eta = \rho \ldots = 0} = 0$$

holds at

$$\Pi = \Pi(u + \varepsilon\, \hat{u}, \lambda_1 + \eta\, \hat{\lambda}_1, \ldots, \lambda_n + \rho\, \hat{\lambda}_n)$$

then $u = u_s$ (i.e. u satisfies all conditions of the bvp) and then the parameters λ_i are exactly the force-terms of $u = u_s$ they replace in the transition from $\Pi(u)$ to the modified functional $\Pi(u, \lambda_1, \lambda_2, \ldots \lambda_n)$.

(For a proof see section 4.12).

The reader might ask why we introduce Lagrange multipliers. To circumvent subsidiary conditions no multipliers would be necessary, for this purpose we have already the basic functional. But Lagrange multipliers offer more:

Firstly, by introducing Lagrange multipliers we avoid the calculation of the force-terms of the function u on the boundary (this always requires differentiation). We, instead, simply introduce these forces as additional independent unknowns. The rule formulated above then guarantees that the λ_i of the variational solution coincide with the quantities they replace.

Secondly, by introducing Lagrange multipliers we obtain a variational principle which with respect to the degree of exactness with which the geometric boundary conditions are approximated stands between the basic principle and the principle of minimum potential energy.

geometric boundary conditions

Basic Principle	Basic Principle + L. M.	Min. Pot. Energy
"none"	relaxed	strict

Consider a simple example. The deflection of a prestressed (N) membrane under a pressure p satisfies the equations

$$- N \Delta w = p \ \text{in} \ \Omega \quad w = 0 \ \text{on} \ \Gamma$$

To approximate the deflection we substitute the function $w = \Sigma \, \delta_i \, \varphi_i$ into the basic functional

$$\Pi (w) = \frac{1}{2} N \int_\Omega \text{grad}^2 \, w \, d\Omega - \int_\Omega p \, w \, d\Omega + \int_\Gamma \frac{\partial w}{\partial n} (w - 0) \, ds$$

and require

$$\frac{\partial \Pi}{\partial \delta_i} = 0 \quad i = 1, 2 \ldots$$

Hence, the vector of the nodal displacements δ_i must satisfy the equation

$$\left[\ K + R \ \right] \left[\ \delta \ \right] = \left[\ f \ \right]$$

where K and R are the matrices

$$K_{ij} = N \int_\Omega \text{grad} \, \varphi_i \, \text{grad} \, \varphi_j \, d\Omega, \quad R_{ij} = \int_\Gamma \left[\frac{\partial \varphi_i}{\partial n} \varphi_j + \varphi_i \frac{\partial \varphi_j}{\partial n} \right] ds$$

and f the vector

$$f_i = \int_\Omega p \, \varphi_i \, d\Omega$$

The Lagrange functional is obtained if we replace the force-term $\partial w / \partial n$ in the basic functional by a boundary function λ

$$\Pi (w, \lambda) = \frac{1}{2} N \int_\Omega \text{grad}^2 \, w \, d\Omega - \int_\Omega p \, w \, d\Omega + \int_\Gamma \lambda (w - 0) \, ds$$

To approximate the unknowns w and λ we introduce the functions

$$w = \sum_{i=1}^{n} \delta_i \varphi_i \qquad \lambda = \sum_{j=1}^{m} \varepsilon_j \psi_j$$

and require

$$\frac{\partial \Pi}{\partial \delta_i} = 0 \quad i = 1, 2 \ldots n, \qquad \frac{\partial \Pi}{\partial \varepsilon_j} = 0 \quad j = 1, 2 \ldots m$$

This leads to

$$\begin{bmatrix} K \end{bmatrix} \begin{bmatrix} L \end{bmatrix} \begin{bmatrix} \delta \end{bmatrix} = \begin{bmatrix} f \end{bmatrix}$$
$$\begin{bmatrix} L^T \end{bmatrix} \quad [\, 0 \,] \quad [\, \varepsilon \,] \qquad [\, 0 \,]$$

where L is the $(n \times m)$ matrix

$$L_{ij} = \int_\Gamma \varphi_i \, \psi_j \, ds$$

The difference with respect to the basic functional is that the boundary condition $w = 0$ appears explicitly as a variational side condition and it is, thus, guaranteed that the function w satisfies the boundary condition at least in the mean, that is the error in deflection on the boundary is orthogonal to every ψ_j.

$$\frac{\partial \Pi}{\partial \varepsilon_j} = \int_\Gamma \psi_j \left(\sum_i \delta_i \varphi_i - 0 \right) ds = 0 \quad j = 1, 2 \ldots m$$

4.12 The Algebra of Structural Mechanics

The reader, certainly, will have noticed that structural mechanics once the hard mathematics is done, once the identities are calculated, becomes a mere play with symbols, that is algebra. For this game to be successful our toys must satisfy certain prerequisites. These prerequisites are the topic of this chapter.

Let D a linear or nonlinear differential operator and assume that integration by parts applied to the integral $(Du, \hat{u})$ has led to the identity

$$G(u, \hat{u}) = \int_\Omega Du \cdot \hat{u} \, d\Omega + \int_\Gamma r(u) \cdot \hat{u} \, ds - E(u, \hat{u}) = 0 \tag{4.37}$$

where, without loss of generality, we assumed that D is of degree 2 and that, therefore, only two boundary operators $\partial^0 = \text{id.}$ and $\partial^1 = r(\ )$ appear in the first identity.

The *essential condition* is that the domain integral $E(u, \hat{u})$ in this equation is the first variation of a functional $F(u)$, that is we must have

$$E(u, \hat{u}) = \frac{d}{d\varepsilon} F(u + \varepsilon \hat{u}) \bigg|_{\varepsilon = 0} = \delta F(u, \hat{u}) \tag{4.38}$$

If this condition is satisfied then the principle of virtual displacements applies

$$G(u, \hat{u}) = \int_\Omega Du \cdot \hat{u} \, d\Omega + \int_\Gamma r(u) \cdot \hat{u} \, ds - \delta F(u, \hat{u}) = 0$$

and if $E(u, \hat{u})$ is symmetric, $E(u, \hat{u}) = E(\hat{u}, u)$, as in linear mechanics, then also the principle of virtual forces

$$G(\hat{u}, u) = \int_\Omega D\hat{u} \cdot u \, d\Omega + \int_\Gamma r(\hat{u}) \cdot u \, ds - \delta F(\hat{u}, u) = 0$$

and Betti's Theorem

$$B(u, \hat{u}) = G(u, \hat{u}) - G(\hat{u}, u) = \int_\Omega Du \cdot \hat{u} \, d\Omega + \int_\Gamma r(u) \cdot \hat{u} \, ds - \int_\Gamma u \cdot r(\hat{u}) \, ds$$
$$- \int_\Omega u \cdot D\hat{u} \, d\Omega = 0$$

The principle of virtual displacements states that the first variation, $\delta F(u, \hat{u})$, of the functional $F(u)$ is equal to the work done by the external forces Du, $r(u)$, acting through the virtual displacements $\hat{u}$.

$$\delta F(u, \hat{u}) = \int_\Omega Du \cdot \hat{u} \, d\Omega + \int_\Gamma r(u) \cdot \hat{u} \, ds$$

and the principle of virtual forces states that the first variation, $\delta F(u, \hat{u})$, of the functional $F(u)$ is equal to the work done by virtual forces $D\hat{u}$, $r(\hat{u})$ acting through the true displacements u.

$$\delta F(u, \hat{u}) = \int_\Omega D\hat{u} \cdot u \, d\Omega + \int_\Gamma r(\hat{u}) \cdot u \, ds$$

Let us turn now to the energy principles of structural mechanics. While for the principles of virtual work to exist one condition, $E(u, \hat{u}) = \delta F(u, \hat{u})$, is sufficient, for energy principles to exist the *bvp*, in addition, must be regular.

We termed a *bvp* as

$$Du = p \ \text{ in } \Omega, \quad r(u)|_{\Gamma_1} = \bar{r}, \quad u|_{\Gamma_2} = \bar{u}$$

regular if on every part of the boundary exactly one of the two conjugated boundary terms which appear in the first identity is prescribed. In other words it must be $\Gamma_1 \cup \Gamma_2 = \Gamma$ and $\Gamma_1 \cap \Gamma_2 = 0$ (with the exception of points ($\mathbb{R}^2$) or curves ($\mathbb{R}^3$) the two parts have in common).

With regard to such *bvps* we may state:

If the *bvp* is regular and if the essential condition, Eq. (4.38), is satisfied then a basic functional exists.

The question then is: how do we find the basic functional?

At the beginning of this chapter our strategy was to transform $G(u, \hat{u})$ into $B(u, \hat{u})$, to substitute the data into $B(u, \hat{u})$ and to go then back again to $G(u, \hat{u})$ which then was $\delta \Pi (u, \hat{u})$.

What we achieved with these steps is that we added (in terms of our model operator D) to the first identity the two identical $(\pm)$ boundary integrals

$$\int_\Gamma r_u(\hat{u}) \cdot u \, ds - \int_\Gamma r_u(\hat{u}) \cdot u \, ds =: N(u, \hat{u}) = 0$$

(This is best seen in the case of Eq. (4.17)).

In other words, the structure of $V(u, \hat{u})$ is

$$V(u, \hat{u}) = (-1) \, G(u, \hat{u}) + N(u, \hat{u}) = \left\{ -\int_\Omega \underline{Du} \cdot \hat{u} \, d\Omega - \int_\Gamma \underline{r(u)} \cdot \hat{u} \, ds \right.$$

$$\left. + \int_\Gamma r_u(\hat{u}) \cdot u \, ds \right\} - \int_\Gamma r_u(\hat{u}) \cdot u \, ds + \underline{E(u, \hat{u})} = 0$$

where the underlined terms are the integrals of the first identity. The curly bracket keeps the two terms of $N(u, \hat{u})$ apart; otherwise they would simply drop out.

The two boundary integrals within the curly bracket are (neglect for the time being their different signs) the complete first variation

$$\delta R(u, \hat{u}) = \int_\Gamma r_u(\hat{u}) \cdot u \, ds + \int_\Gamma r(u) \cdot \hat{u} \, ds \tag{4.39}$$

of the functional

$$R(u) = \int_\Gamma r(u) \cdot u \, ds$$

(Remember that

$$r_u(\hat{u}) = \frac{d}{d\varepsilon} r(u + \varepsilon \hat{u}) \Big|_{\varepsilon = 0}$$

is the Gateaux differential of $r(u)$ in the direction of $\hat{u}$, see section 1.7).

This functional $R(u)$ is the boundary integral in the first identity $G(u, u)$ (note $\hat{u} = u$).

That is, the form $V(u, \hat{u})$ is obtained by adding to $(-1) \, G(u, \hat{u})$ the *missing term* (with a plus sign) of the first variation $\delta R(u, \hat{u})$ of $R(u)$ and subtracting it again (with a minus sign). "Missing" term because $G(u, \hat{u})$ contains only the second integral of Eq. (4.39), not both.

This understood, we can now construct the form $V(u, \hat{u})$ without making use of the second identity. This is a helpful hint because in nonlinear mechanics no second identity exists.

We next demonstrate how with the help of the form $V(u, \hat{u})$ the basic functional of the first, second and mixed bvp associated with the operator D is found.

We start with the first bvp

$$Du = p \quad \text{in } \Omega, \quad u = \bar{u} \quad \text{on } \Gamma$$

Substitution of the data into the form $V(u, \hat{u})$ results in

$$\delta\,\Pi\,(u,\,\hat{u}) = -\int_{\Omega} p\hat{u}\cdot d\Omega - \int_{\Gamma} r(u)\cdot\hat{u}\,ds + \int_{\Gamma} r_u(\hat{u})\cdot(\bar{u}-u)\,ds$$
$$+\,\delta\,F\,(u,\,\hat{u}) = 0$$

which is the first variation of the functional

$$\Pi\,(u) = F(u) - \int_{\Omega} p\cdot u\,d\Omega + \int_{\Gamma} r(u)\cdot(\bar{u}-u)\,ds$$
$$= F(u) - R(u) - \int_{\Omega} p\cdot u\,d\Omega + \int_{\Gamma} r(u)\cdot\bar{u}\,ds \tag{4.40}$$

In the case of the second bvp

$$Du = p \quad \text{in } \Omega, \quad r(u) = \bar{r} \quad \text{on } \Gamma$$

the form $V(u, \hat{u})$ becomes after the same process

$$\delta\,\Pi\,(u,\,\hat{u}) = -\int_{\Omega} p\cdot\hat{u}\,d\Omega - \int_{\Gamma} \bar{r}\cdot\hat{u}\,ds + \delta\,F(u,\,\hat{u}) = 0$$

which is the first variation of the functional

$$\Pi\,(u) = F(u) - \int_{\Omega} p\cdot u\,d\Omega - \int_{\Gamma} \bar{r}\cdot u\,ds$$

In the case of the mixed bvp

$$Du = p \quad \text{in } \Omega, \quad u = \bar{u} \quad \text{on } \Gamma_1, \quad r(u) = \bar{r} \quad \text{on } \Gamma_2$$

the basic functional, naturally, is

$$\Pi\,(u) = F(u) + \int_{\Gamma_1} r(u)\cdot(\bar{u}-u)\,ds - \int_{\Omega} p\cdot u\,d\Omega - \int_{\Gamma_2} \bar{r}\cdot u\,ds$$

and its restriction to the class

$$R_1 = \{u \in C^m | u = \bar{u} \text{ on } \Gamma_1\}$$

is

$$\Pi\,(u)|_{R_1} = \Pi_1\,(u) = F(u) - \int_{\Omega} p\cdot u\,d\Omega - \int_{\Gamma_2} \bar{r}\cdot u\,ds$$

and its restriction to the class

$$R_2 = \{u \in C^{2m} | Du = p, r(u) = \bar{r} \text{ on } \Gamma_2\}$$

is

$$\begin{aligned}
\Pi(u)|_{R_2} = \Pi_2(u) &= -\int_\Omega p \cdot u \, d\Omega + \int_{\Gamma_1} r(u) \cdot (\bar{u} - u) \, ds - \int_{\Gamma_2} \bar{r} \cdot u \, ds + F(u) \\
&= -\int_\Omega Du \cdot u \, d\Omega + \int_{\Gamma_1} r(u) \cdot (\bar{u} - u) \, ds - \int_{\Gamma_2} r(u) \cdot u \, ds + F(u) \\
&= -E(u, u) + \int_{\Gamma_1} r(u) \cdot \bar{u} \, ds + F(u)
\end{aligned}$$

In linear mechanics (formally self-adjoint operators D) the functional $F(u)$ is simply

$$F(u) = \frac{1}{2} E(u, u)$$

and hence

$$\Pi_2(u) = -\frac{1}{2} F(u) + \int_{\Gamma_1} r(u) \cdot \bar{u} \, ds$$

We close this section with a proof of the rule concerning the Lagrange multipliers as stated in section 4.11.

Let $\Pi(u)$ be the basic functional of the first bvp, see Eq. (4.40). If we replace the force-term $r(u)$ by a function λ then the functional becomes

$$\Pi(u, \lambda) = F(u) - \int_\Omega p \cdot u \, d\Omega + \int_\Gamma \lambda \cdot (\bar{u} - u) \, ds$$

Now, let u, λ a stationary point of the functional, i.e.

$$\begin{aligned}
\delta \Pi(u, \lambda; \hat{u}, \hat{\lambda}) &= \delta F(u, \hat{u}) - \int_\Omega p \cdot \hat{u} \, d\Omega + \int_\Gamma \hat{\lambda} \cdot (\bar{u} - u) \, ds - \int_\Gamma \lambda \cdot \hat{u} \, ds \\
&= 0 \quad \forall \, \hat{u}, \hat{\lambda}
\end{aligned}$$

then follows, if we replace $\delta F(u, \hat{u}) = E(u, \hat{u})$ by Eq. (4.37),

$$\int_\Omega (Du - p) \cdot \hat{u} \, d\Omega + \int_\Gamma (r(u) - \lambda) \cdot \hat{u} \, ds + \int_\Gamma \hat{\lambda} \cdot (\bar{u} - u) \, ds = 0 \quad \forall \, \hat{u}, \hat{\lambda}$$

and therewith

$$Du = p \quad \text{in } \Omega, \quad r(u) = \lambda \quad \text{on } \Gamma, \quad u = \bar{u} \quad \text{on } \Gamma$$

The proof is done.

5 Concentrated Forces

In this chapter we shall formulate

the principle of virtual displacements
the principle of virtual forces
Betti's Theorem
the principle "eigenwork = int. energy"

when the structural elements (bars, beams, Kirchhoff plates and elastic plates or bodies) are loaded with concentrated forces.

The difficulty with concentrated forces is that the principles of structural mechanics of continua are integral identities and, therefore, to be applicable require the displacement functions to satisfy certain smoothness conditions. Simply stated we must be sure that the integrals exist and we must know in which sense.

The displacement functions which correspond to concentrated loads are the fundamental solutions and as these functions do not satisfy the necessary smoothness conditions the formulation of the principles above requires a special care.

But before we discuss all this in more detail we better first introduce the fundamental solutions of the single structural elements.

Whenever a bar, a beam, a plate etc. is loaded in its interior with a concentrated force then the displacement is the superposition of a fundamental solution and a regular function $u_R \in C^{2m}$.

$$u[x] = g_0[x] + u_R$$

(If there are disturbances on the boundary then the regular function u_R might not be so regular after all, but this is negligible here).

The fundamental solution $g_0[x]$ is the displacement we observe at a point y of the infinite medium when a concentrated force of magnitude 1 acts at some (distant) point x. Hence, $g_0[x]$ depends on the position of the source point x and of the observation point y.

$$g_0[x] = g_0(y, x)$$

The source point x acts as a parameter, we put it between brackets, and y, the observation point, is the true variable.

5.1 Fundamental Solutions

In the following we list the fundamental solutions of the single structural elements.

5.1.1 The Bar, $-EA\,d^2/dy^2$

The fundamental solution of a bar is

$$g_0(y,\,x)=\frac{1}{EA}\begin{cases}(1-x)\,y & y\leqslant x\\(1-y)\,x & x\leqslant y\end{cases}\tag{5.1}$$

To $g_0[x]$ belongs the normal force

$$N_0[x]=N_0(y,\,x)=EA\,\frac{d}{dy}\,g_0(y,\,x)=\begin{cases}1-x & y\leqslant x\\-x & x\leqslant y\end{cases}$$

If the source point x lies in the interval $[0,1]$ then $g_0[x]$ is just the axial displacement $u(y)$ of a fixed bar loaded at the point x with a concentrated force $P=1$, see Fig. 5.1.

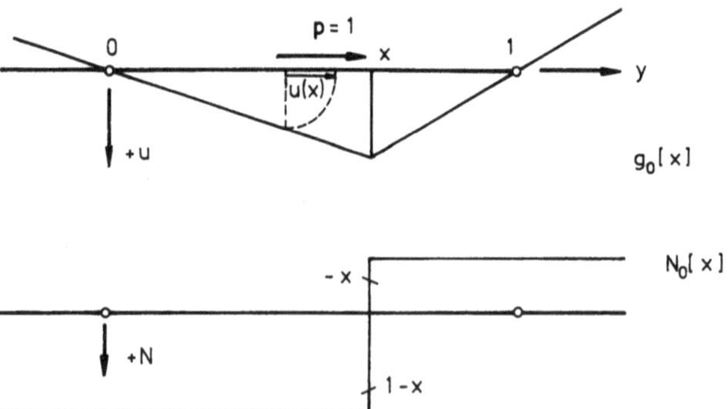

Figure 5.1

As an example consider the bar in Fig. 5.2 which is loaded at $x=l/2$ with a single force P. The displacement $u(y)$ of the bar

$$u(y)=\frac{P}{EA}\begin{cases}y & y\leqslant l/2\\[4pt]\dfrac{l}{2} & l/2\leqslant y\end{cases}$$

is the superposition, $u(y)=\{g_0[l/2]+u_R\}\,P$, of the fundamental solution in Eq. (5.1) and the regular solution $u_R=ly/2EA$.

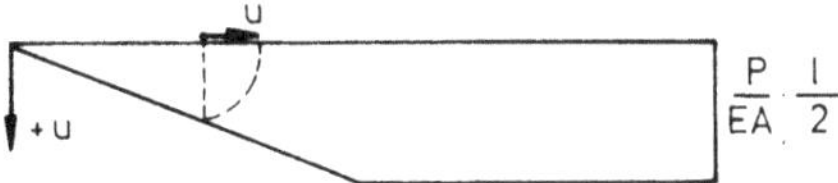

Figure 5.2

5.1.2 The Beam, $EI\,d^4/dy^4$

The fundamental solution of a beam is

$$g_0(y, x) = \frac{1}{6EI} \times \begin{cases} x(1-x)(2-x)\,y - (1-x)\,y^3 & y \leqslant x \\ (y-x)^3 + x(1-x)(2-x)\,y - (1-x)\,y^3 & x \leqslant y \end{cases}$$

$$(5.2)$$

To this function belong the quantities

$$w_0'[x] = \frac{1}{6EI} \times \begin{cases} x(1-x)(2-x) - 3y^2(1-x) & y \leqslant x \\ 3(y-x)^2 + x(1-x)(2-x) - 3y^2(1-x) & x \leqslant y \end{cases}$$

$$M_0[x] = \begin{cases} y(1-x) \\ x(1-y) \end{cases}, \qquad V_0[x] = \begin{cases} 1-x & y \leqslant x \\ -x & x \leqslant y \end{cases}$$

If the source point x lies in the interval $[0,1]$ then the fundamental solution is just the deflection $w(y)$ of a simply supported beam of length 1 loaded with a concentrated force $P = 1$ at x, see Fig. 5.3.

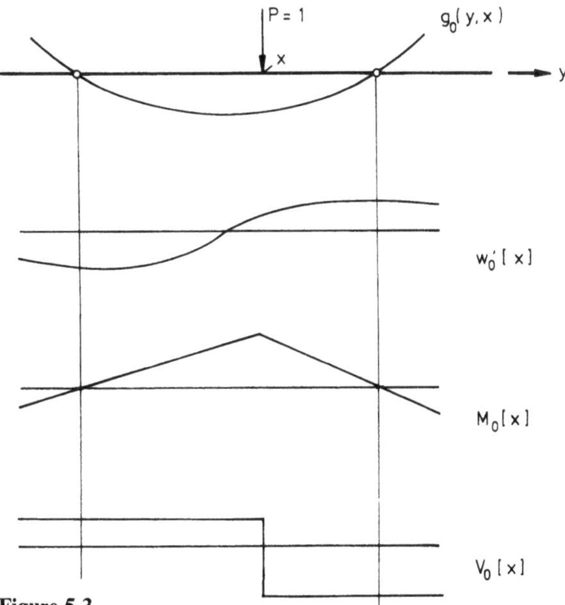

Figure 5.3

As an example consider the beam in Fig. 5.4

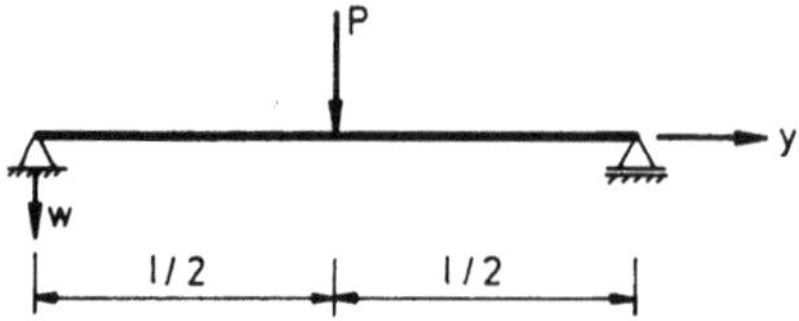

Figure 5.4

whose deflection $w(y) = \{g_0[l/2] + w_R\}\, P$ is a superposition of the fundamental solution $g_0[l/2]$ in Eq. (5.2) and the regular, homogeneous solution

$$w_R(y) = \frac{1}{6EI}\left\{\left(\frac{9}{8}l^2 - l - \frac{l^3}{8}\right)y - \frac{1}{2}(l-1)y^3\right\}$$

5.1.3 The Kirchhoff Plate, $K\,\Delta\Delta\,w$

The fundamental solution of the plate is the function

$$g_0(y, x) = \frac{1}{8\pi K}r^2\ln r \tag{5.3}$$

where $r = |y - x|$ is the distance between the observation point, $y = (y_1, y_2)$, and the source point, $x = (x_1, x_2)$.

If both points lie on the boundary Γ of a plate then we must distinguish between the normal and tangent at x or y. The first pair we call n and t and the second pair v and τ.

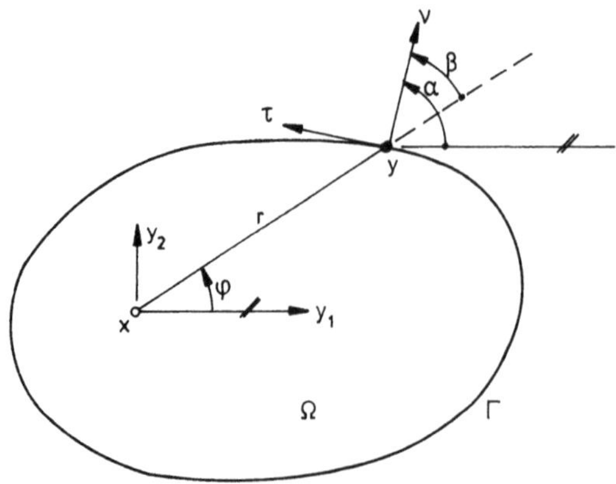

Figure 5.5

The force and displacement terms of the fundamental solution $g_0[x]$ at a boundary point y then are

$$\frac{\partial}{\partial v} g_0[x] = \frac{1}{8\pi K} rr_v(1 + 2\ln r)$$
(5.4)

$$M_v(g_0[x]) = -\frac{1}{8\pi}\{2(1+v)\ln r + (3+v) r_v^2 + (1+3v) r_\tau^2\}$$

$$= -\frac{2}{8\pi}(1+v)(1+\ln r) - \frac{1}{8\pi}(1-v)\cos 2\beta$$
(5.5)

$$V_v(g_0[x]) = -\frac{2}{8\pi r}\{2r_v + (1-v)(r_v - \varkappa r)(r_v^2 - r_\tau^2)\}$$

$$= -\frac{2}{8\pi r}\cos\beta[2 + (1-v)\cos 2\beta] + \frac{2(1-v)}{8\pi R}\cos 2\beta$$
(5.6)

$$M_{v\tau}(g_0[x]) = \frac{1}{8\pi}(1-v)\sin 2\beta$$
(5.7)

where

$$r_v = r_{,y_1} v_1(y) + r_{,y_2} v_2(y) = \cos\varphi\,\cos\alpha + \sin\varphi\,\sin\alpha = \cos\beta$$

and

$$r_\tau = r_{,y_1}\tau_1(y) + r_{,y_2}\tau_2(y) = \cos\varphi(-\sin\alpha) + \sin\varphi\,\cos\alpha = -\sin\beta$$

are the normal and tangential derivatives of the distance $r = |y - x|$ and $\varkappa = 1/R$ is the curvature of the boundary at y. The angle β is defined in Fig. 5.5.

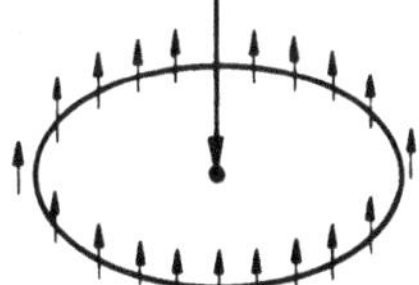

Figure 5.6

To become acquainted with the fundamental solution (5.3) let us calculate the force and displacement terms in the neighborhood of the source point, that is for points y which lie on a circle with radius r and center at x.

On such a circle it is $r_v = 1$, $r_\tau = 0$ and $\beta = 0$, $\alpha = \varphi$. Hence

$$\frac{\partial}{\partial v}(g_0[x]) = \frac{1}{8\pi K} r(1 + 2\ln r), \quad M_v(g_0[x]) = -\frac{(3+v)}{8\pi} - \frac{2}{8\pi}(1+v)\ln r$$

$$V_v(g_0[x]) = -\frac{4}{8\pi r}, \quad M_{v\tau}(g_0[x]) = 0$$
(5.8)

The bending moment M_v and the Kirchhoff-shear V_v have a singularity at the source point x and the functions do not depend on φ, that is the elastic state in the infinite plate is axi-symmetric.

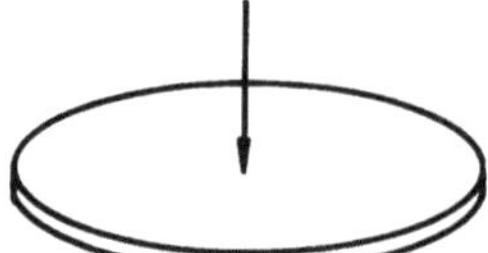

Figure 5.7

This symmetry is also observed in the problem of a circular, simply supported plate of radius R loaded with a concentrated force P at its center, see Fig. 5.7. The deflection of this plate

$$w(y) = \frac{P}{16\pi K}\left\{\frac{3+v}{1+v}R^2\left(1-\frac{r^2}{R^2}\right)-2r^2\ln R\right\}+\frac{P}{8\pi K}r^2\ln r$$

$$= \{w_R + g_0[0]\}\, P \tag{5.9}$$

is the superposition of $g_0[0]$ plus a regular solution.

5.1.4 The Elastic Plate and the Elastic Body

At an internal point of an elastic plate or body we might observe concentrated forces which point in the direction of the x_1, x_2 or x_1, x_2, x_3 -axis. Consequently there exist two or three fundamental solutions $g_0^i[x]$, $i=1,2$ or $i=1,2,3$, (Kelvin solutions).

These solutions are vector-valued functions with two or three components. If we place these vectors side by side then their components form a symmetric matrix, the Somigliana matrix U,

$$U[x] = \left[g_0^1[x], g_0^2[x]\right], \quad U[x] = \left[g_0^1[x], g_0^2[x], g_0^3[x]\right] \tag{5.10}$$

whose elements in two dimensions are

$$U_{ij} = U_{ji} = \frac{1}{8\pi\mu(1-v)}\left[(3-4v)\ln\frac{1}{r}\delta_{ij}+r_{,x_i}r_{,x_j}\right] \tag{5.11a}$$

and in three dimensions

$$U_{ij} = U_{ji} = \frac{1}{16\pi\mu(1-v)\,r}\left[(3-4v)\,\delta_{ij}+r_{,x_i}r_{,x_j}\right] \tag{5.11b}$$

Let us denote by $S_0^i[x] = S_0^i(y, x)$ the stress tensor which belongs to the fundamental solution $g_0^i[x]$. If y is a point on the surface of a volume which contains the source point x, see Fig. 5.8, then the traction vector at y is the product of this stress tensor with the exterior unit normal v at y.

$$S_0^i(y, x)\, v(y) = t_0^i[x]$$

In terms of the traction operator, see Eq. (1.39), this can be expressed as

$$\tau_y(g_0^i[x]) =: t_0^i[x] \tag{5.12}$$

where the subscript y indicates that we differentiate with respect to y.

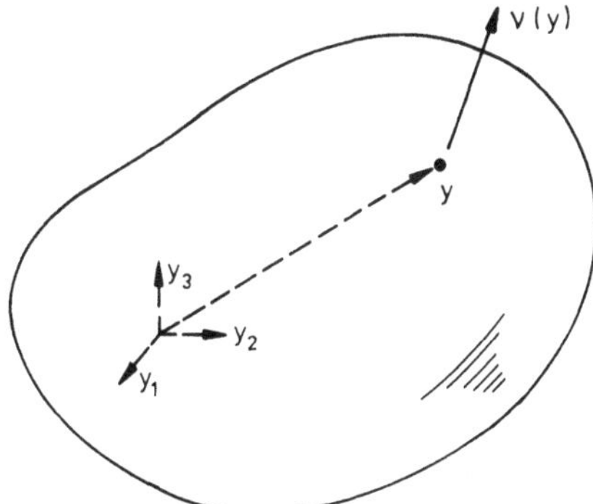

Figure 5.8

The traction vectors of the two and three fundamental solutions are the rows of two unsymmetric matrices $T(y, x)$

$$T[x] = [t_0^1[x], t_0^2[x]]^T, \quad T[x] = [t_0^1[x], t_0^2[x], t_0^3[x]]^T \tag{5.13}$$

whose elements are

$$T_{ij} = -\frac{1}{4\,\alpha\,\pi(1-v)\,r^\alpha}\left[\frac{\partial r}{\partial v}\left\{(1-2v)\,\delta_{ij} + \beta\, r_{,y_i} r_{,y_j}\right\}\right.$$
$$\left. - (1-2v)\left\{r_{,y_i} v_j(y) - r_{,y_j} v_i(y)\right\}\right] \tag{5.14}$$

where $\alpha = 1$, $\beta = 2$ in 2-D and $\alpha = 2$, $\beta = 3$ in 3-D.

To become acquainted with the behaviour of the fundamental solutions we consider an infinite elastic plate when it is loaded at a point x with a force $P = 1$ that points into the direction of the x_1-axis, see Fig. 5.9.

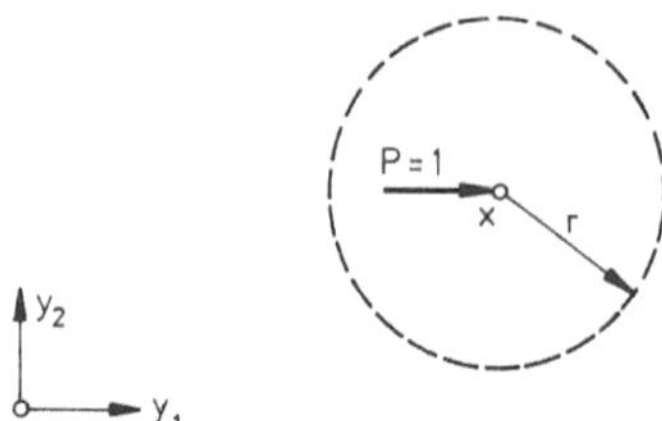

Figure 5.9

The displacement field of the infinite elastic medium under the action of this force is

$$g_0^1[x] = \{U_{11}, U_{12}\}^T$$

and its values on a circle with radius r and center at x, see Fig. 5.9, are (horizontal displacement)

$$U_{11} = \frac{1}{8\pi\mu(1-v)}\left[(3-4v)\ln\frac{1}{r} + \cos^2\varphi\right]$$

and (vertical displacement)

$$U_{12} = \frac{1}{8\pi\mu(1-v)}[\cos\varphi\,\sin\varphi]$$

The horizontal displacement, U_{11}, tends to infinity if r tends to zero while the vertical displacement remains bounded but oscillates considerably and the more so the more we approach the source point x. A small change in the direction φ from which we near x causes a large change in U_{12}. Such an oscillation can also be observed in the horizontal displacement, though it is, compared with the singularity $-\ln r$, negligible.

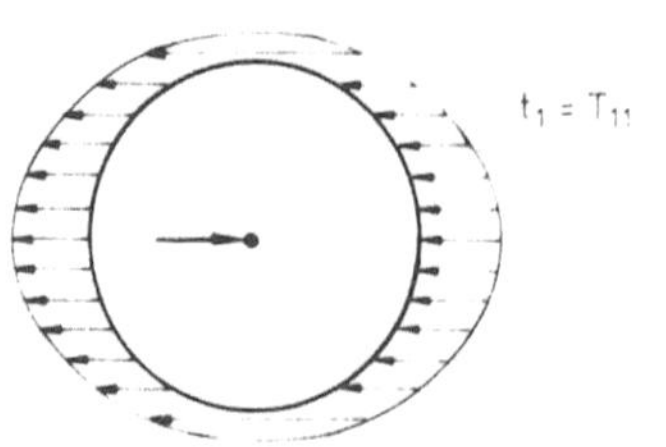

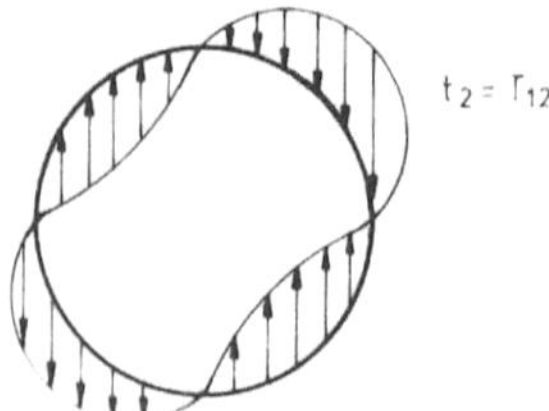

Figure 5.10

The components of the traction vector $t = \{t_1, t_2\} = \{T_{11}, T_{12}\}^T$ on the circle are

$$T_{11} = -\frac{1}{4\pi(1-v)\,r}[(1-2v) + 2\cos^2\varphi],$$

$$T_{12} = -\frac{1}{4\pi(1-v)\,r}[2\cos\varphi\,\sin\varphi]$$

see Fig. 5.10, and we easily verify that they satisfy the equilibrium conditions

$$\oint T_{11}\, ds = -\frac{1}{4\pi(1-v)} \int_0^{2\pi} [(1-2v)+2\cos^2\varphi]\, d\varphi = -1$$

$$\oint T_{12}\, ds = -\frac{2}{4\pi(1-v)} \int_0^{2\pi} \cos\varphi\, \sin\varphi\, d\varphi = 0$$

5.1.5 Summary

All fundamental solutions are symmetric functions

$$g_0(y,\, x) = g_0(x,\, y)$$

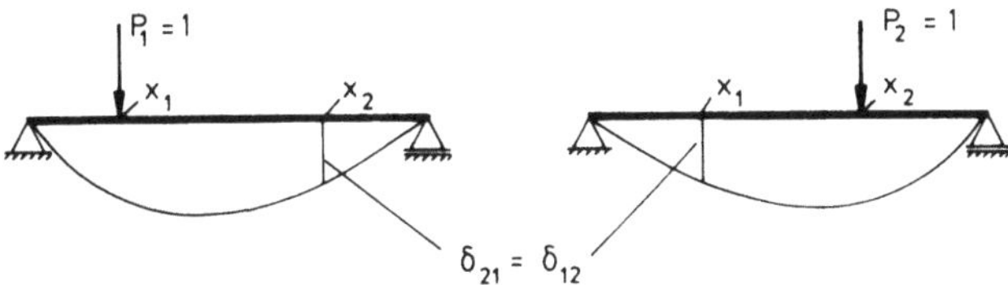

Figure 5.11

this is known as *Maxwell's principle*, see Fig. 5.11.

$$\delta_{21} = g_0(x_2,\, x_1) = g_0(x_1,\, x_2) = \delta_{12}$$

and they are homogeneous solutions of the governing equation

$$D_y g_0(y,\, x) = D_x g_0(y,\, x) = 0$$

at all points y which do not coincide with x and they are infinitely often differentiable at all such points.

(The subscripts y and x are to denote that differentiation is done with respect to y and x. That differentiation with respect to y as to x must have the same effect is obvious since the functions are symmetric). The exception is the source point x. There the solution is no longer infinitely often differentiable, not even $2\,m$-times (what would be sufficient).

Consider, e. g., a beam loaded with a concentrated force. At the point where the single force acts the shear force V, the third derivative, jumps. Hence, the deflection is no longer in $C^4[0,l]$.

On the other hand, the source point is the only exceptional point. At every other point y, it may be as close as possible to x, the fundamental solutions are infinitely smooth.

Or else, the fundamental solutions do belong to $C^\infty(\bar{\Omega} - \{x\})$ but not to $C^{2m}(\bar{\Omega})$.

5.2 Fundamental Solutions and Integral Identities

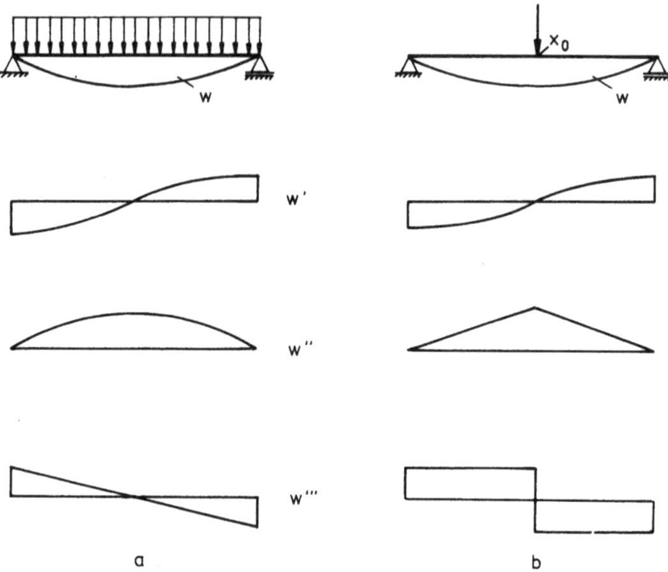

Figure 5.12

To illustrate the difficulties which these properties of fundamental solutions entail consider the beam in Fig. 5.12a.

If the beam is loaded with an evenly distributed force p then its deflection w belongs to $C^4[0,l]$. (This is a simple consequence of the equation $EIw^{IV}=p$). Hence, it is permissible to substitute the deflection w into the first identity, $1/2\,G(w,w)$, the principle "eigenwork = int. energy" applies.

But if a concentrated force acts on the same beam then the shear force jumps and, consequently, the deflection no longer is in $C^4[0,l]$ but only, see Fig. 5.12b, in $C_p^3[0,l]$. Hence, it is not permissible to substitute w into the first identity and we, therefore, face the question: how do we formulate the principle of "eigenwork = int. energy" or other principles if the displacements are no longer in C^{2m} but are worse?.

This problem is, in our case, identical with the problem to define the first and second identity for fundamental solutions because the singularity is caused by concentrated forces.

The reader might object that we formulated these principles already in chapter 3 for frames loaded with concentrated forces and he may ask: why suddenly this break in the routine?

But if he looks back then he will realize that we only treated concentrated forces which acted on bars and beams but not on Kirchhoff plates and elastic plates or bodies.

If the shear force of a beam jumps at x then it is always possible to divide the beam into two parts, the interval to the left and to the right of x. On the left-hand side the deflection belongs to $C^4[0,x]$ and on the right-hand side to $C^4[x,l]$; this is easily verified with the help of Fig. 5.12b.

The beam divided into two parts can be considered a continuous beam and the procedure employed to formulate the basic principles for continuous beams, the summation of zeros, $0+0=0$, works here as well.

But this technique is no longer applicable in two- or three-dimensional domains.

The difference is that the shear force of a beam only jumps at the source point (the singularity is distinct) while the Kirchhoff-shear (the third derivative) of a plate behaves as r^{-1} (the singularity is "blurred").

Consequently, if a cut through x divides the plate in two parts, Ω_l and Ω_r, then the deflection $w[x] = g_0[x] + w_R$ belongs neither to $C^4(\bar{\Omega}_l)$ nor to $C^4(\bar{\Omega}_r)$ and it would, therefore, be an error to simply substitute $w[x]$ into the first identity and to claim that

$$G(w[x], \hat{w})_{\Omega_l} = 0, \qquad G(w[x], \hat{w})_{\Omega_r} = 0$$

The identies apply only if the functions w and $\hat{w}$ are four-times and two-times differentiable and this not only in the interior, in Ω_l and Ω_r, but up to the boundaries, up to the source point x, that is in $\bar{\Omega}_l$ and $\bar{\Omega}_r$.

An instructive example is the function $f(x) = x^{-0.5}$. This function belongs in the open interval $(0, 1)$ to $C^\infty(0, 1)$ but fails to satisfy even the meager conditions of the class $C^0[0, 1]$. The function is infinitely smooth in the interior but discontinuous at $x = 0$. Consequently $x^{-0.5}$ does not belong to $C^0[0, 1]$ and, a fortiori, not to $C^1[0, 1]$. Hence, we may not transform an integral as

$$\int_0^1 x^{-0.5} \cos x \, dx$$

by integration by parts, at least not directly.

But indirectly. Because the function $x^{-0.5}$ belongs to $C^1[\varepsilon, l]$ whenever ε is a small positive quantity we are sure that we may integrate the expression

$$\int_\varepsilon^1 x^{-0.5} \cos x \, dx = [2\sqrt{x} \, \cos x]_\varepsilon^1 + \int_\varepsilon^1 2\sqrt{x} \, \sin x \, dx$$

by parts.

As this equation is valid for every $\varepsilon > 0$ it must be valid in the limit too, that is if ε tends to zero. The integral on the left-hand side tends to

$$\lim_{\varepsilon \to 0} \int_\varepsilon^1 x^{-0.5} \cos x \, dx = \int_0^1 x^{-0.5} \cos x \, dx$$

and the two terms on the right-hand side to

$$\lim_{\varepsilon \to 0} \left\{ [2\sqrt{x} \, \cos x]_\varepsilon^1 + \int_\varepsilon^1 2\sqrt{x} \, \sin x \, dx \right\} = [2\sqrt{x} \, \cos x]_0^1 + \int_0^1 2\sqrt{x} \, \sin x \, dx$$

Hence, it follows

$$\int_0^1 x^{-0.5} \cos x \, dx = [2\sqrt{x} \, \cos x]_0^1 + \int_0^1 2\sqrt{x} \, \sin x \, dx$$

This is the result we would have obtained if we had applied integration by parts regardless of legitimate or not. But this is rather the exception. All the integrands in the sections to come are such badly behaved that a blind application of the integration by parts rule would lead to incorrect results.

This simple example provides also the answer to the question how we extend the first and second identities to fundamental solutions.

The fundamental solutions are in any domain

$$\Omega_\varepsilon = \Omega - N_\varepsilon(x)$$

which does not contain the source point x regular functions

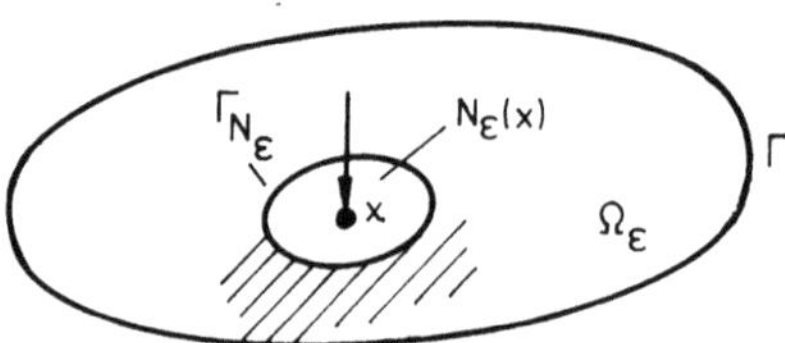

Figure 5.13

that is they belong in any such domain to $C^{2m}(\bar{\Omega}_\varepsilon)$. Hence, it is admissible to formulate the identities in any such domain Ω_ε.

$$G_\varepsilon(g_0[x], \hat{u}) = 0 \quad \forall \hat{u} \in C^m(\bar{\Omega}) \tag{5.15}$$

(The subscript ε indicates that integration is done over the domain Ω_ε and its boundary $\Gamma_\varepsilon = \Gamma \cup \Gamma_{N_\varepsilon}$, see Fig. 5.13).

Because the left-hand side of Eq. (5.15) is zero for all $\varepsilon > 0$, the limit must be zero as well.

$$\lim_{\varepsilon \to 0} G_\varepsilon(g_0[x], \hat{u}) = 0$$

or at full length and in the notation of section 2.10

$$G(g_0[x], \hat{u}) := \lim_{\varepsilon \to 0} \left\{ (Dg_0[x], \hat{u})_{\Omega_\varepsilon} - \sum_{i=1}^{m} (-1)^i \, [\partial^{2m-i} g_0[x], \partial^{i-1} \hat{u}]_{\Gamma_\varepsilon} \right.$$
$$\left. - E(g_0[x], \hat{u})_{\Omega_\varepsilon} \right\} = 0 \tag{5.16}$$

The meaning of $G(g_0[x], \hat{u})$ is now exactly this limit.
Similarly we understand $B(g_0[x], \hat{u})$ to be the limit

$$B(g_0[x], \hat{u}) := \lim_{\varepsilon \to 0} \left\{ (Dg_0[x], \hat{u})_{\Omega_\varepsilon} - \sum_{i=1}^{2m} (-1)^i \, [\partial^{2m-i} g_0[x], \partial^{i-1} \hat{u}]_{\Gamma_\varepsilon} \right.$$
$$\left. - (g_0[x], D\hat{u})_{\Omega_\varepsilon} \right\} = 0 \tag{5.17}$$

With these equations the definition is done but the job not yet, namely the calculation of these limits. This is not a trivial task.

While the sum of all the integrals must be zero in the limit (because this is true for every $\varepsilon > 0$) the limit of the single integral is yet unknown. The difficulty in the evaluation of Eq. (5.16) or (5.17) lies, exactly, in the calculation of these single limits.

When we calculate in the following now these limits and, therewith, extend the principles of chapter 2 to fundamental solutions then we do not consider the fundamental solutions alone but we add to every fundamental solution $g_0[x]$ from the beginning a regular function u_R from $C^{2m}(\bar{\Omega})$. These regular functions are considered to be homogeneous solutions of the governing equation

$$Du_R = 0$$

Such functions are, as the examples in section 5.2 show, natural companions of the fundamental solutions. They match the fundamental solutions with the boundary conditions of the domain.

Fig. 5.14 shows a cut along a diameter of the plate in Fig. 5.7. The two curves are the fundamental solution $g_0[x]$ and the regular solution w_R, see Eq. (5.9), and their superposition the deflection of the circular plate.

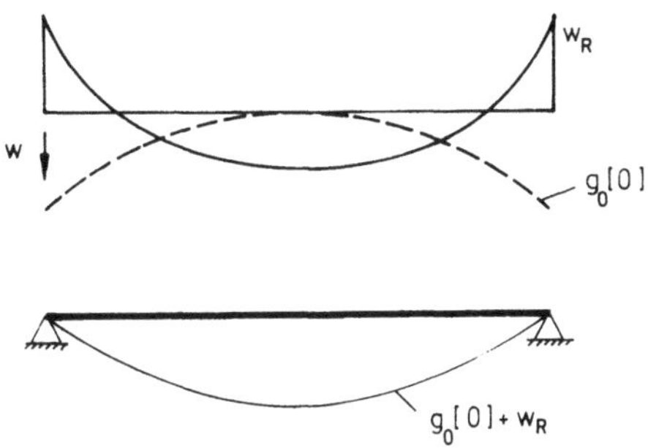

Figure 5.14

In the following we shall denote by

$$u[x] = u(y,x) = g_0(y,x) + u_R(y)$$

or by

$$w[x] = w(y,x) = g_0(y,x) + w_R(y)$$

the sum of fundamental solution plus regular solution and we shall also call these composite functions fundamental solutions (for the sake of simplicity).

To be complete we must consider a variety of combinations. The fundamental solution can be the first or the second argument in an identity, it can appear twice and then either the source points coincide or they do not. All these possibilities make a long list.

$$G(u[x], \hat{u}), \quad G(\hat{u}, u[x]), \quad B(u[x], \hat{u}), \quad G(u[x], u[x])$$

$$G(u[x^1], u[x^2]), \quad B(u[x^1], u[x^2]), \quad x^1 \neq x^2,$$

and our task in the next section is it to formulate these equations for bars, beams, Kirchhoff plates, elastic plates and bodies. To reduce, somewhat, the number of equations we do not list $G(\hat{u}, u[x]) = 0$, i.e. the first identity when the fundamental solution is the second argument. The reason is that this expression looks just as in chapter 2, as if $\{\hat{u}, u[x]\}$ were a couple of regular functions. No extra term, $c(x)\,\hat{u}(x)$, appears as in the other equations.

5.3 Results

To simplify the notation we mention the source point x only once, on the left-hand side, in $G(u[x], \hat{u})$, $B(u[x], \hat{u})$ etc. but no more so on the right-hand side where we write, e.g., simply u or N instead of $u[x]$ or $N[x]$.

If the magnitude of the concentrated force P is not unity, $P \neq 1$, then the displacement is

$$Pu[x] = P\{g_0[x] + u_R\}$$

where $u[x]$ is the displacement under the action of the force $P = 1$. The form $G(Pu[x], \hat{u}) = 0$ is obtained by multiplying $G(u[x], \hat{u}) = 0$ with P. The same rule applies to $G(\hat{u}, Pu[x])$ and $B(Pu[x], \hat{u})$ and to the forms

$$G(Pu[x^1], \hat{P}\hat{u}[x^2]) = 0, \quad G(Pu[x], Pu[x]) = 0$$

as well which are obtained by multiplying

$$G(u[x^1], \hat{u}[x^2]) = 0, \quad G(u[x], u[x]) = 0$$

with $P\hat{P}$ and PP resp.

5.3.1 Bars

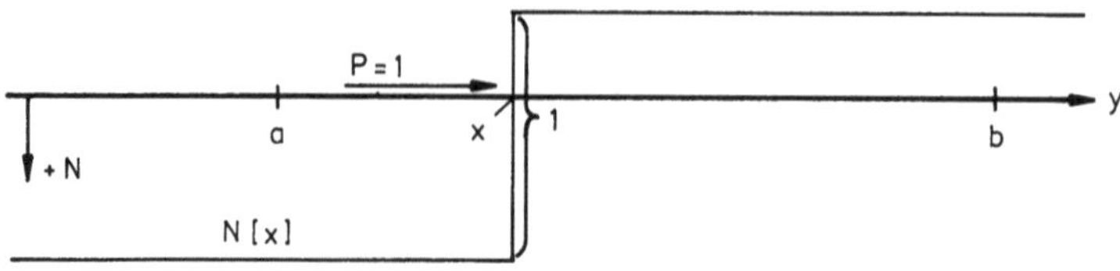

Figure 5.15

Let $x \in [a, b]$ and $u \in C^1[a, b]$. It is for every $\varepsilon > 0$

$$G_\varepsilon(u[x], \hat{u}) = \int_a^{x-\varepsilon} -EAu''\hat{u}\,dy + \int_{x+\varepsilon}^b -EAu''\hat{u}\,dy + [N\hat{u}]_a^{x-\varepsilon} + [N\hat{u}]_{x+\varepsilon}^b$$

$$- \int_a^{x-\varepsilon} \frac{N\hat{N}}{EA}\,dy - \int_{x+\varepsilon}^b \frac{N\hat{N}}{EA}\,dy = 0 \tag{5.18}$$

The function $u[x]$ is still continuous in $[a, b]$ but its first derivative jumps at x, see Fig. 5.15.

Furthermore, it is in Ω_ε

$$-EA\frac{d^2}{dy^2}u(y, x) = -EAg_0''[x] - EAu_R'' = 0$$

Substituting these results into Eq. (5.18) we obtain the following theorem.

Theorem 5.1

$$p:\ \hat{u} \in C^1[a, b],\ x \in \mathbb{R}^1$$

$$q:\ G(u[x], \hat{u}) = c(x)\,\hat{u}(x) + [N\hat{u}]_a^b - \int_a^b \frac{N\hat{N}}{EA}\,dy = 0$$

Similarly the next theorem is obtained.

Theorem 5.2

$$p:\ \hat{u} \in C^2[a, b],\ x \in \mathbb{R}^1$$

$$q:\ B(u[x], \hat{u}) = c(x)\,\hat{u}(x) + [N\hat{u}]_a^b - [u\hat{N}]_a^b - \int_a^b u(-EA\hat{u}'')\,dy = 0$$

In both theorems the function

$$c(x) = \begin{cases} 1 & \text{if } \ x \in (a, b) \\ 0 & \text{else} \end{cases} \tag{5.19}$$

is the characteristic function of the open interval (a, b), see Fig. 5.16.

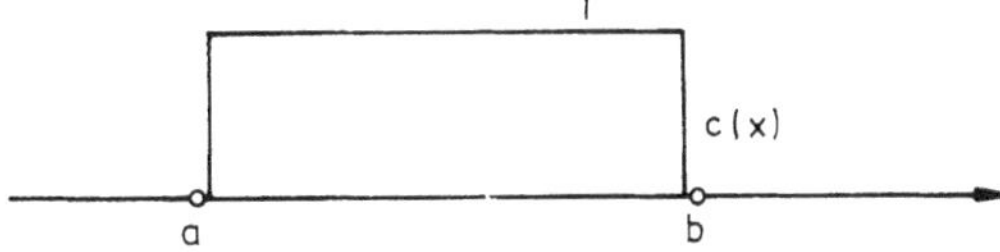

Figure 5.16

Note that the end points of the interval, a and b, do not belong to (a,b) and, consequently, $c(a) = c(b) = 0$.

The reason for the introduction of this function $c(x)$ is the following:

If the source point x lies outside of (a, b) then $u[x] = u(y, x)$ belongs, as a function of y, to $C^2[a, b]$ and, therefore, as demonstrated in chapter 2

$$G(u[x], \hat{u}) = \qquad [N\hat{u}]_a^b - \int_a^b \frac{N\hat{N}}{EA}\,dy = 0$$

If x lies in (a, b) then, as demonstrated above,

$$G(u[x], \hat{u}) = \hat{u}(x) + [N\hat{u}]_a^b - \int_a^b \frac{N\hat{N}}{EA}\,dy = 0$$

The two expressions differ only in the term $\hat{u}(x)$. The introduction of $c(x)$ is a convenient way to represent these two equations by one single equation alone.

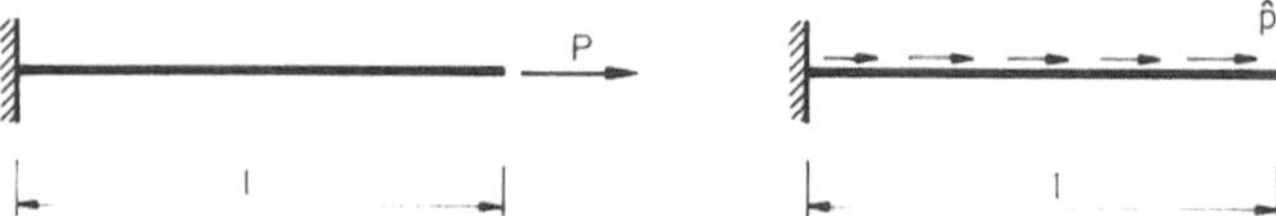

Figure 5.17

An illustrative application of Theorem 5.2, Betti's principle, provides the bar in Fig. 5.17 which we once load with a single force and once with evenly distributed forces $\hat{p}$. The corresponding displacements are

$$u[l] = \frac{y}{EA} \quad (P=1), \quad \hat{u}(y) = \frac{\hat{p}}{EA}\left(yl - \frac{y^2}{2}\right)$$

Substituting these two displacements into the second identity renders

$$B(Pu[l], \hat{u}) = P\Big\{ c(l)\,\hat{u}(l) + N(l)\,\hat{u}(l) - N(0)\,\hat{u}(0) - u(l)\,\hat{N}(l)$$

$$+ u(0)\,\hat{N}(0) - \int_0^l u\hat{p}\,dy \Big\} = P\hat{u}(l) - \int_0^l \frac{Py}{EA}\,\hat{p}\,dy = 0$$

The first term in the last equation is the work of the force P acting through $\hat{u}(l)$ and the second term the work of the distributed forces $\hat{p}$ acting through the displacements $Pu[l]$. Simple calculation confirms that, indeed, the reciprocal work of the two displacements is equal.

$$P\hat{u}(l) - \int_0^l \frac{Py}{EA}\,\hat{p}\,dy = P\left\{ \frac{\hat{p}l^2}{2EA} - \frac{\hat{p}l^2}{2EA} \right\} = 0$$

5.3.2 Beams

The deflection $w[x] = g_0[x] + w_R$ is a homogeneous solution in $\Omega - N_\varepsilon(x)$

$$EI \frac{d^4}{dy^4} w(y,x) = EI g_0^{IV}(y,x) + EI w_R^{IV}(y) = 0$$

while its third derivative jumps at x, see Fig. 5.18,

$$\lim_{y \to x_-} V[x] - \lim_{y \to x_+} V[x] = 1$$

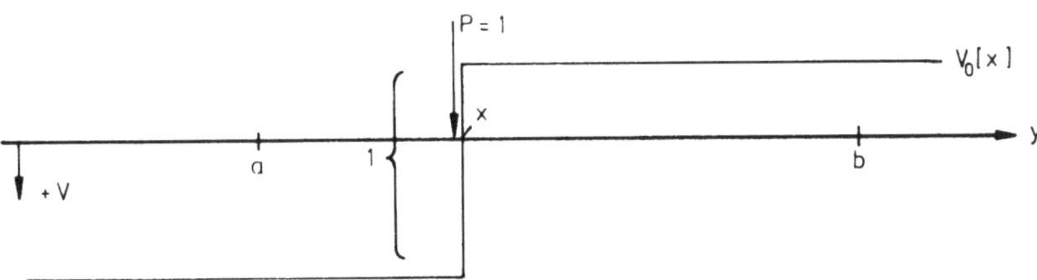

Figure 5.18

These properties render the following results.

Theorem 5.3

$$p: \quad \hat{w} \in C^2[a,b]$$

$$q: \quad G(w[x], \hat{w}) = c(x)\,\hat{w}(x) + [V\hat{w} - M\hat{w}']_a^b - \int_a^b \frac{M\hat{M}}{EI}\,dy = 0 \qquad (5.20)$$

Theorem 5.4

$$p: \quad \hat{w} \in C^4[a,b]$$

$$q: \quad B(w[x], \hat{w}) = c(x)\,\hat{w}(x) + [V\hat{w} - M\hat{w}']_b^a$$

$$- [w\hat{V} - w'\hat{M}]_a^b - \int_a^b w\,EI\hat{w}^{IV}\,dy = 0$$

The function $c(x)$ has the same meaning as before, see Eq. (5.19).

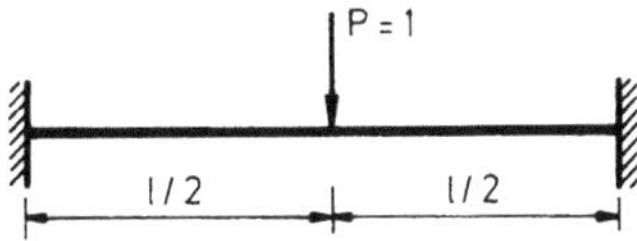

Figure 5.19

A chance to apply these theorems offers the beam in Fig. 5.19 which is loaded with a concentrated force $P = 1$ at $x = l/2$. This force causes the deflection

$$w\left[\frac{l}{2}\right] = \frac{ly^2}{48\,EI}\left(3 - \frac{4y}{l}\right) \qquad y \leqslant \frac{l}{2}$$

The principle of virtual displacements, Theorem 5.3, formulated with the virtual displacement $\hat{w} = 1$

$$G\left(w\left[\frac{l}{2}\right], 1\right) = c\left(\frac{l}{2}\right) + V(l) - V(0) = 1 + V(l) - V(0) = 0 \tag{5.21}$$

is just the equilibrium condition with regard to the vertical forces $V(0) = 1/2$, $V(l) = -1/2$ of the deflection $w[l/2]$.

Next, assume that the same beam is loaded with distributed forces $\hat{p}$, as in Fig. 5.20. The corresponding deflection is

$$\hat{w}(y) = \frac{\hat{p}\,l^4}{24\,EI}\,(\eta^2 - 2\eta^3 + \eta^4), \qquad \eta = \frac{y}{l}$$

If we let this deflection act on the first beam as a virtual displacement then the principle of virtual displacements formulates

$$G\left(w\left[\frac{l}{2}\right], \hat{w}\right) = c\left(\frac{l}{2}\right)\hat{w}\left(\frac{l}{2}\right) + [V\hat{w} - M\hat{w}']_0^l$$

$$- \int_0^l \frac{M\hat{M}}{EI}\,dy = \hat{w}\left(\frac{l}{2}\right) - \int_0^l \frac{M\hat{M}}{EI}\,dy = 0$$

which is just the well known result

$$\hat{w}\left(\frac{l}{2}\right) = \int_0^l \frac{M\hat{M}}{EI}\,dy$$

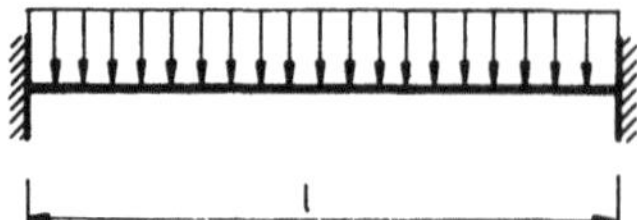

Figure 5.20

5.3.3 The Kirchhoff Plate

The derivation of the same results in two and three dimensions is of a rather technical nature and we, therefore, state only the results.

Theorem 5.5

p: $\hat{w} \in C^2(\bar{\Omega})$, $x \in \mathbb{R}^2$, Ω is a bounded domain with a smooth or piecewise smooth boundary (n corner points x^i, $i = 1, 2 \ldots n$).

$$c(x) = \begin{cases} 1 & x \in \Omega \\ \Delta\varphi/2\pi & x \in \Gamma \\ 0 & x \in \Omega^c \end{cases} \tag{5.22}$$

$\Delta\varphi = \Delta\varphi_i$ is the internal angle of the boundary point x, see section 1.3.

$$q:\ G(w[x], \hat{w}) = c(x)\,\hat{w}(x) + \int_\Gamma \left(V_\nu \hat{w} - M_\nu \frac{\partial\hat{w}}{\partial\nu} \right) ds_y + [[M_{\nu\tau}\hat{w}]]$$
$$- E(w, \hat{w}) = 0 \tag{5.23}$$

The following rule A applies:

If the point x is a corner point, $x = x^k$, then the contribution of x^k to the sum $[[M_{\nu\tau}[x]\,\hat{w}]]$ is neglected. In such a case

$$[[M_{\nu\tau}[x]\,\hat{w}]] = \sum_{\substack{i=1 \\ i \neq k}}^n \{M_{\nu\tau}(\hat{w})\,(y^i_+) - M_{\nu\tau}(\hat{w})\,(y^i_-)\}$$

Theorem 5.6

p: $\hat{w} \in C^4(\bar{\Omega})$

$$q:\ B(w[x], \hat{w}) = c(x)\,\hat{w}(x) + \int_\Gamma \left(V_\nu \hat{w} - M_\nu \frac{\partial\hat{w}}{\partial\nu} \right) ds_y + [[M_{\nu\tau}\hat{w}]]$$

$$- \int_\Gamma \left(w\hat{V}_\nu - \frac{\partial w}{\partial\nu}\,\hat{M}_\nu \right) ds_y - [[w\,\hat{M}_{\nu\tau}]] - \int_\Omega w\,K\Delta\Delta\hat{w}\,d\Omega_y = 0 \tag{5.24}$$

If x is a boundary point then rule A applies, see Theorem 5.5. For a proof of this theorem we refer to $[S1]$.

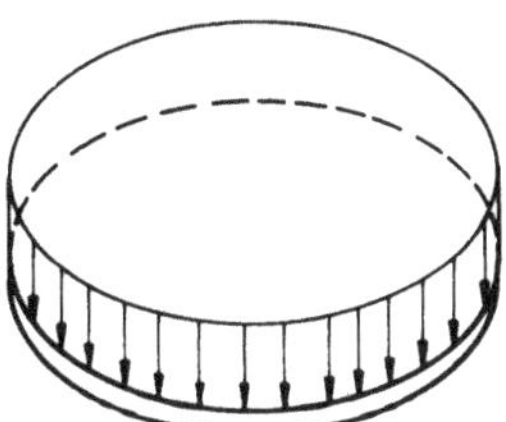

Figure 5.21

As an application of these results consider the simply supported and uniformly loaded plate with radius R in Fig. 5.21. The deflection of this plate

$$\hat{w}(y) = \frac{\hat{p}}{64K}\left\{\frac{5+v}{1+v}R^4 - 2\frac{3+v}{1+v}R^2 r^2 + r^4\right\}, \quad r = |y - \mathbf{0}|$$

has its maximum at the center

$$\hat{w}(\mathbf{0}) = \frac{\hat{p}}{64K}\left\{\frac{5+v}{1+v}R^4\right\} \tag{5.25}$$

Assume we let this deflection $\hat{w}$ act as a virtual displacement on the plate in Fig. 5.7, the plate loaded with a concentrated force. Then according to the principle of virtual displacements

$$G(w[\mathbf{0}], \hat{w}) = c(\mathbf{0})\,\hat{w}(\mathbf{0}) + \int_\Gamma \left(V_v \hat{w} - M_v \frac{\partial \hat{w}}{\partial v}\right) ds_y + [[M_{v\tau}\hat{w}]] - E(w, \hat{w})$$

$$= \hat{w}(\mathbf{0}) - E(w, \hat{w}) = 0$$

the work done by the concentrated load $P = 1$ acting through the deflection $\hat{w}(\mathbf{0})$ is stored as virtual strain energy.

When we balance the reciprocal external work of the two deflections then we formulate the second identity or Betti's principle

$$B(w[\mathbf{0}], \hat{w}) = \hat{w}(\mathbf{0}) - \int_\Omega w\hat{p}\, d\Omega = 0 \tag{5.26}$$

Let us check this equation by computing the domain integral

$$\int_\Omega w\hat{p}\, d\Omega = \frac{2\pi\hat{p}}{16\pi K}\int_0^R\left[\frac{3+v}{1+v}R^2\left(1 - \frac{r^2}{R^2}\right) - 2r^2 \ln R + r^2 \ln r^2\right]r\, dr$$

$$= \frac{\hat{p}}{8K}\left\{\frac{3+v}{1+v}\frac{R^4}{4} - \frac{1}{2}R^4 \ln R + \frac{1}{2}R^4 \ln R - \frac{R^4}{8}\right\}$$

$$= \frac{\hat{p}}{64K}\left\{\frac{5+v}{1+v}R^4\right\}$$

The result is just $\hat{w}(\mathbf{0})$, see Eq. (5.25), hence, Eq. (5.26) is correct.

5.3.4 Elastic Plates and Bodies

The two and three vector-valued fundamental solutions $g_0^i[x]$ have components $g_{0j}^i[x]$ and each component possesses a characteristic function $c_{ij}(x)$.

All these functions together form a (2×2) or (3×3) matrix $C(x)$ which is the unit matrix in the interior and the zero matrix in the complement, that is outside the plate or the body.

$$C(x) = \begin{cases} I, & x \in \Omega \\ \dot{C}(x), & x \in \Gamma \\ \mathbf{0}, & x \in \Omega^c \end{cases}, \quad c_{ij}(x) = \begin{cases} \delta_{ij} & x \in \Omega \\ \dot{c}_{ij}(x), & x \in \Gamma \\ 0 & x \in \Omega^c \end{cases} \tag{5.27}$$

The form of the boundary matrix $\dot{C}(x)$ depends on the dimension. In 2-D, the boundary is a curve and the value of $\dot{C}(x)$ is

$$\dot{C}(x) = \frac{\Delta\varphi}{2\pi}\, I + \frac{1}{4\pi(1-v)}\,\{J(\varphi_1) - J(\varphi_2)\}$$

where $J(\varphi)$ is the symmetric matrix

$$J(\varphi) = \begin{bmatrix} \dfrac{1}{2}\sin 2\varphi & \sin^2\varphi \\[2ex] \text{sym.} & -\dfrac{1}{2}\sin 2\varphi \end{bmatrix}$$

and the angles φ_1 and φ_2 the angles between the x_1-axis and the tangents t_1 and t_2, see Fig. 1.4; $\Delta\varphi = \varphi_1 - \varphi_2$ is the internal angle of the boundary point x in question.

At smooth boundary points ($\Delta\varphi = \pi$) the boundary matrix $\dot{C}(x)$ is simply

$$\dot{C}(x) = \frac{1}{2}\, I$$

1. Smooth boundary point

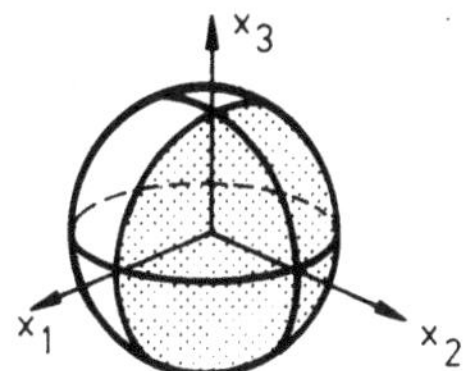

$$\dot{C} = \frac{1}{8\pi} \begin{vmatrix} 4\pi & 0 & 0 \\ 0 & 4\pi & 0 \\ 0 & 0 & 4\pi \end{vmatrix}$$

2. Edge

$$\dot{C} = \frac{1}{8\pi} \begin{vmatrix} 2\pi & 0 & 0 \\ 0 & 2\pi & \dfrac{2}{1-v} \\ 0 & \dfrac{2}{1-v} & 2\pi \end{vmatrix}$$

3. Corner

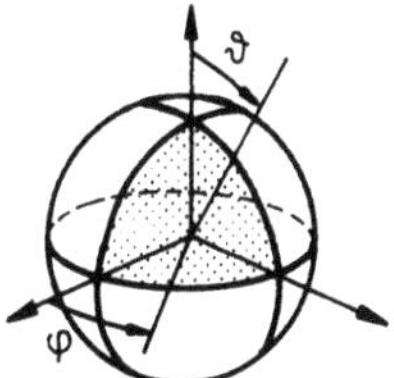

$$\dot{C} = \frac{1}{8\pi} \begin{vmatrix} \pi & \dfrac{1}{1-v} & \dfrac{1}{1-v} \\ \dfrac{1}{1-v} & \pi & \dfrac{1}{1-v} \\ \dfrac{1}{1-v} & \dfrac{1}{1-v} & \pi \end{vmatrix}$$

Figure 5.22

In 3-*D*, the boundary is a surface, and the boundary matrix $\dot{C}(x)$ the integral $(a = 1 - 2v)$

$$\dot{C}(x) = \frac{1}{8\pi(1-v)} \int_{A_i} \begin{bmatrix} a + 3r^2,_1 & 3r,_1 r,_2 & 3r,_1 r,_3 \\ & a + 3r^2,_2 & 3r,_2 r,_3 \\ \text{sym.} & & a + 3r^2,_3 \end{bmatrix} \sin\vartheta\, d\vartheta\, d\varphi \qquad (5.28)$$

where integration is done over that part, A_i, of the unit sphere which determines the internal angle of the boundary point x, see Eq. (1.4).

Note that at boundary points which have the same internal angle (the angle is the size of A_i) $\dot{C}(x)$ must not necessarily have the same value. The four corners of a cube have the same size, the same angle, but the location of A_i on the unit sphere (global coordinate system!) differs from point to point. As a consequence the signs of the single entries in $\dot{C}(x)$ change from point to point.

Fig. 5.22 shows three different boundary points and their matrices $\dot{C}(x)$: a smooth point, where A_i is one half of the unit sphere, a point on an edge, where A_i is one quarter of the sphere and a corner point, where A_i is one eighth of the sphere.

We are now in a position to formulate the single results for fundamental solutions as $u_0^i[x] = g_0^i[x] + u_R$ where u_R is a homogeneous displacement field from $C^2(\bar{\Omega})$.

Theorem 5.7

$$p: \hat{u} \in C^1(\bar{\Omega}),\ i = 1, 2\,(\mathbb{R}^2)\,;\ i = 1, 2, 3\,(\mathbb{R}^3)$$

$$q: G(u^i[x], \hat{u}) = c_{ij}(x)\,\hat{u}_j(x) + \int_\Gamma \tau(u^i) \cdot \hat{u}\, ds_y - E(u^i, \hat{u}) = 0$$

Note that the $\hat{u}_j$ are the components of the vector $\hat{u} = \{\hat{u}_j\}$ and summation has to be done with respect to j in the expression $c_{ij}\hat{u}_j$.

Theorem 5.8

$$p: \hat{u} \in C^2(\bar{\Omega}),\quad i = 1, 2\,(\mathbb{R}^2)\,;\quad i = 1, 2, 3\,(\mathbb{R}^3)$$

$$q: B(u^i[x], \hat{u}) = c_{ij}(x)\,\hat{u}_j(x) + \int_\Gamma \tau(u^i) \cdot \hat{u}\, ds_y - \int_\Gamma u^i \cdot \tau(\hat{u})\, ds_y$$

$$- \int_\Omega u^i \cdot (-L\hat{u})\, d\Omega_y = 0 \qquad (5.29)$$

If we let $u^i[x] = g_0^i[x]$, that is if we drop u_R, and if we formulate this equation with the two or three fundamental solutions $g^i[x]$ consecutively then we obtain the so-called Somigliana formula.

$$C(x)\,\hat{u}(x) + \int_\Gamma \left(T(y, x)\,\hat{u}(y) - U(y, x)\,\tau(\hat{u})\,(y) \right) ds_y$$

$$- \int_\Omega U(y, x)\,(-L\hat{u}(y))\, d\Omega_y = 0 \qquad (5.30)$$

The matrices U and T are defined in Eqs. (5.10) and 5.13)

5.4 Summary

We summarize in this section our results in the notation of section 2.10.

Imagine $u[x]$ to be the displacement of a structure and $\hat{u}$ to be a virtual displacement, Eq. (5.31), or $\hat{u}$ to be the displacement under the action of complementary virtual loads, Eq. (5.32), or $\hat{u}$ to be the displacement of the structure under a second system of loads, Eq. (5.33).

Principle of virtual displacements

$$p: \ \hat{u} \in C^m(\bar{\Omega})$$

$$q: \ G(u[x], \hat{u}) = c(x)\,\hat{u}(x) - \sum_{i=1}^{m} (-1)^i \left[\partial^{2m-i} u, \partial^{i-1} \hat{u}\right] - E(u, \hat{u}) = 0$$

$$(5.31)$$

Principle of virtual forces

$$p: \ \hat{u} \in C^{2m}(\bar{\Omega})$$

$$q: \ G(\hat{u}, u[x]) = (D\hat{u}, u) - \sum_{i=1}^{m} (-1)^i \left[\partial^{2m-1} \hat{u}, \partial^{i-1} u\right] - E(\hat{u}, u) = 0$$

$$(5.32)$$

Betti's principle

$$p: \ \hat{u} \in C^{2m}(\bar{\Omega})$$

$$q: \ B(u[x], \hat{u}) = c(x)\,\hat{u}(x) - \sum_{i=1}^{2m} (-1)^i \left[\partial^{2m-i} u, \partial^{i-1} \hat{u}\right] - (u, D\hat{u}) = 0$$

$$(5.33)$$

If D is the operator of elastic plates and bodies, then $u[x]$ has to be replaced by $u^i[x]$ and $c(x)\,u(x)$ by $c_{ij}\,u_j(x)$. If $D = K\Delta\Delta$ the work, $[[\ldots\ldots]]$, of the corner forces must be added.

5.5 An Extension

These principles can, immediately, be extended to pairs of functions as

$$\{u[x^1], u[x^2]\} = \{g_0[x^1] + u_R, g_0[x^2] + \hat{u}_R\}, \quad x^1 \neq x^2$$

that is when both functions are fundamental solutions which have their source at different points, $x^1 \neq x^2$. The point x^1 at which the first function has its singularity is a smooth point with respect to the second function and vice versa.

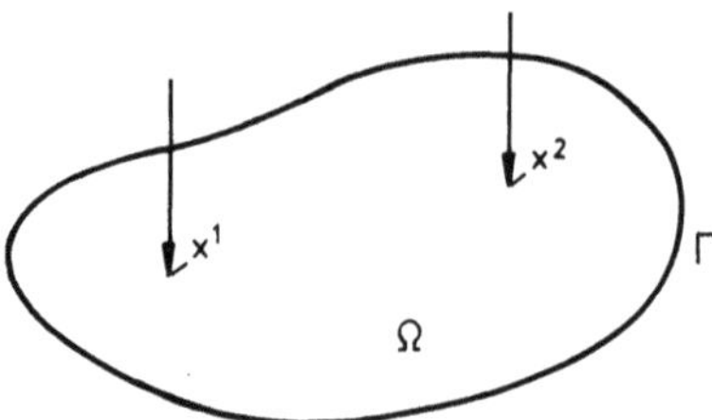

Figure 5.23

Theorem 5.9

$$p: x^1 \neq x^2$$

$$q: G(u[x^1], u[x^2]) = c(x^1)\, u(x^1, x^2)$$

$$- \sum_{i=1}^{m} (-1)^i \left[\partial^{2m-1} u[x^1],\, \partial^{i-1} u[x^2] \right] - E(u[x^1], u[x^2]) = 0$$

Theorem 5.10

$$p: x^1 \neq x^2$$

$$q: B(u[x^1], u[x^2]) = c(x^1)\, u(x^1, x^2)$$

$$- \sum_{i=1}^{2m} (-1)^i \left[\partial^{2m-i} u[x^1],\, \partial^{i-1} u[x^2] \right] - c(x^2)\, u(x^2, x^1) = 0$$

5.6 Theorems "eigenwork = int. energy"

A theorem "eigenwork = int. energy", typically, has the form

$$\frac{1}{2} G(u, u) = 0 \qquad u \in C^{2m}$$

In the case of functions as $u[x] = g_0[x] + u_R$ it becomes

$$\frac{1}{2} G(u[x], u[x]) = 0$$

This is a generalization of Theorem 5.9. Now the two points x^1 and x^2 coincide, both functions have their singularities at the same point, the singularities add up, and consequently the resulting singularity is worse than in the case of one singularity alone.

An exception from this rule are again the fundamental solutions of bars and beams because their singularities are distinct, the functions simply jump, but otherwise are well behaved. An extension of the foregoing results to the case $x^1 = x^2$ poses, therefore, no serious problems in one dimension.

Theorem 5.11 *"eigenwork = int. energy" for a bar loaded with a concentrated force*

$$p: \; u[x] = g_0[x] + u_R, \; u_R \in C^2[a, b]$$

$$q: \; \frac{1}{2} G(u[x], u[x]) = \frac{1}{2} c(x) \, u(x, x) + \frac{1}{2} [Nu]_a^b - \frac{1}{2} \int_a^b \frac{N^2}{EA} \, dy = 0$$

Theorem 5.12 *"eigenwork = int. energy" for a beam loaded with a concentrated force*

$$p: \; w[x] = g_0[x] + w_R, \; w_R \in C^4[a, b]$$

$$q: \; \frac{1}{2} G(w[x], w[x]) = \frac{1}{2} c(x) \, w(x, x) + \frac{1}{2} [Vw - Mw']_a^b - \frac{1}{2} \int_a^b \frac{M^2}{EI} \, dy = 0$$

$$(5.34)$$

The extension is also easy in the case of a Kirchhoff plate. The only (minor) difficulty poses the limit of the integral

$$\lim_{\varepsilon \to 0} E(w[x], w[x])_{\Omega_\varepsilon} \tag{5.35}$$

because the strains behave as $\ln r$ but due to $d\Omega = r \, dr \, d\varphi$ and

$$\lim_{r \to 0} r \ln^2 r = 0$$

the limit of the internal energy is bounded, that is the internal energy under the action of a concentrated load is finite.

Theorem 5.13 *"eigenwork = int. energy" for a Kirchhoff plate loaded with a concentrated force*

$$p: \; w[x] = g_0[x] + w_R, \; w_R \in C^4(\bar{\Omega})$$

$$q: \; \frac{1}{2} G(w[x], w[x]) = \frac{1}{2} c(x) \, w(x, x) + \frac{1}{2} \int_\Gamma \left(V_\nu w - M_\nu \frac{\partial w}{\partial \nu} \right) ds_y$$

$$+ \frac{1}{2} [[M_{\nu\tau} w]] - \frac{1}{2} E(w, w) = 0$$

In case x is a boundary point rule A applies, see Theorem 5.5.

Elastic plates and bodies pose more difficulties. Their internal energy is infinite, $E(g_0^i[x], g_0^j[x]) = \infty$, because the fundamental solutions behave as

$$g_0^i[x] = 0(\ln r) \; (\mathbb{R}^2), \quad g_0^i[x] = 0(r^{-1}) \; (\mathbb{R}^3)$$

The strains are the first derivatives of g_0^i, hence, in two dimensions

$$\varepsilon_{kl}(g_0^i) = 0(r^{-1})$$

and consequently the strain energy density (the integrand) behaves as

$$\varepsilon_{kl}(g_0^i)\, C^{klmn}\, \varepsilon_{mn}(g_0^i) = 0\,(r^{-2})$$

But in the plane where $d\Omega = r\,dr\,d\varphi$ an integral as (let Ω the unit circle)

$$\int_\Omega r^{-2}\, d\Omega = \int_0^{2\pi}\int_0^1 r^{-2}\, r\,d\varphi\,dr = 2\pi \int_0^1 r^{-1}\, dr = \infty$$

does not exist and, hence, also the internal energy is infinite. In agreement with this the displacement at the source point in the direction of the single load is infinite, too. The situation is the same in three dimensions.

It seems then that the formulation of the first identity for pairs as $\{g_0^i[x], g_0^j[x]\}$ is impossible because their strain energy is infinite. But this is not true. The singularities simply drop out.

Recall that the first identity $G(g_0^i[x], g_0^j[x])$ is defined as the limit of a sequence of regular integrals.

$$\lim_{\varepsilon \to 0} G_\varepsilon(g_0^i[x], g_0^j[x]) = G(g_0^i[x], g_0^j[x])$$

One of these regular integrals is the strain energy in the domain Ω_ε. This integral splits into the strain energy in the domain $\Omega_1 = \Omega - N_1(x)$ and the strain energy in the remainder, the ring shaped domain $\varepsilon < r < 1$, the domain in the vicinity of the singularity.

$$E(g_0^i, g_0^j)_{\Omega_\varepsilon} = E(g_0^i, g_0^j)_{\Omega_1} + E(g_0^i, g_0^j)_{(\varepsilon, 1)}$$

The second term is the critical, the singular term. This term tends to infinity when the radius ε of the hole shrinks to zero. But, luckily, this singular term is counter-balanced by an identical singular term with opposite sign, the integral of the tractions over the circle (or sphere) which forms the boundary of the hole $N_\varepsilon(x)$. Hence, in the limit the two singularities cancel each other.

$$\lim_{\varepsilon \to 0} \{E(g_0^i, g_0^j)_{(\varepsilon, 1)} - \int_{\Gamma N_\varepsilon(x)} g_0^i \cdot \tau_y(g_0^j)\, ds_y\} = 0$$

Naturally, this is not sheer luck. It must be so.

The identity $G_\varepsilon(g_0^i[x], g_0^j[x])$ is zero for every $\varepsilon > 0$ and, hence, also its limit. But this is only possible if the singularities drop out.

The results so obtained are the subject of the following theorems. To simplify matters somewhat we assumed that, if the source point x lies on the boundary, then its unit neighborhood, $|y - x| \leqslant 1$, $y \in \Gamma$, on Γ is not curved. This is, because the scale is relative, no real restriction. We can choose the unit distance arbitrarily small.

We first list the results for two dimensions, Theorems 5.14 and 5.15, and then for three dimensions, Theorems 5.16 and 5.17.

Theorem 5.14

$$p:\ \Omega \subset \mathbb{R}^2,\ i,j \in \{1,2\},\ \boldsymbol{x} \in \Omega\ \ (\textit{internal point})$$

$$q:\ G(\boldsymbol{g}_0^i[\boldsymbol{x}],\boldsymbol{g}_0^j[\boldsymbol{x}]) = \int_\Gamma \boldsymbol{t}_0^i \cdot \boldsymbol{g}_0^j\, ds_y + M\delta_{ij} - E(\boldsymbol{g}_0^i,\boldsymbol{g}_0^j)_{\Omega_1} = 0$$

where

$$M = \frac{3-2v}{32\pi\mu(1-v)^2} \tag{5.36}$$

Theorem 5.15

$$p:\ \Omega \subset \mathbb{R}^2,\ i,j \in \{1,2\},\ \boldsymbol{x} \in \Gamma\ (\textit{boundary point with angles}\ \varphi_1,\varphi_2)$$

$$q:\ G(\boldsymbol{g}_0^i[\boldsymbol{x}],\boldsymbol{g}_0^j[\boldsymbol{x}]) = \int_{\Gamma_1'(x)} \boldsymbol{t}_0^i \cdot \boldsymbol{g}_0^j\, ds_y + M_{ij}(\varphi_1,\varphi_2) - E(\boldsymbol{g}_0^i,\boldsymbol{g}_0^j)_{\Omega_1} = 0$$

where

$$M_{ij} = \frac{1}{c}\sum_{k=1}^{2}\int_{\varphi_1}^{\varphi_2} \{(1-2v)\,\delta_{ik} + 2r_{,i}\,r_{,k}\}\,r_{,j}\,r_{,k}\ d\varphi$$

$$r_{,1} = \cos\varphi,\quad r_{,2} = \sin\varphi,\quad c = 32\,\pi^2\mu(1-v^2)$$

Theorem 5.16

$$p:\ \Omega \subset \mathbb{R}^3,\ i,j \in \{1,2,3\},\ \boldsymbol{x} \in \Omega\ (\textit{internal point})$$

$$q:\ G(\boldsymbol{g}_0^i[\boldsymbol{x}],\boldsymbol{g}_0^j[\boldsymbol{x}]) = \int_\Gamma \boldsymbol{t}_0^i \cdot \boldsymbol{g}_0^j\, ds_y - E(\boldsymbol{g}_0^i,\boldsymbol{g}_0^j)_{\Omega_1} - N\delta_{ij} = 0$$

where

$$N = \frac{12v^2 - 22v + 11}{24\pi\mu(1-v)^2} \tag{5.37}$$

Theorem 5.17

$$p:\ \Omega \subset \mathbb{R}^3,\ i,j \in \{1,2,3\},\ \boldsymbol{x} \in \Gamma\ (\textit{boundary point})$$

$$q:\ G(\boldsymbol{g}_0^i[\boldsymbol{x}],\boldsymbol{g}_0^j[\boldsymbol{x}]) = \int_{\Gamma_1'(x)} \boldsymbol{t}_0^i \cdot \boldsymbol{g}_0^j\, ds + Q_{ij}(\boldsymbol{x}) - E(\boldsymbol{g}_0^i,\boldsymbol{g}_0^j)_{\Omega_1} - K_{ij}(\boldsymbol{x}) = 0$$

If x is a smooth boundary point then $Q_{ij}(x) = 0$ and $K_{ij}(x) = \frac{1}{2}N$ with N as in Eq. (5.37).

The meaning of $K_{ij}(x)$ and $Q_{ij}(x)$ is the following: The integrand in the energy integral $E(g_0^i, g_0^j)$ is of the form

$$C^{qrst} \varepsilon_{qr}(g_0^i)\, \varepsilon_{st}(g_0^j) = \frac{1}{r^4}\, k_{ij}(\varphi, \vartheta)$$

and the term $K_{ij}(x)$ is the integral

$$K_{ij}(x) = \int\limits_{\Gamma_{N_1(x)}} k_{ij}(\varphi, \vartheta)\, ds$$

of small $k_{ij}(y)$ over the boundary of the unit neighborhood $N_1(x)$.

Analogously, the integrand in the boundary integral splits into a factor which depends on the distance and a factor which depends on the direction

$$t_0^i[x] \cdot g_0^j[x] = -\frac{1}{r^3}\, q_{ij}(\varphi, \vartheta)$$

The integral of the latter

$$Q_{ij}(x) = \oint\limits_{\gamma} q_{ij}(\varphi, \vartheta)\, ds$$

over the closed curve γ (this curve γ forms when the unit sphere with center at x cuts the surface) is $Q_{ij}(x)$.

5.7 The Characteristic Functions

When we first encountered characteristic functions they were mere tools to simplify the notation. Now, in two and three dimensions, they have developed into functions in their own right and they, therefore, deserve our attention.

5.7.1 Their Origin

What we call the characteristic function of a Kirchhoff plate is in reality the limit of an integral, namely

$$\lim_{\varepsilon \to 0} \int\limits_{\Gamma_{N_\varepsilon(x)}} V_\nu(y, x)\, ds_y = \begin{cases} 1, & x \in \Omega \\ \Delta\varphi/2\pi, & x \in \Gamma \\ 0, & x \in \Omega^c \end{cases}$$

where $\Gamma_{N_\varepsilon}(x)$ is the boundary of the neighborhood $N_\varepsilon(x)$ of the source point x, see

Fig. 5.24. Else, the function $c(x)$ is the limit of the work done by the Kirchhoff-shear acting through the virtual displacement $\hat{w}=1$ when the radius of the circle (or segment) $\Gamma_{N_\varepsilon}(x)$ tends to zero.

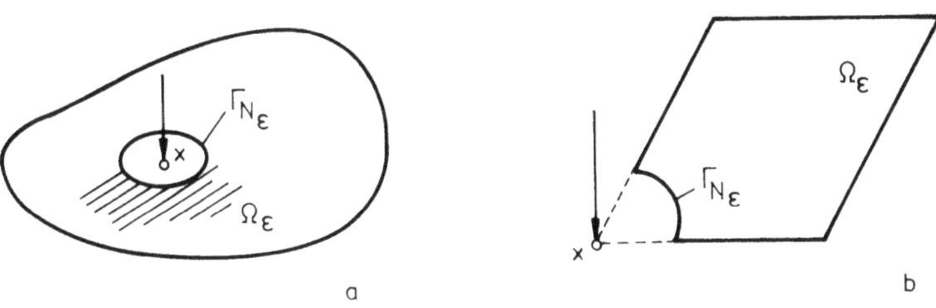

Figure 5.24

The value of the Kirchhoff-shear on $\Gamma_{N_\varepsilon}(x)$ is, see Eq. (5.8)

$$V_\nu(y,x)=\frac{1}{2\pi\varepsilon}, \quad \varepsilon=|y-x|$$

(Contrary to Eq. (5.8) the normal ν now points to x and not away from it, hence the different sign). In case $\Gamma_{N_\varepsilon}(x)$ is the full circle as in Fig. 5.24a, we obtain

$$\lim_{\varepsilon\to 0}\int_{\Gamma_{N_\varepsilon}(x)} V_\nu(y,x)\,ds_y=\lim_{\varepsilon\to 0}\int_0^{2\pi}\frac{1}{2\pi\varepsilon}\,\varepsilon\,d\varphi=1$$

and in case $\Gamma_{N_\varepsilon}(x)$ is only a quarter of a circle as in Fig. 5.24b then

$$\lim_{\varepsilon\to 0}\int_{\Gamma_{N_\varepsilon}(x)} V_\nu(y,x)\,ds_y=\lim_{\varepsilon\to 0}\int_0^{\pi/2}\frac{1}{2\pi\varepsilon}\,\varepsilon\,d\varphi=\frac{1}{4}$$

What we call the characteristic function of an elastic plate or body is the limit of the following integral

$$\lim_{\varepsilon\to 0}\int_{\Gamma_{N_\varepsilon}(x)} t_{0j}^i(y,x)\,ds_y=\begin{cases}\delta_{ij}, & x\in\Omega \\ \dot{c}_{ij}, & x\in\Gamma \\ 0, & x\in\Omega^c\end{cases}=c_{ij}(x)$$

where t_{0j}^i is the $j-th$ component of the traction vector t_0. In other words, $c_{ij}(x)$ is the limit of the work done by the component t_{0j}^i acting through the displacement component $\hat{u}_j=1$ when the radius ε of $\Gamma_{N_\varepsilon}(x)$ tends to zero.

5.7.2 A Mechanical Interpretation

Let a plate Ω with a smooth boundary be embedded into the infinite plate and assume a concentrated force $P=1$ acts at an interior point, see Fig. 5.25a. The equilibrium condition then requires that the concentrated force plus the integral of the Kirchhoff-shear over the boundary Γ is zero

$$1 + \int_\Gamma V_v(y, x)\, ds_y = 0 \qquad\qquad \text{if } x \in \Omega$$

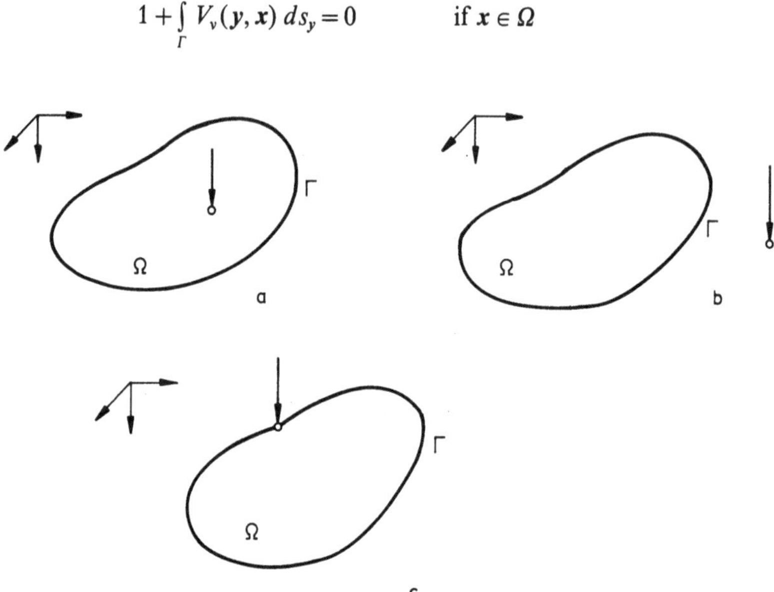

Figure 5.25

The same condition requires the integral of the Kirchhoff-shear alone to be zero if we place the concentrated force outside the plate Ω, as in Fig. 5.25 b.

$$\int_\Gamma V_v(y, x)\, ds_y = 0 \qquad\qquad \text{if } x \in \Omega^c$$

If the concentrated force acts on the boundary then the integrand is singular (the source point, the singular point, lies in the integration path) and the integral can only be understood in the sense of a Cauchy principal value. This value is

$$\int_\Gamma V_v(y, x)\, ds_y = -\frac{1}{2} \qquad\qquad \text{if } x \in \Gamma$$

If there are corners then the sum $[[M_{vt}]]$ of the corner forces, too, must be considered in the equilibrium conditions and, hence, the equations in all generality look like

$$-\int_\Gamma V_v(y, x)\, ds_y - [[M_{vt}[x]]] = \begin{cases} 1,\, x \in \Omega \\ \Delta\varphi/2\pi,\, x \in \Gamma \\ 0,\, x \in \Omega^c \end{cases} = c(x) \qquad (5.38)$$

Else, the function $c(x)$ is the integral of the Kirchhoff-shear over the boundary Γ (+ corner forces).

5.7.3 Integral Representation of $c(x)$

The same result, Eq. (5.38), is obtained if we choose in Eq. (5.24) for $\hat{w}$ the virtual displacement $\hat{w} = 1$.

$$B(g[x], 1) = c(x) + \int_\Gamma V_v(y, x)\, ds_y + \left[\left[M_{vt}[x]\right]\right] = 0$$

because this equation

$$c(x) = -\int_\Gamma V_v(y, x)\, ds_y - \left[\left[M_{vt}[x]\right]\right]$$

is just Eq. (5.38). We, thus, learn that the characteristic function of a Kirchhoff plate can be represented by a boundary integral.

The same holds true for the characteristic function $C(x)$ of an elastic plate or body. If we substitute into Eq. (5.30) for $\hat{u}$ the rigid-body movements $\hat{u} = \{1, 1\}$ and $\hat{u} = \{1, 1, 1\}$ then the result is an integral representation of the matrix $C(x)$.

$$C(x) = -\int_\Gamma T(y, x)\, ds_y$$

5.8 An Alternative

A fundamental solution $g_0[x]$ of a structural element, say a beam, can be considered the limit of a sequence of regular functions w_ε which are the deflection of the beam under the action of distributed forces p_ε which tend to the concentrated load $P = 1$ located at the source point x.

As all w_ε correspond to distributed forces they are all in $C^4[0, l]$ and, hence, one might ask: could we not formulate the principles of this chapter by substituting the function w_ε into, say, the first identity and letting then ε tend to zero? And if so, is the result the same? In other words, do we have

$$\lim_{\varepsilon \to 0} G(w_\varepsilon, w_\varepsilon) = \lim_{\varepsilon \to 0} G_\varepsilon(g_0[x], g_0[x]) \quad ?$$

Answer: we may assume so as we want to show in the following.

(The difficulty to prove this seemingly simple conjecture might be a hint as to why we did not opt for this alternative).

Consider the beam in Fig. 5.26 loaded with a series of distributed forces $p_\varepsilon(y, x)$

$$p_\varepsilon(y, x) = \begin{cases} 0 & \text{if } |y - x| \geqslant \varepsilon \\ \dfrac{1}{2\varepsilon}\left(1 + \cos\dfrac{\pi}{\varepsilon}(y - x)\right) & \text{if } |y - x| < \varepsilon \end{cases}$$

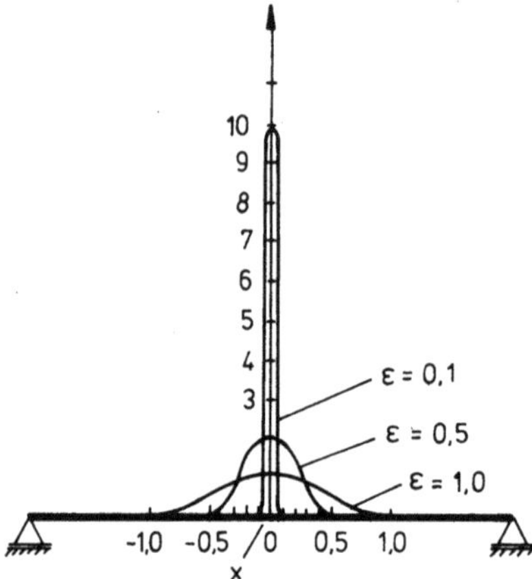

Figure 5.26

which are zero outside an ε-neighborhood of x and whose integral is always unity

$$\int_0^1 p_\varepsilon(y,x)\,dy = 1$$

When ε becomes smaller and smaller then these distributed loads resemble more and more a concentrated force of magnitude 1 acting at x.

Now let $\hat{w} \in C^2[0,l]$ a virtual displacement, then the work of the distributed force p_ε acting through this displacement is

$$\int_0^1 p_\varepsilon(y,x)\,\hat{w}(y)\,dy = \int_0^1 p_\varepsilon(y,x)\,(\hat{w}(y)-\hat{w}(x))\,dy + \int_0^1 p_\varepsilon(y,x)\,dy \ \hat{w}(x)$$

$$= \int_{x-\varepsilon}^{x+\varepsilon} p_\varepsilon(y,x)\,(\hat{w}(y)-\hat{w}(x))\,dy + \hat{w}(x)$$

Because of $p_\varepsilon = 0\,(\varepsilon^{-1})$ and $\hat{w}(y)-\hat{w}(x)=0(\varepsilon)$ (this is a consequence of $\hat{w} \in C^2[0,l]$) the integral in the last equation tends to zero and we, thus, obtain

$$\lim_{\varepsilon \to 0} \int_0^1 p_\varepsilon(y,x)\,\hat{w}(y)\,dy = \hat{w}(x) \tag{5.39}$$

that is, the work done by the distributed force $p_\varepsilon(y,x)$ acting through a virtual displacement $\hat{w}(y)$ tends to $\hat{w}(x)\cdot 1$.

Next, consider the deflection $w_\varepsilon(y)$ of the beam itself under the action of the distributed load $p_\varepsilon(y,x)$. The function $w_\varepsilon(y)$ satisfies the equation

$$EIw_\varepsilon^{IV}(y) = p_\varepsilon(y,x)$$

and the boundary conditions

$$w_\varepsilon(0) = w_\varepsilon(l) = w_\varepsilon''(0) = w_\varepsilon''(l) = 0$$

Each load p_ε is smooth, $p_\varepsilon \in C[0,1]$, hence each w_ε belongs to $C^4[0,1]$ and, consequently, it is admissible to formulate the first identity with any virtual deflection $\hat{w}$ from $C^2[0,l]$

$$G(w_\varepsilon, \hat{w}) = 0$$

The left-hand side is zero for every $\varepsilon > 0$ and, therefore, also the limit

$$\lim_{\varepsilon \to 0} G(w_\varepsilon, \hat{w}) = \lim_{\varepsilon \to 0} \left\{ \int_0^1 EIw_\varepsilon^{IV} \hat{w}\, dy + [V_\varepsilon \hat{w} - M_\varepsilon \hat{w}']_0^1 - \int_0^1 \frac{M_\varepsilon \hat{M}}{EI}\, dy \right\} = 0$$

The limit of the first integral is, because of $EIw_\varepsilon^{IV} = p_\varepsilon$, just $\hat{w}(x)$, see Eq. (5.39). We may, furthermore, assume that the solutions $w_\varepsilon(y)$ tend to the fundamental solution $g_0(y, x)$ in Eq. (5.2) and the work done by their end forces to the work done by the end forces of the fundamental solution, that is we may assume

$$\lim_{\varepsilon \to 0} [V_\varepsilon \hat{w} - M_\varepsilon \hat{w}']_0^1 = [V[x]\, \hat{w} - M[x]\, \hat{w}']_0^1$$

In addition, we may assume that

$$\lim_{\varepsilon \to 0} \int_0^1 \frac{M_\varepsilon \hat{M}}{EI}\, dy = \int_0^1 \frac{M[x]\, \hat{M}}{EI}\, dy$$

Hence, it follows

$$\lim_{\varepsilon \to 0} G(w_\varepsilon, \hat{w}) = \hat{w}(x) + [V[x]\, \hat{w} - M[x]\, \hat{w}']_0^1 - \int_0^1 \frac{M[x]\, \hat{M}}{EI}\, dy = 0$$

which is exactly the Eq. (5.20) derived in section 5.3.

But note: though the result is the same, there is a difference in the genesis of the term $\hat{w}(x)$. In section 5.3 $\hat{w}(x)$ was the limit of a "boundary integral", here it is the limit of a domain integral

$$\lim_{\varepsilon \to 0} \int_0^1 EIw_\varepsilon^{IV} \hat{w}\, dy = \hat{w}(x)$$

This domain integral is the work done by the distributed load $EIw_\varepsilon^{IV} = p_\varepsilon$ acting through the virtual displacement $\hat{w}(y)$. Naturally, we expect this work to be $\hat{w}(x) \cdot 1$ if EIw^{IV} represents a concentrated load $P = 1$ located at x

$$\int_0^1 EIw^{IV} \hat{w}\, dy = \hat{w}(x) \cdot 1 \tag{5.40}$$

But this "natural result" cannot be obtained, mathematically, other than by limits. There is no function (in the proper sense) which corresponds to a concentrated force. If we substitute into Eq. (5.40) a function EIw^{IV} which has almost everywhere the value 0 but at x the value 1 then the result is not $\hat{w}(x)$ but zero.

This is the point at which Dirac's delta-function, the function with the properties

$$\text{(i)}\quad \delta(y-x)=0 \qquad y\neq x$$

$$\text{(ii)}\quad \int_a^b \delta(y-x)\,\hat{w}(y)\,dy = \hat{w}(x) \quad \forall \hat{w}\in C^1[a,b] \quad x\in(a,b)$$

comes in. Due to (ii) the virtual work of this function acting through a virtual displacement $\hat{w}$ is just $1\cdot\hat{w}(x)$, i.e. the work of a concentrated force. In this sense the function $\delta(y-x)$ represents a concentrated force.

With the help of this function the differential equation for a beam loaded with a concentrated force can be written as

$$EIw^{IV}(y)=\delta(y-x)$$

and if we now apply (formally) the first identity to the deflection w and a displacement $\hat{w}$ then this results in

$$G(w,\hat{w})=\int_0^1 \delta(y-x)\,\hat{w}(y)\,dy + [V\hat{w}-M\hat{w}']_0^1 - \int_0^1 \frac{M\hat{M}}{EI}\,dy$$

$$=\hat{w}(x) \qquad\qquad + [V\hat{w}-M\hat{w}']_0^1 - \int_0^1 \frac{M\hat{M}}{EI}\,dy$$

which coincides with Eq. (5.20).

5.9 Castigliano's Theorem

This chapter on concentrated loads would be incomplete without Castigliano's Theorem. With the identities defined for fundamental solutions the mathematics is done and the derivation of Castigliano's Theorem now a simple exercise.

Let the beam in Fig. 5.27 be our model problem. The deflection w is the sum

$$w=\sum_i w[x^i]\,P_i$$

of three single deflections w_i each of which is the deflection of the beam under the action of the force $P_i=1$ alone. Hence, each of these functions satisfies the homogeneous boundary conditions

$$w(0)=w(l)=M(0)=M(l)=0$$

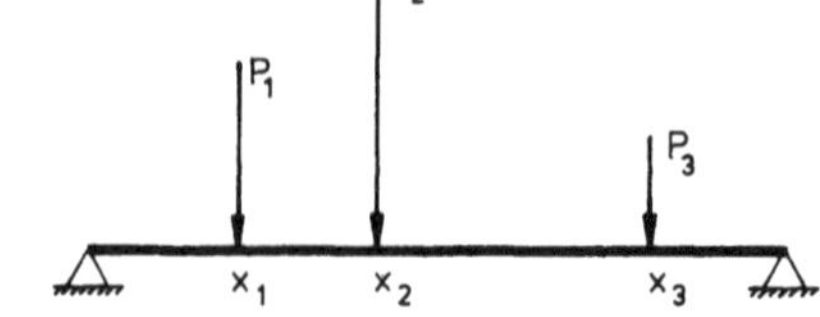

Figure 5.27

The energy balance, "eigenwork = int. energy" for the beam reads

$$\frac{1}{2} G(w, w) = \frac{1}{2} G\left(\sum_i w[x^i]\, P_i, \sum_j w[x^j]\, P_j \right)$$

$$= \frac{1}{2} \sum_{i,j=1}^{3} G(w[x^i], w[x^j])\, P_i P_j = 0$$

For any two indices $i, j \in \{1, 2, 3\}$ holds

$$G(w[x^i], w[x^j]) = w(x^i, x^j) - \int_0^l \frac{M[x^i]\, M[x^j]}{EI}\, dy = 0$$

Hence,

$$\frac{1}{2} G(w, w) = \frac{1}{2} \sum_{i,j=1}^{3} \left\{ w(x^i, x^j) - \int_0^l \frac{M[x^i]\, M[x^j]}{EI}\, dy \right\} P_i P_j = 0$$

or with the notation

$$\delta_i = \sum_{j=1}^{3} w(x^i, x^j)\, P_j, \quad M = \sum_{j=1}^{3} M[x^j]\, P_j \tag{5.41}$$

simply

$$\frac{1}{2} G(w, w) = \frac{1}{2} \sum_{i=1}^{3} P_i \delta_i - \frac{1}{2} \int_0^l \frac{M^2}{EI}\, dy = 0$$

Rearranging the equation and differentiating on both sides with respect to an arbitrary parameter P_k we obtain due to

$$\frac{\partial}{\partial P_k} (\Sigma\, P_i \delta_i) = \sum_{i=1}^{3} \left(\frac{\partial P_i}{\partial P_k} \delta_i + P_i \frac{\partial \delta_i}{\partial P_k} \right) = \delta_k + \sum_{i=1}^{3} P_i w(x^i, x^k) = 2\delta_k$$

the result

$$\delta_k = \frac{\partial}{\partial P_k} \frac{1}{2} \int_0^l \frac{M^2}{EI}\, dy \tag{5.42}$$

This is

Castigliano's Second Theorem

The derivation of the strain energy with respect to any force P_k is equal to the displacement δ_k at k in the direction of P_k.

The derivation of Castigliano's First Theorem starts with the observation that there exists a relation as

$$W\,p = \delta$$

between the vectors $\delta = \{\delta_i\}$ and $p = \{P_i\}$ where the matrix W has the elements $W_{ij} = w(x^{i}, x^{j})$. Its inverse, W^{-1}, exists because the *bvp* is uniquely solvable and, hence, the vector p can be expressed in terms of the vector δ

$$p = W^{-1}\,\delta$$

The inverse is, furthermore, symmetric (Betti), hence, the energy balance formulates

$$\frac{1}{2}\,G(w,w) = \frac{1}{2}\,p^{T}\,\delta - \frac{1}{2}\int_{0}^{l}\frac{M^{2}}{EI}\,dy$$

$$= \frac{1}{2}\,\delta^{T}\,W^{-1}\,\delta - \frac{1}{2}\int_{0}^{l}\frac{M^{2}}{EI}\,dy = 0$$

and if we differentiate this expression with respect to δ_k then because of

$$\frac{\partial}{\partial \delta_k}(\delta^{T}\,W^{-1}\,\delta) = \frac{\partial}{\partial \delta_k}\sum_{i,j=1}^{3} W_{ij}^{(-1)}\,\delta_i\delta_j$$

$$= \sum_{j=1}^{3}(W_{kj}^{(-1)}\,\delta_j + W_{jk}^{(-1)}\,\delta_j) = 2\sum_{j=1}^{3}W_{kj}^{(-1)}\,\delta_j = 2P_k$$

the result is

$$P_k = \frac{\partial}{\partial \delta_k}\frac{1}{2}\int_{0}^{l}\frac{M^{2}}{EI}\,dy$$

This is

Castigliano's First Theorem

The derivation of the strain energy with respect to any displacement at point k is equal to the force P_k at k in the direction of δ_k.

The two theorems are, essentially, equivalent statements and we shall, therefore, in the following put no emphasis on the distinction between the first and the second theorem and simply speak of Castigliano's Theorem.

The validity of Castigliano's Theorem requires that the internal energy is finite and that the conjugated quantity (e. g. the displacement if the concentrated load is a single force) is bounded. Only then does it make sense to speak of the derivation of the energy with respect to the conjugated quantity. As not all displacement fields corresponding to concentrated loads comply with these two (equivalent) conditions Castigliano's Theorem cannot be considered a universal law of mechanics.

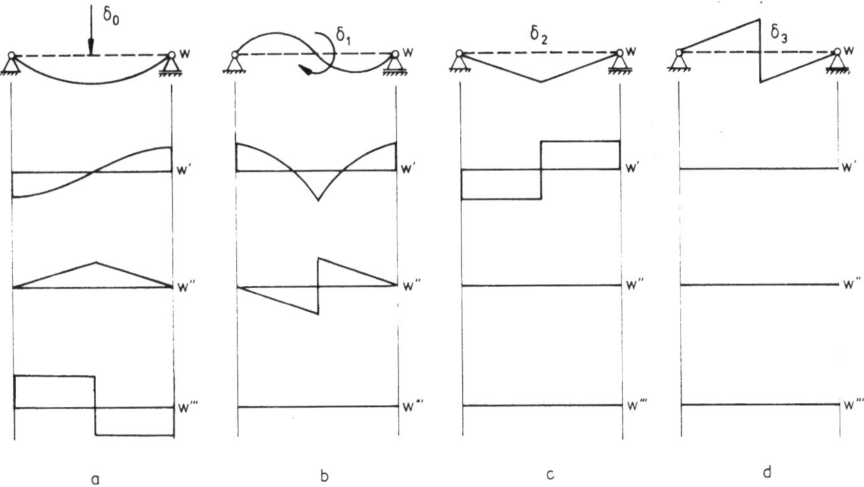

Figure 5.28

It is Sobolev's Embedding Theorem which asserts that these two conditions, finite energy and bounded conjugated quantity, depend on the degree, m, of the strain energy, the dimension, n, of the continuum and the degree, i, of the singularity.

To explain: the differential operators which govern the displacements are of even order, $2m$, (order in this book always means the maximum degree of the derivatives) and the highest derivative in the strain energy, therefore, of order m. Hence, we call m the order of the strain energy.

The Sobolev space with the same index, the space $H^m(\Omega)$, is called the energy space because the functions in $H^m(\Omega)$ are the functions with finite energy.

$$u \in H^m(\Omega) \quad \rightarrow \quad \frac{1}{2} E(u,u) < \infty$$

The classification, i, of the singularities of an operator of degree $2m = 4$ is explained in Fig. 5.28.

The simplest singularity is a concentrated force, Fig. 5.28a. The deflection corresponding to this load case is a (relatively) smooth function. A jump in the derivatives occurs very late, at level 3, the level of the shear force. The worst singularity is a jump in deflection, Fig. 5.28d, because this means that already the 0-th derivative is discontinuous.

Finally, bars and beams are continua of dimension $n = 1$, plates continua of dimension $n = 2$ and elastic bodies continua of dimension $n = 3$.

We may then state:

Theorem 5.18

Castigliano's Theorem applies if and only if the internal energy is finite and the conjugated quantity bounded. This is the case if the three numbers satisfy the inequality

$$m - i > n/2.$$

To start the proof let us first assume that the structural element is loaded not with a singularity $\delta_i(y-x)$ but with a distributed force p and let us, for simplicity, assume that all the boundary conditions are of displacement type and homogeneous ("clamped beam"). That is the displacement u satisfies the equations

$$Du = p \ \text{ in } \Omega, \quad \partial^j u = 0 \quad j = 0, 1, 2 \ldots m-1 \text{ on } \Gamma$$

We are looking for a solution of this bvp in the Sobolev space $H_0^m(\Omega)$, the completion of the class $C_0^m(\bar{\Omega})$ with respect to the norm $\|u\|_m$.

As the question now centers on the existence of a solution we better switch in the following from equilibrium in a pointwise sense to equilibrium in a variational sense (weak formulation) because then the problem is easily resolved with Hilbert-space methods.

Recall, if the potential energy

$$\Pi_1(u) = \frac{1}{2} E(u, u) - \int_\Omega pu \, d\Omega$$

of the structural element has a minimum at u then the first variation of $\Pi_1(u)$ at u with respect to all virtual displacements $\hat{u}$ vanishes

$$\delta \Pi_1(u, \hat{u}) = E(u, \hat{u}) - \int_\Omega p\hat{u} \, d\Omega = 0 \quad \forall \, \hat{u} \in H_0^m(\Omega)$$

So the question is: *does this so-called weak problem have a solution in $H_0^m(\Omega)$?*

The answer depends on the properties of the strain energy, $E(u, u)$, and the character of the load.

With respect to the operators D introduced in section 1.9 it is well known that their strain energies $E(u, u)$ are symmetric and continuous

$$|E(u, \hat{u})| < c_1 \|u\|_m \|\hat{u}\|_m \quad \forall \, u, \hat{u} \in H_0^m(\Omega)$$

as well as coercive

$$\frac{1}{2} E(u, u) > c_2 \|u\|_m^2 \quad \forall \, u \in H_0^m(\Omega)$$

bilinear forms on the space $H_0^m(\Omega)$ in question.

In the case of elastic bodies, $\boldsymbol{u} = \{u_1, u_2, u_3\}$, and Reissner plates, $\boldsymbol{u} = \{\varphi_1, \varphi_2, w\}$, the pertinent energy space for which these inequalities hold true is the triple product

$$(H_0^1(\Omega))^3 = H_0^1(\Omega) \times H_0^1(\Omega) \times H_0^1(\Omega)$$

This is demonstrated for elastic bodies in $[F_1]$. In the case of Reissner plates the proof of the coerciveness

$$\frac{1}{2} E(\boldsymbol{u}, \boldsymbol{u}) = \frac{1}{2} K(1-v) \int_{\Omega} \left(\varphi_{1,1}^2 + \frac{1}{2}(\varphi_{1,2} + \varphi_{2,1})^2 + \varphi_{2,2}^2 \right.$$

$$\left. + \frac{v}{1-v}(\varphi_{1,1} + \varphi_{2,2})^2 + \frac{\bar{\lambda}^2}{2} \{(\varphi_1 + w_{,1})^2 + (\varphi_2 + w_{,2})^2\} \right) d\Omega$$

$$> c_3 (\|\varphi_1\|_1^2 + \|\varphi_2\|_1^2 + \|w\|_1^2)$$

is done with the inequalities of Poincaré and Young.

In the case of shells $\boldsymbol{u} = \{u_1, u_2, u_3\}$, see chapter 8, the energy space is the product

$$H_0^1(\Omega) \times H_0^1(\Omega) \times H_0^2(\Omega)$$

The proof of the coerciveness can be found, e. g., in [B & C].

The question "when does the weak problem have a solution?" is now answered by the famous

Lax-Milgram Theorem

Let $E(u, u)$ be a continuous, symmetric and coercive bilinear functional on a Hilbert space V, that is a pair of positive constants a, b exists such that

$$|E(u, \hat{u})| < a\|u\|_V \|\hat{u}\|_V \quad \forall\, u, \hat{u} \in V$$

$$E(u, u) > b\|u\|_V^2 \quad \forall\, u \in V$$

and assume that the load p belongs to the dual V', that is the load p is a continuous functional on V, this is the case if there exists a constant c such that

$$|(p, u)| = |\int_{\Omega} p u\, d\Omega| < c\|u\|_V \quad \forall\, u \in V,$$

then the weak problem has one and only one solution.

The Sobolev spaces $H_0^m(\Omega)$ (and also the products thereof) are Hilbert spaces and because the strain energy $E(u, u)$ satisfies, as seen above, the necessary conditions the question which remains is: when is the load a continuous functional on $H_0^m(\Omega)$, that is when holds

$$|\int_{\Omega} p u\, d\Omega| < c\|u\|_m \quad \forall\, u \in H_0^m(\Omega)?$$

This is certainly the case if p is in $L_2(\Omega) = H^0(\Omega)$ (if p^2 has finite area (volume)) because then, according to Schwarz' inequality

$$|\int_{\Omega} p u\, d\Omega| \leqslant \left(\int_{\Omega} p^2 d\Omega \right)^{1/2} \left(\int_{\Omega} u^2 d\Omega \right)^{1/2} \leqslant c_4 \|u\|_0 \leqslant c_5 \|u\|_m$$

Hence, all load cases with continuous or piecewise continuous loads (only those occur in practice) have a solution in $H_0^m(\Omega)$, are load cases with finite energy.

If p is continuous and if Γ is smooth then these solutions are our classical, our $C^{2m}(\Omega)$, solutions. If the boundary has corners then singularities in the force-terms, the terms $\partial^i u$, $m \leqslant i \leqslant 2m-1$, might occur on the boundary but these are not so excessive as to render the energy infinite.

The cardinal question now is: what happens if we replace the distributed load p by a concentrated singularity $\delta_i(y-x)$?

For the corresponding weak formulation

$$E(u,\hat{u}) - \int_\Omega \delta_i \hat{u} \, d\Omega = E(u,\hat{u}) - \partial^i \hat{u}(x) = 0 \quad \forall \, \hat{u} \in H_0^m(\Omega)$$

to have a solution u in $H_0^m(\Omega)$ we must require that the functional

$$P(u) = \int_\Omega \delta_i(y-x) \, u(y) \, d\Omega_y = \partial^i u(x)$$

is a bounded linear functional on $H_0^m(\Omega)$, that is that δ_i belongs to $H^{-m}(\Omega)$, see section 6.8. The outcome of this question depends on

Sobolev's Embedding Theorem

If $\Omega \subset \mathbb{R}^n$ *is a bounded domain which satisfies the cone hypothesis (any "reasonable" domain satisfies this condition) and if the index m of the Sobolev space $H^m \Omega$) exceeds $n/2$*

$$m > n/2$$

then $H^m(\Omega) \subset C(\bar{\Omega})$ *and, furthermore, there exists a constant* c_6, *depending only on the domain Ω and the index m of the Sobolev space such that*

$$\max_{x \in \bar{\Omega}} |u(x)| \leqslant c_6 \|u\|_m$$

Consider now an arbitrary function in $H_0^m(\Omega)$. Because u is in $H_0^m(\Omega)$ the term $\partial^i u$ belongs to $H_0^{m-i}(\Omega)$ and if $m - i > n/2$ then the Embedding Theorem asserts that

$$\max_{x \in \bar{\Omega}} |\partial^i u(x)| \leqslant c_6 \|\partial^i u\|_{m-i}$$

But as obviously

$$\|\partial^i u\|_{m-i} \leqslant c_7 \|u\|_m$$

we may continue and state that

$$\max_{x \in \bar{\Omega}} |\partial^i u(x)| \leqslant c_6 \|\partial^i u\|_{m-i} \leqslant c_6 c_7 \|u\|_m$$

We have, thus, demonstrated that Dirac's function δ_i is a bounded continuous functional on $H_0^m(\Omega)$ as long as the index m of the Sobolev space minus the index i of the delta-function exceeds $n/2$. That is as long as the inequality

$$m - i > n/2$$

is satisfied.

In the following table this inequality is evaluated for operators of degree $2m = 2$ (these operators allow only two singularities) and for operators of degree $2m = 4$ (these operators allow four singularities).

		bars, beams $n = 1$	plates $n = 2$	bodies $n = 3$	
$m = 1$					
	$1 - 0 > n/2$	yes	no	no	(1)
	$1 - 1 > n/2$	no	no	no	(2)
$m = 2$					
	$2 - 0 > n/2$	yes	yes	yes	(3)
	$2 - 1 > n/2$	yes	no	no	(4)
	$2 - 2 > n/2$	no	no	no	(5)
	$2 - 3 > n/2$	no	no	no	(6)

If the answer is "yes" then the energy is finite and the conjugated quantity is bounded, Castigliano's Theorem applies. If "no" then the energy and the conjugated quantity are both infinite, Castigliano's Theorem does not apply.

Consider, e.g. the results in row (1). The energy of a bar, $n=1$, loaded with a concentrated force is finite but the energy is infinite if the same load acts on an elastic plate, a Reissner plate, a membrane or on an elastic body.

With operators of degree 4 ($m=2$) the situation improves, somewhat. The energy corresponding to a concentrated force is finite, in all dimensions. Now it is the couple which causes a state of stress with infinite energy in two dimensions (Kirchhoff plate) but finite energy in one dimension (beam).

If the singularity is a jump in displacement or rotation then the internal energy is always infinite. This is what we must expect if we "mistreat" an elastic medium. The forces needed to contort a beam are infinite.

The simple problem of a prestressed membrane perhaps best illustrates what happens when the energy becomes infinite.

The first identity of the Laplace operator Δ which governs the deflection of a prestressed membrane ($N=1$) is, see Eq. (1.40),

$$G(w, \hat{w}) = \int_\Omega -\Delta w \, \hat{w} \, d\Omega + \int_\Gamma \frac{\partial w}{\partial n} \hat{w} \, ds - \int_\Omega \operatorname{grad} w \cdot \operatorname{grad} \hat{w} \, d\Omega = 0$$

We learn from this equation that the force-term in a membrane is the normal derivative $\partial w / \partial n$. This is easily understood if we study Fig. 5.29. The greater the load the greater the tensile forces (the normal derivative) needed to keep the membrane in place (needed to follow the deflection).

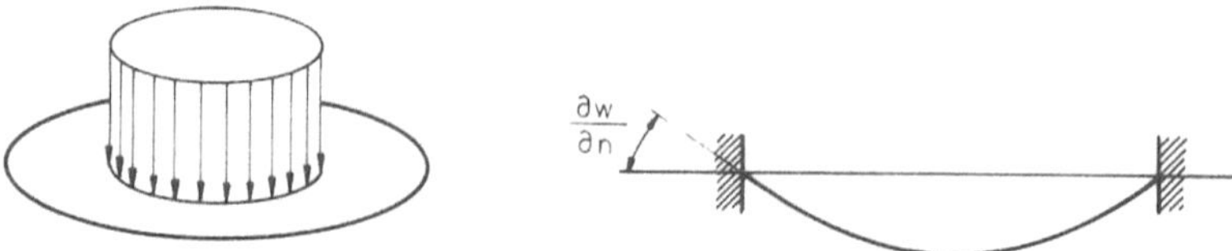

Figure 5.29

Suppose now the unit circle is loaded with a concentrated force $P = 2\pi$ at its center $x = 0$, see Fig. 5.30a. The corresponding *bvp*

$$-N \Delta w = \delta_0 (y - 0) \, 2\pi \quad \text{in } \Omega \quad w = 0 \ \text{ on } \Gamma, \quad N = 1$$

has the solution $w = -\ln r$.

The internal energy is defined as

$$\frac{1}{2} E(w, w) = \frac{1}{2} \int_\Omega \operatorname{grad} w \cdot \operatorname{grad} w \, d\Omega = \frac{1}{2} \int_\Omega (w^2{}_{,1} + w^2{}_{,2}) \, d\Omega$$

Hence, in the ring $\varepsilon < r < 1$ the value of the internal energy is

$$\frac{1}{2}\,E(w,w)_{\Omega_\varepsilon} = \frac{1}{2}\int_\varepsilon^1 \int_0^{2\pi} \frac{1}{r^2}\,(\cos^2\varphi + \sin^2\varphi)\,r\,dr\,d\varphi$$

$$= \pi \int_\varepsilon^1 \frac{1}{r}\,dr = \pi\,\ln\frac{1}{\varepsilon}$$

and consequently the internal energy tends to infinity when we close the ring

$$\lim_{\varepsilon\to 0}\frac{1}{2}\,E(w,w)_{\Omega_\varepsilon} = \lim_{\varepsilon\to 0}\pi\,\ln\frac{1}{\varepsilon} = +\infty$$

In agreement with this the displacement at the center is unbounded, $w(0) = -\ln 0$, and, therefore, Castigliano's Theorem does not apply.

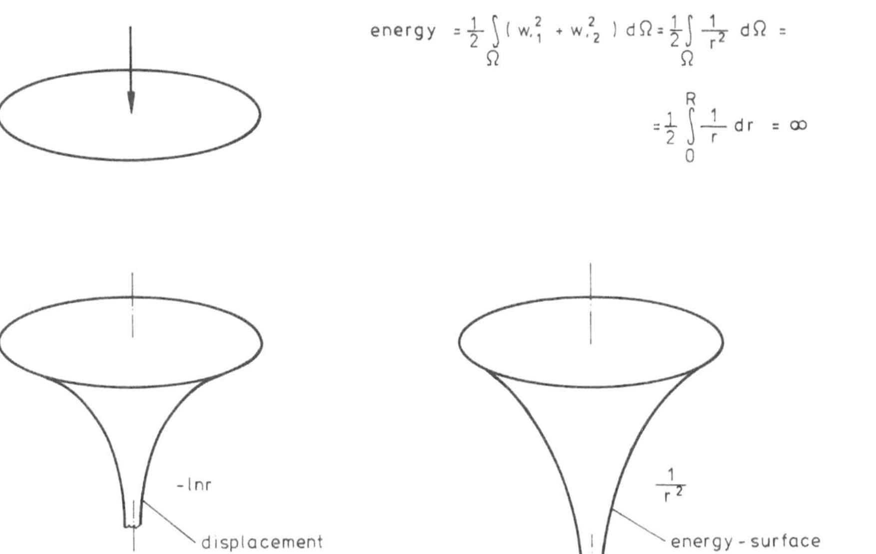

Figure 5.30

In other words, infinite energy means that the square of the strains, ε^2, no longer can be measured, the volume under the energy-surface has no measure, it is infinite.

Illustrative examples provide also the four singularities of a beam, see Fig. 5.28.

The energy of a beam is the integral of M^2/EI, hence, the energy is finite if the area of the M-diagram is finite. Clearly then, the energy under the action of δ_0 and δ_1 is finite. But what about δ_2 and δ_3? No M-diagram exists.

In this case we expand the deflection, e. g. the function with the bend in Fig. 5.28 c, into a Fourier series

$$w(x) = \frac{\pi}{2} - \frac{4}{\pi}\left(\cos 1x + \frac{1}{3^2}\cos 3x + \frac{1}{5^2}\cos 5x + \ldots\right)$$

and we remember that the integral of the square of a Fourier series is just the sum of the coefficients squared. Hence, the energy, the square of the second derivative, becomes

$$\frac{1}{2}\int_0^l \frac{M^2}{EI}\,dx = \frac{1}{2}\,EI\,\frac{16}{\pi^2}\{1+1+1+\ldots\}$$

and this, evidently, is infinite.

A good insight into the physics behind Sobolev's Embedding Theorem provides, we think, the following comparative study of a membrane, a Kirchhoff plate and an elastic body, each loaded with a concentrated force $P=1$.

The normal derivative, $\partial w/\partial n$ ($=$ force per unit length) of a membrane loaded with a concentrated force $P=1$ must satisfy the equation

$$\lim_{\varepsilon\to 0}\int_{\Gamma_{N_\varepsilon}(x)}\frac{\partial w}{\partial v}\,ds_y = \lim_{\varepsilon\to 0}\int_0^{2\pi}\frac{\partial w}{\partial v}\,\varepsilon\,d\varphi = 1 \tag{5.43}$$

the Kirchhoff-shear of a plate the equation

$$\lim_{\varepsilon\to 0}\int_{\Gamma_{N_\varepsilon}(x)} V_v\,ds_y = \lim_{\varepsilon\to 0}\int_0^{2\pi} V_v\,\varepsilon\,d\varphi = 1 \tag{5.44}$$

and the traction vector $\tau(u)$ of an elastic body loaded with a concentrated force $P=e_1=\{1,0,0\}$ at x the equation

$$\lim_{\varepsilon\to 0}\int_{\Gamma_{N_\varepsilon}(x)}\tau(u)\,ds_y = \lim_{\varepsilon\to 0}\int_0^\pi\int_0^{2\pi}\tau(u)\,\varepsilon^2\sin\vartheta\,d\varphi\,d\vartheta = e_1 \tag{5.45}$$

In two dimensions $\Gamma_{N_\varepsilon}(x)$ is a circle whose measure (circumference) is

$$\mathrm{mes}\,\Gamma_{N_\varepsilon}(x) = \int_{\Gamma_{N_\varepsilon}(x)} ds = 2\pi\varepsilon$$

and in three dimensions a sphere

$$\mathrm{mes}\,\Gamma_{N_\varepsilon}(x) = \int_{\Gamma_{N_\varepsilon}(x)} ds = 4\pi\varepsilon$$

Consequently for Eqs. (5.43) and (5.44) to hold the membrane forces and the Kirchhoff-shear must tend to infinity as ε^{-1} to balance the shrinking size, $2\pi\varepsilon$, of $\Gamma_{N_\varepsilon}(x)$ when ε tends to zero, see Fig. 5.31.

$$\frac{\partial w}{\partial v} = 0(\varepsilon^{-1}), \quad V_v = 0(\varepsilon^{-1})$$

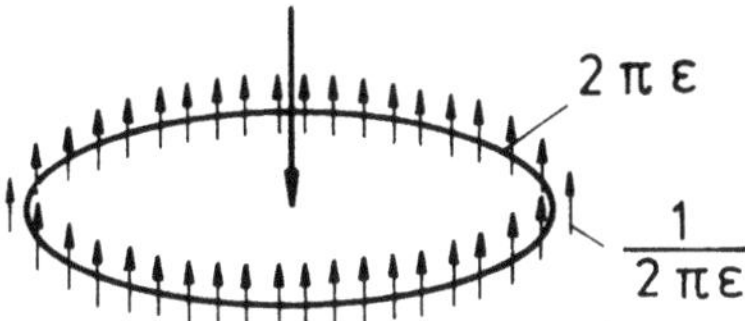

Figure 5.31

Similarly for Eq. (5.45) to hold the traction vector must behave as

$$\tau(\boldsymbol{u}) = 0(\varepsilon^{-2})$$

We, thus, come to understand why the number n appears in Sobolev's inequality. The measure of $\Gamma_{N_\varepsilon}(\boldsymbol{x})$ and, therefore, the dimension, n, of the continuum determines how fast the stresses must tend to infinity. The importance of the second number, m, in Sobolev's inequality is understood if we consider the following:

The forces $\partial w/\partial n$ of a membrane are essentially the first derivatives of the deflection w and, therefore, the deflection w the integral of the forces. If we place a concentrated force on a membrane then the normal derivative behaves as

$$\frac{\partial w}{\partial n} = 0(r^{-1})$$

and, hence, the deflection as

$$w = 0(\ln r)$$

Or consider an elastic body loaded with a concentrated force. The stresses are the first derivatives of the displacement field

$$\tau(\boldsymbol{u}) \cong \nabla \boldsymbol{u}$$

and, hence, $\boldsymbol{u}$ the integral of the stresses. The stresses in the vicinity of a concentrated force behave as r^{-2} and, therefore, the displacements as r^{-1},

$$\boldsymbol{u} = 0(r^{-1})$$

The force and displacement terms of equations of second degree ($m = 1$) are only one differentiation apart while they lie three differentiations apart if the operator is of fourth degree ($m = 2$).

Integrating r^{-1} once we still have a singular function, namely $\ln r$, but integrating it thrice we obtain a well behaved function

$$\iiint \frac{1}{r}\, dr\, dr\, dr = \frac{1}{2} r^2 \left(\ln r - \frac{3}{2} \right)$$

Note that

$$\lim_{r\to 0} r \ln r = 0$$

The following table illustrates these facts. The heading 3, 2, 1, 0 indicates the level of differentiation. The Kirchhoff-shear ($\simeq$ 3rd derivative) has its place in column 3, the deflection in column 0, etc.

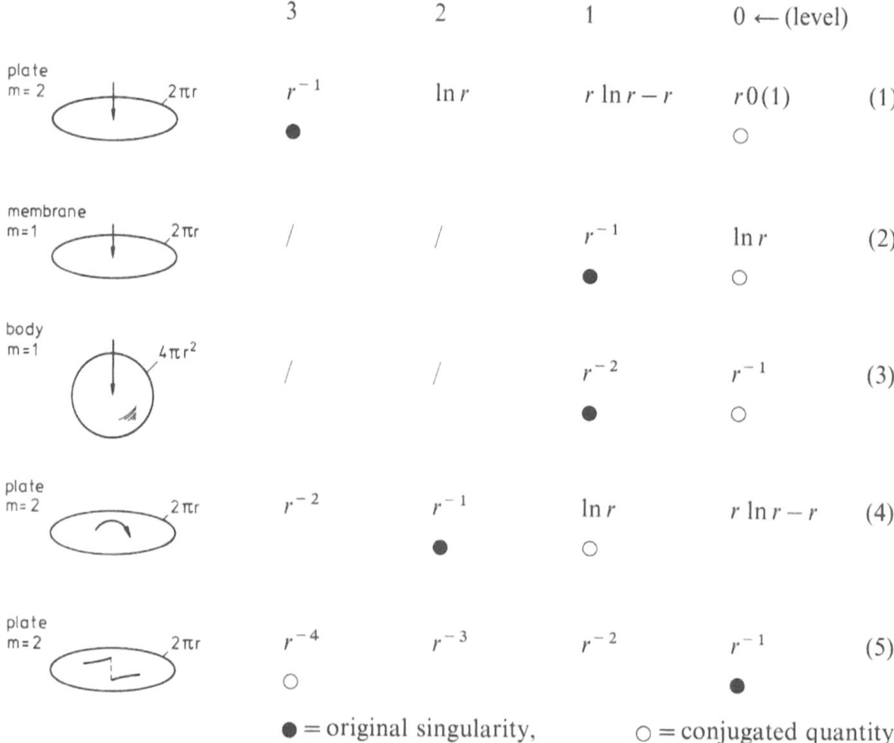

The third number which appears in Sobolev's inequality is i, the degree or the level of the singularity. If i is low, say $i = 0$, then the singularity is a force, i. e. a function on level 3 (in case $m = 2$). The conjugated quantity is the displacement, i. e. a function on level 0 and consequently the distance between the two levels is at its maximum. There are enough integrations in between to dampen the singularity r^{-1} (see the first line).

If i increases then the place of the singularity r^{-1} moves to the right, e. g. from level 3 to level 2 (we load the plate with a couple, see line 4) while the place of the conjugated quantity moves, at the same time, to the left, from level 0 to level 1. The distance becomes smaller, only one integration separates the two conjugated quantities and this is no longer sufficient to transform the singularity r^{-1} into a well behaved function. We only obtain $\ln r$ for the normal derivative, the conjugated quantity.

If the level of the singularity is lower than the level of the conjugated quantity — this happens, e. g., if we force upon the plate a jump in deflection, see line 5, — then we must perform differentiations instead of integrations to calculate the conjugated quantity, the Kirchhoff-shear. This is, with respect to Castigliano's Theorem, ill luck because then the behaviour of the conjugated quantity is even worse than the behaviour of the original singularity which caused the disturbance.

5.10 Castigliano's Generalized Theorem

Often engineers refer to Castigliano's Theorem when they calculate stiffness matrices. But this must not be taken literally because, in general, the calculation of stiffness matrices has little to do with Castigliano's Theorem as we shall explain in this section.

Let us choose a simple bar as our model problem, that is an elastic element whose displacement satisfies the equation

$$-EA\,u'' = p$$

In finite element analysis the function u has the form

$$u(x) = \sum_i \delta_i\,\varphi_i(x) = \boldsymbol{\delta}^T\boldsymbol{\varphi}$$

Substituting u into the first identity, see Eq. (2.14), we obtain

$$\frac{1}{2}\,\boldsymbol{\delta}^T K\,\boldsymbol{\delta} = \frac{1}{2}\,\boldsymbol{\delta}^T(A+B)\,\boldsymbol{\delta}$$

where

$$K_{ij} = \int_0^l EA\,\varphi_i'\varphi_j'\,dx, \quad A_{ij} = \int_0^l -EA\,\varphi_i''\varphi_j\,dx, \quad B_{ij} = [EA\,\varphi_i'\varphi_j]_0^l$$

According to Betti's law (the second identity)

$$\int_0^l -EA\,\varphi_i''\varphi_j\,dx + [N_i\varphi_j]_0^l = \int_0^l \varphi_i(-EA\,\varphi_j'')\,dx + [\varphi_i N_j]_0^l$$

the sum of the two matrices A and B is symmetric

$$(A+B)^T = A+B$$

and, hence, the derivation of the internal energy with respect to a parameter δ_k yields

$$\frac{\partial}{\delta\,\delta_k}\,F(u) = \sum_j (A_{kj} + B_{kj})\,\delta_j =: f_k \tag{5.46}$$

In general the term f_k is not a genuine nodal force but only an "equivalent nodal force" because, in general, the functions φ_i are no fundamental solutions.

Two sets of functions φ_i may illustrate the difference. First we choose the splines in Fig. 5.32a

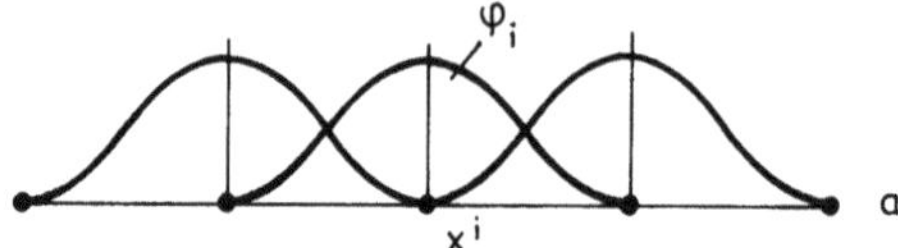

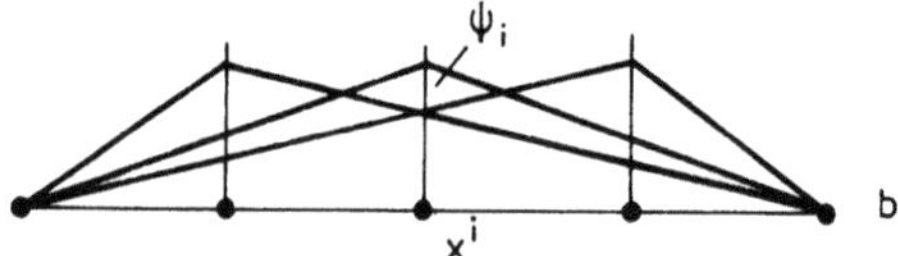

Figure 5.32

The end displacement and end forces are zero

$$\varphi_i(0) = \varphi_i'(0) = \varphi_i(l) = \varphi_i'(l) = 0$$

hence, the matrix B is zero and Eq. (5.46) simplifies to

$$\frac{\partial}{\delta \delta_k} F(u) = \sum_j A_{kj}\delta_j = f_k$$

The second derivatives of the splines, $-EA\,\varphi_i''$, the loads, are piecewise continuous functions, therefore, no concentrated force acts on the truss element and consequently the number f_k,

$$f_k = \sum_j A_{kj}\delta_j = \sum_j A_{jk}\delta_j = \sum_j \int_0^l -EA\,\varphi_j''\varphi_k\,dx\,\delta_j$$

is not a single force but the work done by the distributed load $-EA\,\varphi_j''\delta_j$ acting through the displacement φ_k.

Consider now the function

$$u(x) = \sum_i f_i\psi_i(x) = f^T\,\psi \tag{5.47}$$

where the functions ψ_i are fundamental solutions, i. e. the displacements of the bar (both ends fixed) if a concentrated load, $f_i = 1$, acts at the node x^i, see Fig. 5.32b.

$$-EA\,\psi_i'' = \delta_0(x - x^i) \qquad \psi_i(0) = \psi_i(l) = 0$$

On substituting the function (5.47) into the first identity we obtain

$$\frac{1}{2} f^T K f = \frac{1}{2} f^T (A + B) f$$

where

$$K_{ij} = \int_0^l EA\, \psi_i' \psi_j'\, dx, \quad A_{ij} = \int_0^l -EA\, \psi_i'' \psi_j\, dx = \psi_j(x^i),$$

$$B_{ij} = [EA\, \psi_i' \psi_j']_0^l = 0$$

Therefore,

$$\frac{\partial}{\partial \delta_k} F(u) = \sum_j A_{kj} f_j = \sum_j \psi_j(x^k) f_j = \delta_k$$

or if we formulate this equation, Castigliano's second Theorem, with every component f_k

$$A f = \delta$$

As the symmetric matrix A is positive definite

$$f^T A f = \int_0^l \frac{N^2}{EA}\, dx > 0 \qquad \forall f \neq 0$$

the force-terms f_i in Eq. (5.47) can be replaced by displacement-terms giving

$$u(x) = f^T \psi = \delta^T A^{(-1)} \psi = \delta^T \varphi$$

The derivation of the strain energy with respect to one of these terms δ_k yields

$$\frac{\partial}{\partial \delta_k} F(u) = f_k$$

which is in agreement with Castigliano's first Theorem.

Obviously then, for Castigliano's Theorem (the first and the second) to hold the functions φ_i must be linear combinations, $\varphi = A^{-1} \psi$, of fundamental solutions, i.e. fundamental solutions (every linear combination of such functions is again a fundamental solution).

The functions φ_i used in the calculation of stiffness matrices for bars and beams are homogeneous solutions and as such represent either rigid-body movements or they are fundamental solutions which correspond to end actions. As only the latter are actually used every φ_i is a fundamental solution and, hence, the necessary condition satisfied.

One dimensional problems are the only problems where the derivation of stiffness matrices can be considered a genuine application of Castigliano's Theorem.

In finite element analysis of plates and bodies the functions φ_i are, in general, no fundamental solutions of the governing equation. To refer to Castigliano's Theorem to justify the equation

$$\frac{\partial}{\partial \delta_k} F(u) = f_k$$

is hence, strictly speaking, illegitimate.

But legitimate or not, the engineer considers this equation meaningful and, hence, there should be a mathematical basis for it. This basis is a theorem we call Castigliano's generalized Theorem. It rests on the following observations.

Common to all our problems is the existence of a functional $F(u)$, termed internal energy, and the fact that the first variation of this functional

$$\delta F(u, \hat{u}) = \frac{d}{d\varepsilon} F(u + \varepsilon \hat{u})\bigg|_{\varepsilon = 0}$$

satisfies an equation as

$$\delta F(u, \hat{u}) = \int_{\Omega} Du \cdot \hat{u} \, d\Omega + \int_{\Gamma} r(u) \cdot \hat{u} \, ds \tag{5.48}$$

where D and $r(\ )$ are linear or nonlinear differential operators (additional boundary operators appear if the degree of D is 4). Eq. (5.48), which is just the first identity of the operator D, see section 4.12, states that the first variation of the internal energy in the direction of $\hat{u}$ is equal to the work done by the external forces, Du and $r(u)$, acting through the virtual displacement $\hat{u}$.

In the case of the bar, e. g., the operators D and $r(\ )$ are

$$D = -EA \, d^2/dx^2, \quad r(\ ) = EA \, d/dx$$

The second, important, observation is the following:
Let:

$$u = \sum_i \delta_i \varphi_i$$

with arbitrary (but sufficiently smooth) functions φ_i and parameters δ_i then

$$\frac{\partial}{\partial \delta_k} F(u) = \delta F(u, \varphi_k) \tag{5.49}$$

which means that the derivation of $F(u)$ at $u = \sum \delta_i \varphi_i$ with respect to δ_k is equal to the first variation of $F(u)$ in the direction of φ_k.

This equation which establishes the equivalence between derivation and variation is valid in the linear theory and in the nonlinear theory (proof by actual trial).

Combining now the two Eqs. (5.48) and (5.49) we obtain

Castigliano's Generalized First Theorem

$$\frac{\partial}{\partial \delta_k} F(u) = \int_\Omega Du \cdot \varphi_k \, d\Omega + \int_\Gamma r(u) \cdot \varphi_k \, ds$$

The derivation of $F(u)$ at $u = \Sigma \, \delta_i \varphi_i$ with respect to a parameter δ_k is equal to the work done by the external forces Du and $r(u)$ acting through φ_k that is equal to the "equivalent nodal force"

$$(Du, \varphi_k) + [r(u), \varphi_k]$$

If the first variation is symmetric (formally self-adjoint operators) then we may interchange u and φ_k in Eq. (5.49) to obtain

Castigliano's Generalized Second Theorem

$$\frac{\partial}{\partial f_k} F(u) = \int_\Omega D\psi_k \cdot u \, d\Omega + \int_\Gamma r(\psi_k) \cdot u \, ds$$

$$\left(\text{We replaced } u = \sum_i \delta_i \varphi_i \quad \text{by } u = \sum_i f_i \psi_i \right).$$

5.11 Concentrated Forces or Disturbances on the Boundary

If concentrated forces act on the boundary and not in the interior then the fundamental solutions in section 5.2 no longer apply (at least in two and three dimensions) because they do not satisfy the boundary conditions of a free boundary (the fundamental solutions of bars and beams apply at free ends as well).

Fundamental solutions which satisfy these conditions are the half-space solutions. Their formulas can be found in the literature.

Disturbances in the displacement field occur (sometimes) at points where the boundary conditions change or (possibly) at reentrant corners. By disturbance it is meant that the field u no longer is smooth, no longer belongs to $C^{2m}(\bar{\Omega})$.

Such a disturbance bears in principle the same consequences for the formulation of the identities as the singularities of the fundamental solutions. It is only that the functions which solve such problems are in general relatively well behaved, better at least than the fundamental solutions.

At the reentrant corner ($\Delta\varphi = 3/2\pi$) of a prestressed membrane the deflection behaves as, s. [L 1].

$$w = r^{\frac{2}{3}} \sin\left(\frac{2}{3}\varphi\right)$$

Consequently the tensile forces

$$\frac{\partial w}{\partial n} = r^{-\frac{1}{3}} \sin\left(\frac{2}{3}\varphi\right) + \ldots$$

become infinite at the corner $(r=0)$ but the deflection still has finite energy, $E(w,w) < \infty$, and no extra term $c(x)\,w(x)$ appears in the first identity, as in the case of the fundamental solution $\ln r$.

6 Influence Functions

The equations formulated in chapter 5 find many applications. Not because concentrated loads occur so often, they are rather fictitious, abstract quantities, but because concentrated loads are useful in the calculation of single displacements. This method is known as "the dummy-unit-load method".

The idea is to place a single load at the point at which we want to measure the displacement and to let the force act in the direction of the unknown displacement. According to the principle of virtual displacements the external work, $1 \cdot \hat{\delta}$, is equal to the virtual strain energy $E(u, \hat{u})$, that is $\hat{\delta}$ can be found by quadrature.

The mathematical basis of this dummy-unit-load method and its consequences is the topic of this chapter.

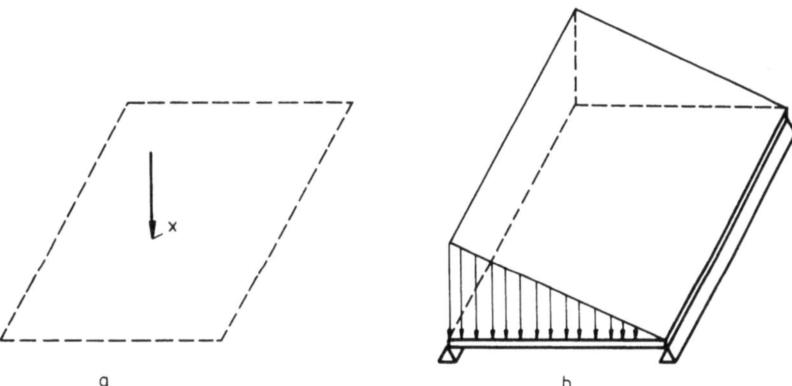

Figure 6.1

6.1 Integral Representations

In chapter 5 we formulated the first and second identity for pairs of functions as $\{w[x], \hat{w}\}$,

$$G(w[x], \hat{w}) = 0, \quad B(w[x], \hat{w}) = 0 \tag{6.1}$$

where the function $w[x] = g_0[x] + w_R$ was the sum of a fundamental solution and a regular, homogeneous solution, $D w_R = 0$. The function w_R could be zero.

What it means, mechanically, if we formulate the identities $G(g_0[x], \hat{w}) = 0$ and $B(g_0[x], \hat{w}) = 0$ with the fundamental solutions alone, that is if we let $w_R = 0$, explains Fig. 6.1.

Let the rectangular plate on the right have the deflection $\hat{w}$ and the infinite plate on the left, loaded with a concentrated force $P = 1$ at x, the deflection $g_0[x] = (8\pi K)^{-1} r^2 \ln r$.

When we separate the rectangular plate from its supports the support reactions become external forces and, similarly, when we cut the infinite plate along the dashed line (this part coincides in shape and location with the rectangular plate) then there, too, the internal actions become external forces.

The two functions $g_0[x]$ and $\hat{w}$ represent now two isolated self-contained elastic states in two identical domains.

If we formulate with these two states the principle of virtual displacements, that is if we let the deflection $\hat{w}$ of the plate on the right act as a virtual displacement on the plate on the left, then we just formulate $G(g_0[x], \hat{w}) = 0$.

Or if we apply Betti's Theorem, that is if we balance the reciprocal external work of the two states then we just formulate $B(g_0[x], \hat{w}) = 0$.

In mathematical terms these two identities

$$G(g_0[x], \hat{w}) = 0, \qquad B(g_0[x], \hat{w}) = 0$$

(we write in the following simply w instead of $\hat{w}$) are, up to the quantity $c(x)\,w(x)$, the sum of domain and boundary integrals.

If we choose x in Ω and if we place the term $c(x)\,w(x) = w(x)$ on one side alone then the first identity, see Eq. (5.23), becomes

$$w(x) = -\int_\Gamma \left(V_\nu[x]\,w - M_\nu[x]\,\frac{\partial w}{\partial \nu} \right) ds_y - [[M_{\nu\tau}[x]\,w]] + E(g_0[x], w) \tag{6.2}$$

and the second identity, see Eq. (5.24),

$$w(x) = -\int_\Gamma \left(V_\nu[x]\,w - M_\nu[x]\,\frac{\partial w}{\partial \nu} + \frac{\partial}{\partial \nu}\,g_0[x]\,M_\nu - g_0[x]\,V_\nu \right) ds_y$$

$$- [[M_{\nu\tau}[x]\,w - g_0[x]\,M_{\nu\tau}]] + \int_\Omega g_0[x]\,K\Delta\Delta w\,d\Omega_y \tag{6.3}$$

These equations are integral representations of the deflection $w(x)$.

The first of these equations expresses the influence the boundary values w and $\partial w/\partial n$ and the strains $\varepsilon_{ij}(w)$ in the interior, that is the derivatives of w which appear in the strain energy $E(g_0[x], w)$, have on $w(x)$.

The second equation expresses the influence (all) the displacement- and force-terms on the boundary and the load $K\Delta\Delta w$ in the domain have on $w(x)$.

But if an evaluation of these data is sufficient to calculate the deflection w of a Kirchhoff plate then, obviously, so we may conclude, these data determine w uniquely.

This conclusion applies to all structural elements, as the following theorem explains.

Theorem 6.1

The displacement of a structural element is uniquely determined by
a) *the displacement-terms on the boundary and the strains $\varepsilon(w)$ in the interior ("the bending-moment diagram"), or, alternatively, by*
b) *the displacement- and force-terms on the boundary and the load p in the domain.*

Hence, if we displace a plate as the one in Fig. 6.2

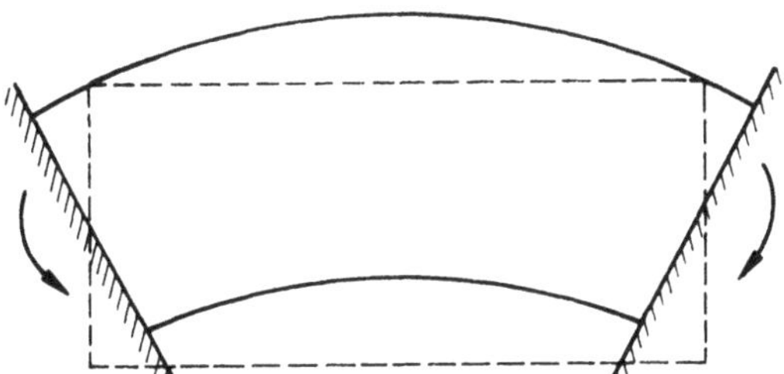

Figure 6.2

and if we measure at the same time the boundary forces ($=$ the traction vector t) and the boundary displacements u then the displacements and stresses at any internal point can be calculated with the Somigliana formula, see Eq. (5.30).

But one word of caution: our interpretation of the second integral representation of the Kirchhoff plate, Eq. (6.3), as an expansion of $w(x)$ with regard to the influence

$$-\int_\Gamma V_\nu[x]\, w\, ds_y, \qquad \int_\Gamma M_\nu[x]\, \frac{\partial w}{\partial v}\, ds_y, \qquad -\int_\Gamma \frac{\partial g}{\partial v}[x]\, M_\nu\, ds_y, \qquad \text{etc.}$$

of the quantities

$$w, \frac{\partial w}{\partial v}, M_\nu, F_k, V_\nu \quad (\text{on } \Gamma), \qquad p \quad (\text{in } \Omega)$$

on $w(x)$ should not lead us to believe that we can substitute arbitrary boundary functions (numbers) w, $\partial w/\partial v$, M_ν, V_ν, F_k (corner forces) into Eq. (6.3) and assume that the function, so constructed, has these functions as boundary values.

This is only true if the boundary functions satisfy a set of compatibility equations, namely two integral equations on the boundary, see section 6.3.2, Theorem 6.6.

Often we are more interested in integral representations ($=$ influence functions) of moments or shear forces than of displacements and we, therefore, must find means to derive integral representations of the moment M_n and the Kirchhoff-shear V_n and, to be complete, of the normal derivative $\partial w/\partial n$ as well.

The first idea will be to simply calculate these quantities by differentiating the integral representation of $w(x)$.

But if we differentiate $w(x)$ at an internal point then we differentiate the domain integrals with respect to their parameter $x = (x_1, x_2)$ and if we differentiate $w(x)$ at a boundary point then we differentiate the boundary integrals, too.

But these differentiations cannot be performed by simply differentiating under the integral signs. This would require, because x lies in the domain of integration, the integrand to be well behaved. This is not the case.

To circumvent this difficulty we simply proceed as in chapter 5. We formulate the identities G and B in the domain Ω_ε, the domain minus a small neighborhood $N_\varepsilon(x)$ of the source point, the point at which we want to differentiate, apply to these equations the operators $\partial w/\partial n$ or $M_n(\ )$ or $V_n(\ )$— now the differentiation under the integral sign is admissible because x does not lie in Ω—and we then let the radius ε of the hole shrink to zero.

Assume we want to calculate the influence function of the normal derivative of a Kirchhoff plate. To this end we pick an arbitrary internal point x and consider the second identity of the Kirchhoff plate in the domain $\Omega_\varepsilon = \Omega - N_\varepsilon(x)$.

$$B_\varepsilon(g_0[x], w) = 0$$

This identity contains, e.g., the two integrals

$$\int_{\Omega_\varepsilon} g_0[x]\, K\Delta\Delta w\, d\Omega_y \qquad \int_{\Gamma_\varepsilon} V_\nu(g_0[x])\, w\, ds_y$$

In the latter integral $V_\nu(g_0[x])$ is the Kirchhoff-shear of the fundamental solution $g_0[x]$ at the boundary point y. The index ν indicates that V depends on the normal ν at y.

If we apply to these integrals the operator

$$\frac{\partial}{\partial n_x} = \frac{\partial}{\partial x_1}\, n_1 + \frac{\partial}{\partial x_2}\, n_2$$

then the result is in the case of the domain integral

$$\frac{\partial}{\partial n_x} \int_{\Omega_\varepsilon} g_0[x]\, K\Delta\Delta w\, d\Omega_y = \int_{\Omega_\varepsilon} \frac{\partial}{\partial n_x}\, g_0[x]\, K\Delta\Delta w\, d\Omega_y$$
$$= \int_{\Omega_\varepsilon} g_1[x]\, K\Delta\Delta w\, d\Omega_y$$

and in the case of the boundary integral

$$\frac{\partial}{\partial n_x} \int_{\Gamma_\varepsilon} V_\nu(g_0[x])\, w\, ds_y = \int_{\Gamma_\varepsilon} \frac{\partial}{\partial n_x}\, V_\nu(g_0[x])\, w\, ds_y = \int_{\Gamma_\varepsilon} V_\nu(g_1[x])\, w\, ds_y$$

Where we introduced, for convenience,

$$g_1(y, x) := \frac{\partial}{\partial n_x}\, g_0(y, x)$$

The other integrals in the identity B_ε are treated analogously and the full equation

$$\frac{\partial}{\partial n_x} B_\varepsilon(g_0[x], w) = 0$$

can then, if we consequently employ the notation $g_1(y, x)$, be written as

$$\frac{\partial}{\partial n_x} B_\varepsilon(g_0[x], w) = B_\varepsilon(g_1[x], w)$$

$$= \int_{\Gamma_\varepsilon} \left(V_v(g_1[x]) \, w - M_v(g_1[x]) \frac{\partial w}{\partial v} + \frac{\partial}{\partial v} g_1[x] M_v - g_1[x] V_v \right) ds_y$$

$$- \int_{\Omega_\varepsilon} g_1[x] K \Delta\Delta w \, d\Omega_y - \left[\left[M_{v\tau}(g_1[x]) \, w - g_1[x] M_{v\tau} \right] \right] = 0$$

If we let the size of the hole tend to zero the result is, (see Theorem 6.3),

$$\lim_{\varepsilon \to 0} B_\varepsilon(g_1[x], w) = \frac{\partial w}{\partial n}(x) + \int_{\Gamma} \left(V_v(g_1[x]) \, w \right.$$

$$\left. - M_v(g_1[x]) \frac{\partial w}{\partial v} + \frac{\partial}{\partial v} g_1[x] M_v - g_1[x] V_v \right) ds_y$$

$$- \int_{\Omega} g_1[x] K \Delta\Delta w \, d\Omega_y - \left[\left[M_{v\tau}(g_1[x]) \, w \right. \right.$$

$$\left. \left. - g_1[x] M_{v\tau} \right] \right] = 0$$

The single term $\partial w(x)/\partial n$ which appears in this equation, naturally, is not the derivative with respect to a normal on the boundary but the directional derivative at the internal point x with respect to the vector n.

When we calculate influence functions for M_n or V_n then we proceed analogously. We first apply the operators $M_n(\ )$ and $V_n(\ )$ to the second identity in the domain $\Omega_\varepsilon = \Omega - N_\varepsilon(x)$

$$M_n(B_\varepsilon(g_0[x], w)) = B_\varepsilon(g_2[x], w) \qquad g_2(y, x) := M_n(g_0(y, x))$$

$$V_n(B_\varepsilon(g_0[x], w)) = B_\varepsilon(g_3[x], w) \qquad g_3(y, x) := V_n(g_0(y, x))$$

and then let ε shrink to zero. In the resulting equations appear as before

$$\lim_{\varepsilon \to 0} B_\varepsilon(g_i[x], w) = \begin{cases} w(x) + \ldots i = 0 \\ \dfrac{\partial w}{\partial n}(x) + \ldots i = 1 \\ M_n(x) + \ldots i = 2 \\ V_n(x) + \ldots i = 3 \end{cases}$$

single terms $M_n(w)(x)$ and $V_n(w)(x)$ which after we place them on one side alone turn the zero sums $B(g_2[x], w)$ and $B(g_3[x], w)$ into integral representations of the moment $M_n(w)(x)$ and the Kirchhoff-shear $V_n(w)(x)$.

As our notation already indicates there is no essential difference between the derivation of the integral representation of $w(x)$ or, say, $V_n(x)$. The single terms $w(x)$, $\partial w(x)/\partial n$, $M_n(w)(x)$ and $V_n(w)(x)$ are always the limit of the work done by the couple {primary quantity, conjugated singular quantity} acting on the shrinking boundary $\Gamma_{N_\varepsilon}(x)$ of the hole $N_\varepsilon(x)$.

$$\lim_{\varepsilon \to 0} \int_{\Gamma_{N_\varepsilon(x)}} V_v(g_0[x]) \, w \, ds_y = w(x) \tag{6.4a}$$

$$\lim_{\varepsilon \to 0} \int_{\Gamma_{N_\varepsilon(x)}} M_v(g_1[x]) \frac{\partial w}{\partial v} \, ds_y = \frac{\partial w}{\partial n}(x) \tag{6.4b}$$

$$\lim_{\varepsilon \to 0} \int_{\Gamma_{N_\varepsilon(x)}} \frac{\partial}{\partial v} g_2[x] \, M_v(w) \, ds_y = M_n(w)(x) \tag{6.4c}$$

$$\lim_{\varepsilon \to 0} \int_{\Gamma_{N_\varepsilon(x)}} g_3[x] \, V_v(w) \, ds_y = V_n(w)(x) \tag{6.4d}$$

The functions $g_i[x]$, first introduced as convenient shorthand notations, are functions in their own right. They, too, are fundamental solutions. They represent the deflection of the plate under the action of a "point source"

$$g_0[x] = \text{concentrated force}$$

$$g_1[x] = \text{couple}$$

$$g_2[x] = \text{jump in the slope}$$

$$g_3[x] = \text{jump in the deflection}$$

In the notation

$$\partial^0 = \text{id.} \quad \partial^1 = \frac{\partial}{\partial n} \quad \partial^2 = M_n(\ ) \quad \partial^3 = V_n(\ )$$

of section 2.10 the definition of the single functions $g_i[x]$ reads

$$g_i[x] := \partial_x^i g_0[x]$$

What we demonstrated here with the Kirchhoff plate applies to all structural elements.

Associated with an operator D of such a structural element are $2m$ boundary operators $\partial^0, \partial^1, \partial^2, \ldots \partial^{2m-1}$ and, therefore, $2m$ fundamental solutions

$$g_i[x] = \partial_x^i g_0[x] \qquad i = 0, 1, \ldots 2m-1$$

which render via

$$\lim_{\varepsilon \to 0} G_\varepsilon (g_i [x], w) = 0, \qquad \lim_{\varepsilon \to 0} B_\varepsilon (g_i [x], w) = 0$$

integral representations of the functions $\partial^i w$

Bars	u	N
Beams	w, w'	M, V
Kirchhoff plates	$w, \partial w/\partial n$	M_n, V_n
Elastic plates and bodies	$\boldsymbol{u}$	$\tau(\boldsymbol{u})$

Integral representations of displacement-terms (left-hand column) result either by way of the first or the second identity. Both is possible. Integral representations of force-terms (right-hand column) only by way of the second identity. The reason for this is that to obtain with G or B and an appropriate fundamental solution an integral representation of $\partial^i u$ the function $\partial^i u$ must appear in one of the boundary integrals. If we check the identities then we see that B, the second identity, contains all the boundary values

$$\partial^i u \quad 0 \leqslant i \leqslant 2m - 1$$

of a function u but G, the first identity, only the lower values (if u is the second argument in G), i.e. only the displacement-terms. We formulate this as a theorem.

Theorem 6.2

Integral representations of displacement-terms can be obtained with the principle of virtual displacements (the first identity) or Betti's principle (the second identity) but integral representations of force-terms only with Betti's principle.

In the sections to come we shall have a closer look at the formulation of all these equations. For completeness we shall repeat the results from chapter 5, the results we obtained with the zero-th fundamental solution $(i = 0)$.

We start with the formulation of the fundamental solutions.

6.1.1 The Bar, $-EA \, d^2/dy^2$

The fundamental solutions of a bar are, see Eq. (5.1),

$$g_0[x] = \frac{1}{EA} \begin{cases} (1-x)y \\ (1-y)x \end{cases}, \quad g_1[x] = EA \frac{d}{dx} g_0[x] = \begin{cases} -y & y \leqslant x \\ (1-y)x \leqslant y \end{cases} \tag{6.5}$$

The function $g_1[x]$ is the axial displacement of an infinite bar whose displacement jumps at x by an amount of 1, see Fig. 6.3.

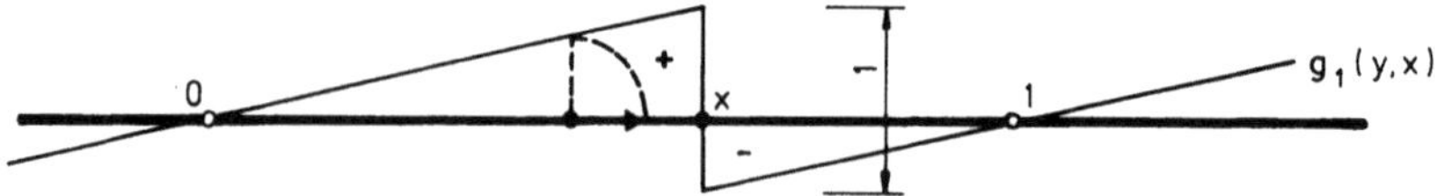

Figure 6.3

The fundamental solutions in Eq. (6.5) render the following results

$$p : u \in C^1\,[a, b],\ i = 0$$

$$q : G(g_i[x], u) = c(x)\,\partial^i u(x) + [N_i u]_b^a - \int_a^b \frac{N_i N}{EA}\,dy = 0 \tag{6.6}$$

and

$$p : u \in C^2\,[a, b],\ i = 0,1$$

$$q : B(g_i[x], u) = c(x)\,\partial^i u(x) + [N_i u - g_i N]_a^b - \int_a^b g_i\,(-EA\,u'')\,dy = 0 \tag{6.7}$$

where $c(x)$ is the characteristic function of the interval (a, b), see Eq. (5.19), and $N_i = EA\,d/dy\,g_i(y, x)$ the normal force associated with the displacement $g_i(y, x)$.

6.1.2 The Beam, $EI\,d^4/dy^4$

The fundamental solutions of a beam are

$$g_0[x] = \frac{1}{6\,EI} \times \begin{cases} \alpha(x)\,y - (1 - x)\,y^3 & y \leqslant x \\ (y - x)^3 + \alpha(x)\,y - (1 - x)\,y^3 & x \leqslant y \end{cases} \tag{6.8}$$

where $\alpha(x) = x\,(1 - x)\,(2 - x)$

$$g_1[x] = \frac{d}{dx}\,g_0[x] = \frac{1}{6\,EI} \times \begin{cases} \alpha'(x)\,y + y^3 \\ -3\,(y - x)^2 + \alpha'(x)\,y + y^3 \end{cases} \tag{6.9}$$

$$g_2[x] = -EI\,\frac{d^2}{dx^2}\,g_0[x] = \begin{cases} (1 - x)\,y \\ x\,(1 - y) \end{cases} \tag{6.10}$$

$$g_3[x] = -EI\,\frac{d^3}{dx^3}\,g_0[x] = \begin{cases} -y \\ 1 - y \end{cases} \tag{6.11}$$

We then have:

$$p : w \in C^2\,[a, b],\ i = 0,1$$

$$q : G(g_i[x], w) = c(x)\,\partial^i w(x) + [V_i w - M_i w']_a^b - \int_a^b \frac{M_i M}{EI}\,dy = 0 \tag{6.12}$$

and

$$p : w \in C^4 [a, b], \ i = 0, 1, 2, 3$$

$$q : B(g_i [x], w) = c(x)\, \partial^i w(x) + [V_i w - M_i w']_a^b - [g_i V - g_i' M]_a^b$$
$$- \int_a^b g_i \, EI w^{IV} \, dy = 0$$

where

$$M_i = - EI \frac{d^2}{dy^2} g_i [x], \qquad V_i = - EI \frac{d^3}{dy^3} g_i [x]$$

The function $c(x)$ is the same as in Eq. (5.19).

6.1.3 The Kirchhoff Plate

There are four fundamental solutions

$$g_0 [x] = \frac{1}{8 \pi K} r^2 \ln r, \quad g_1 [x] = \frac{\partial}{\partial n_x} g_0 [x] = r(1 + 2 \ln r)\, r_n \frac{1}{8 \pi K}$$

$$g_2 [x] = M_n(g_0 [x]) = - \frac{1}{8 \pi} [2(1 + v) \ln r + (3 + v)\, r_n^2 + (1 + 3v)\, r_t^2] \tag{6.13}$$

$$g_3 [x] = V_n(g_0 [x]) = - \frac{2}{8 \pi r} [2 r_n + (1 - v)(r_n - \varkappa r)(r_n^2 - r_t^2)]$$

The functions

$$r_n = \frac{\partial r}{\partial n_x} = r_{,x_1} n_1 + r_{,x_2} n_2, \qquad r_t = \frac{\partial r}{\partial t_x} = r_{,x_1} t_1 + r_{,x_2} t_2$$

in these equations are the normal and tangential derivative of the distance $r = |y - x|$ with respect to x.

The term $\varkappa$ is the curvature at x. If x is not a boundary point but an interior point then we are free to attribute to $\varkappa$ whatever value we wish, i.e. to imagine x to lie on an interior curve of arbitrary shape.

The single fundamental solutions render the following results.

$$p : w \in C^2 (\bar{\Omega}), \quad i = 0, 1$$

$$q : G(g_i [x], w) = c^{(i)}(x)\, \partial^i w(x) + \int_\Gamma \left(V_{v_i} w - M_{v_i} \frac{\partial w}{\partial v} \right) ds_y + [[M_{v \tau_i} w]]$$
$$- E(g_i, w) = 0$$

and

$$p: w \in C^4(\bar{\Omega}), \; i = 0, 1, 2, 3$$

$$q: B(g_i[\mathbf{x}], w) = c^{(i)}(\mathbf{x}) \, \partial^i \, w(\mathbf{x})$$

$$+ \int_\Gamma \left(V_{\nu_i} w - M_{\nu_i} \frac{\partial w}{\partial \nu} + \frac{\partial}{\partial \nu} g_i M_\nu - g_i V_\nu \right) ds_y$$

$$+ \left[[M_{\nu\tau_i} w - g_i M_{\nu\tau}] \right] - \int_\Omega g_i K \Delta\Delta w \, d\Omega_y = 0 \qquad (6.14)$$

where

$$V_{\nu_i} = V_\nu(g_i[\mathbf{x}]), \quad M_{\nu_i} = M_\nu(g_i[\mathbf{x}]), \quad M_{\nu\tau_i} = M_{\nu\tau}(g_i[\mathbf{x}])$$

We shall repeat the equation $B(g_i[\mathbf{x}], w) = 0$, $i = 1$, separately in Theorem 6.3 and we shall then give a precise definition of the function $c^{(1)}(\mathbf{x}) \, \partial^1 \, w(\mathbf{x})$.

The characteristic function $c^{(0)}(\mathbf{x})$ is the function $c(\mathbf{x})$ in Eq. (5.22). The formula of $c^{(1)}(\mathbf{x}) \, \partial^1 \, w(\mathbf{x})$ is given in Eq. (6.17), see Theorem 6.3.

The values of the functions $c^{(2)}$ and $c^{(3)}$ are

$$c^{(i)}(\mathbf{x}) = \begin{cases} 1 & \mathbf{x} \in \Omega \\ ? & \mathbf{x} \in \Gamma \\ 0 & \mathbf{x} \in \Omega^c \end{cases} \qquad i = 2, 3$$

The question mark indicates that their boundary values are, as far as we know, still unknown (though we may assume them to be 1/2 at smooth points) because the calculation of the limit

$$\lim_{\varepsilon \to 0} B_\varepsilon(g_i[\mathbf{x}], w) \quad \mathbf{x} \in \Gamma$$

in the case of $i = 2$ and $i = 3$ has yet to be done.

6.1.4 Elastic Plates and Bodies

The fundamental solutions of elastic plates and bodies form the columns of the symmetric Somigliana matrices, see Eq. (5.10),

$$U_{(2 \times 2)} = [g_0^1, g_0^2], \quad U_{(3 \times 3)} = [g_0^1, g_0^2, g_0^3]$$

Hence, we obtain the fundamental solution $g_1^i[\mathbf{x}]$ if we apply the operator $\tau(\;)$, see Eq. (1.39), to the single columns of the matrices U,

$$\tau_x(U) = [\tau_x(g_0^1), \tau_x(g_0^2)] = [g_1^1, g_1^2] =: T^*(y, x)$$

The resulting matrices $T^*(y, x)$ have the elements

$$T_{ij}^* = -\frac{1}{4\alpha\pi(1-v)\,r^\alpha} \times$$

$$\times \left[\frac{\partial r}{\partial n_x}\left\{(1-2v)\,\delta_{ij} + \beta r_{,x_i} r_{,x_j}\right\} - (1-2v)\left\{r_{,x_j} n_i(x) - r_{,x_i} n_j(x)\right\}\right]$$

where $\alpha=1$, $\beta=2$ in $2-D$ and $\alpha=2$, $\beta=3$ in $3-D$. The two matrices T^* are not symmetric.

It then holds:

$$p: u \in C^1(\bar{\Omega}),\ i=0,\ j=1,2\ (\mathbb{R}^2)\ \text{or}\ j=1,2,3\ (\mathbb{R}^3)$$

$$q: G(g_i^j[x], u) = c_{jk}^{(i)}(x)\,u_k(x) + \int_\Gamma \tau(g_i^j)\cdot u\,ds - E(g_i^j, u) = 0 \tag{6.15}$$

(sum on k)
and

$$p: u \in C^2(\bar{\Omega}),\ i=0\ \text{or}\ 1$$

$$q: B(g_i^j[x], u) = c_{jk}^{(i)}(x)\,\partial^{(i)} u_k(x) + \int_\Gamma \left(\tau(g_i^j)\cdot u - g_i^j\cdot\tau(u)\right)ds_y$$

$$- \int_\Omega g_i^j\cdot(-Lu)\,d\Omega_y = 0 \tag{6.16}$$

The equations (6.15), $i=0$, and (6.16), $i=0$, were already formulated in section 5.3.4. The functions $c_{jk}^{(0)}(x)$ are, therefore, identical with the functions $c_{jk}(x)$ in Eq. (5.27). The functions $c_{jk}^{(1)}(x)$ have the values

$$c_{jk}^{(1)}(x) = \begin{cases} \delta_{jk}, & x \in \Omega \\ ?, & x \in \Gamma \\ 0, & x \in \Omega^c \end{cases}$$

Their boundary values are unknown (though we should have $c_{jk}^{(1)}(x)=1/2$ at smooth points) because the calculation of the limits

$$\lim_{\varepsilon\to 0} B(g_1^j[x], u) = 0, \qquad x \in \Gamma$$

has yet to be done if x is a point on the boundary.

6.1.5 The Integral Representation of $\partial w(x)/\partial n$

We dedicate this section to the study of the influence function for $\partial w(x)/\partial n$ alone because this function deserves some additional comments.

To start, we first repeat the equation $B(g_1[x], w) = 0$, i.e. the integral representation of the normal derivative $\partial w / \partial n$ at full length.

Theorem 6.3

$$p : w \in C^4(\bar{\Omega})$$

$$q : c_1(x)\, w,_1(x) + c_2(x)\, w,_2(x) = \int_\Gamma \Bigg[g_1(y, x)\, V_\nu(w)(y)$$

$$- \frac{\partial}{\partial \nu} g_1(y, x)\, M_\nu(w)(y) - V_\nu(g_1(y, x))\, [w(y) - w(x)]$$

$$+ M_\nu(g_1(y, x))\, \frac{\partial w}{\partial \nu}(y) \Bigg]\, ds_y + \int_\Omega g_1(y, x)\, K\Delta\Delta w(y)\, d\Omega_y$$

$$+ \big[[g_1(y, x)\, M_{\nu\tau}(w)(y) - M_{\nu\tau}(g_1(y, x))\, \{w(y) - w(x)\}]\big] \tag{6.17}$$

The two functions $c_i(x)$ which appear on the left-hand side are defined as

$$c_i(x) = \begin{cases} n_i, & x \in \Omega \\ \dot{c}_i(x), & x \in \Gamma \\ 0, & x \in \Omega^c \end{cases}$$

Their boundary values are

$$\dot{c}_1(x) = \frac{\Delta\varphi}{2\pi}\, n_1 + \frac{\nu}{2\pi} \left[\frac{1}{2} \sin 2\varphi\, n_1 + \sin^2\varphi\, n_2 \right]_{\varphi_2}^{\varphi_1} \tag{6.18a}$$

$$\dot{c}_2(x) = \frac{\Delta\varphi}{2\pi}\, n_2 + \frac{\nu}{2\pi} \left[\sin^2\varphi\, n_1 - \frac{1}{2} \sin 2\varphi\, n_2 \right]_{\varphi_2}^{\varphi_1} \tag{6.18b}$$

The meaning of φ_1, φ_2 and $\Delta\varphi$ is explained in Fig. 6.4. If x is a corner point then, in addition, rule A applies, see Theorem 5.5.

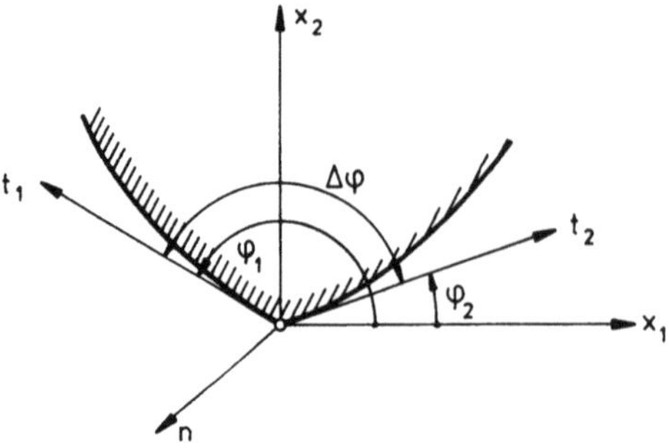

Figure 6.4

The equations of the kernels in Eq. (6.17) are

$$g_1(y, x) = r(1 + 2 \ln r) r_n \frac{1}{8 \pi K}$$

$$\frac{\partial}{\partial v} g_1[x] = [2(r_v r_n + r_t r_\tau) \ln r + 3 r_v r_n + r_t r_\tau] \frac{1}{8 \pi K}$$

$$M_v(g_1[x]) = -\frac{2}{r}[(1 + v) r_n + 2(1 - v) r_v r_\tau r_t] \frac{1}{8 \pi}$$

$$V_v(g_1[x]) = -\frac{2}{r^2} \{[3 - v - 2(1 - v) r_\tau^2](r_\tau r_t - r_v r_n)$$

$$+ 4(1 - v)(r_v - \varkappa r) r_v r_\tau r_t\} \frac{1}{8 \pi}$$

$$M_{v\tau}(g_1[x]) = -\frac{2(1 - v)}{r}[\sin \varphi\, n_1 - \cos \varphi\, n_2] \cos 2\beta \frac{1}{8 \pi}$$

where $\beta = \alpha - \varphi$, see Fig. 5.5.

In formulating Eq. (6.17) we made use of a small trick.

Eq. (6.17) is not the limit of $B_\varepsilon(g_1[x], w(y))$ but of $B_\varepsilon(g_1[x], w(y) - w(x))$. This is admissible because the two functions

$$w(y) \quad \text{and} \quad w(y) - w(x)$$

have the same normal derivative. The advantage of the second approach is that the strongly singular kernel $V_v(g_1(y, x)) = 0(|y - x|^{-2})$ is now coupled with $w(y) - w(x)$ instead of $w(y)$. Because $w(y) - w(x)$ behaves as $0(|y - x|)$ the resulting singularity is of order $0(|y - x|^{-1})$ and, consequently, the difficulty to compute the singular integral decreases.

The two functions in Eq. (6.18)

$$[\sin 2\varphi]_{\varphi_2}^{\varphi_1} = \sin 2\varphi_1 - \sin 2\varphi_2$$

$$[\sin^2 \varphi]_{\varphi_2}^{\varphi_1} = \sin^2 \varphi_1 - \sin^2 \varphi_2$$

are zero if $|\varphi_1 - \varphi_2| = \pi$. So that at smooth boundary points the functions $\dot{c}_i(x)$ simplify to

$$\dot{c}_i(x) = \frac{1}{2} n_i(x)$$

Hence, at such points (we usually let the vector n coincide with the vector n of the exterior unit normal at x) the right-hand side of Eq. (6.17) renders the normal derivative

$$\dot{c}_1(x) w,_1(x) + \dot{c}_2(x) w,_2(x) = \frac{1}{2} \frac{\partial w}{\partial n}(x)$$

at x. If x is a corner point then $\dot c_i(x) \neq n_i$, $i = 1, 2$ whatever the vector n and, hence, the right-hand side is not the directional derivative with respect to n

$$\dot c_1(x)\, w,_1(x) + \dot c_2(x)\, w,_2(x) \neq \frac{1}{2}\frac{\partial w}{\partial n}(x)$$

The question then is: how do we calculate the two normal derivatives at the corner of a plate? — We do so by first calculating the directional derivatives $w,_1(x)$ and $w,_2(x)$ at the corner. To this end we formulate Eq. (6.17) with two different vectors, n^a and n^b, i.e. we once substitute $n = n^a$ and once $n = n^b$ into the fundamental solution $g_1[x]$. This renders a system of two equations for two unknowns.

$$\dot c_1^a(x)\, w,_1(x) + \dot c_2^a(x)\, w,_2(x) = r^a \tag{6.19a}$$

$$\dot c_1^b(x)\, w,_1(x) + \dot c_2^b(x)\, w,_2(x) = r^b \tag{6.19b}$$

(the letter r indicates the right-hand side and the superscripts a and b refer to n^a and n^b). If this system is solved for $w,_1(x)$ and $w,_2(x)$ then the directional derivative $\partial w/\partial n = w,_1 n_1 + w,_2 n_2$ at x with respect to arbitrary vectors n can be calculated.

The Eq. (6.17) bears a close resemblance with the Somigliana formula, the integral representation of the displacement field $u(x)$ of an elastic plate, see Eq. (5.30).

Let x be a boundary point of the Kirchhoff plate then with the help of the unit matrix I and the matrix

$$J(\varphi) = \begin{bmatrix} 1/2 \sin 2\varphi & \sin^2 \varphi \\ \sin^2 \varphi & -1/2 \sin 2\varphi \end{bmatrix}$$

the left-hand side of Eq. (6.17) can be written as

$$\dot c_1(x)\, w,_1(x) + \dot c_2(x)\, w,_2(x) = \mathrm{grad}^T w(x) \left\{ \frac{\Delta \varphi}{2\pi} I + \frac{v}{2\pi} [J(\varphi_1) - J(\varphi_2)] \right\} n \tag{6.20}$$

while the left-hand side of the Somigliana formula at a boundary point reads, see Eq. (5.27)

$$\left\{ \frac{\Delta \varphi}{2\pi} I + \frac{1}{4\pi(1-v)} [J(\varphi_1) - J(\varphi_2)] \right\} u(x) \tag{6.21}$$

The similarity between the two expressions, Eq. (6.20) and (6.21), is evident and it confirms the analogy between the elastic plate and the Kirchhoff plate observed in the literature. Such analogous expressions are, e. g., ($F =$ Airy's stress function)

$$\varepsilon_{11} = \lambda(F,_{22} + \mu F,_{11}) \qquad \text{elastic plate}$$

$$- M_{22} = K(w,_{22} + v\, w,_{11}) \qquad \text{Kirchhoff plate}$$

6.2 Green's Function

At the beginning of this chapter we calculated the deflection $w(x)$ of a rectangular plate by letting the deflection w act as a virtual displacement on a rectangular segment of the infinite Kirchhoff plate loaded with a concentrated force at x, see Fig. 6.1.

But instead of the infinite plate we could also choose a finite plate as in Fig. 6.5 whose boundary conditions coincide with those of the plate on the right.

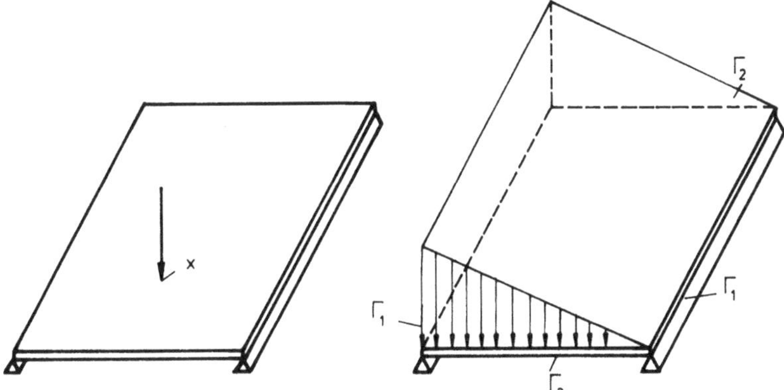

Figure 6.5

The deflection of the finite plate, $G_0[x] = g_0[x] + w_R$, differs from the deflection $g_0[x]$ of the infinite plate by a term w_R which matches the boundary values of $g_0[x]$ with those of the supports.

If we now formulate the identities in the domain Ω

$$G(G_0[x], w) = 0, \quad B(G_0[x], w) = 0$$

then due to the fact that both $G_0[x]$ and w satisfy the same homogeneous boundary conditions, of two conjugated quantities on the boundary one is always zero so that the work done on the boundary is zero as well

$$\int_{\Gamma_2} V_v[\overset{\circ}{x}]\, w\, ds_y, \quad \int_{\Gamma} M_v[\overset{\circ}{x}]\, \frac{\partial w}{\partial v}\, ds_y, \quad \text{etc.}$$

and the identities, therefore, simplify to

$$G(G_0[x], w) = w(x) - E(G_0[x], w) = 0$$

$$B(G_0[x], w) = w(x) - \int_{\Omega} G_0[x]\, p\, d\Omega_y = 0$$

Which means that

$$w(x) = E(G_0[x], w), \quad w(x) = \int_{\Omega} G_0[x]\, p\, d\Omega_y$$

These are the simplest integral representations obtainable for the deflection $w(x)$ of a plate.

The function $G_0[x]$ which achieves this simple result is called a Green's function.

Such functions $G_i[x]$ can be employed for integral representations of higher derivatives $\partial^1 w = \partial w/\partial n$, $\partial^2 w = M_n$, $\partial^3 w = V_n$ as well. The result, any time, is very simple

$$G(G_i[x], w) = \partial^i w(x) - E(G_i[x], w) = 0$$

$$B(G_i[x], w) = \partial^i w(x) - \int_\Omega G_i[x] K \Delta \Delta w \, d\Omega_y = 0$$

The difference between integral representations formulated with the simple fundamental solutions $g_i[x]$ and those formulated with Green's functions $G_i[x] = g_i[x] + w_R$ is that the latter only contain data which appear in the formulation of the bvp which, hence, are a priori known. This is the advantage a Green's function $G_i[x] = g_i[x] + w_R$ compared with a simple fundamental solution $g_i[x]$ has.

If we want to calculate with the simple fundamental solution $g_i[x]$ alone the deflection of the plate in Fig. 6.5b then we must know the values of the normal derivative $\partial w/\partial n$ on the whole boundary, the values of the Kirchhoff-shear (plus corner forces) along the supports and the deflection w along the free boundary.

But as these data do not appear in the formulation of the bvp we, first, must solve the compatibility conditions of section 6.2 for these unknown functions, i. e. we must, first, solve two integral equations before we can really use Eq. (6.3).

All these difficulties disappear if we use a Green's function because the (conjugated) boundary values of a Green's function $G_i[x]$ are zero where the boundary values of the real deflection are unknown. Hence, the unknown quantities do not appear in the identity, the pertinent boundary integrals are zero.

For a function $G_i[x] = g_i[x] + w_R$ to be Green's function of an bvp as

$$Dw = p \text{ in } \Omega, \quad \partial^\lambda w = f_\lambda, \; \lambda \in \Lambda, \text{ on } \Gamma \tag{6.22}$$

the regular part w_R must be a homogeneous solution of

$$Dw_R = 0$$

and it must satisfy the boundary conditions

$$\partial^\lambda w_R = -\partial^\lambda g_i[x] \; \lambda \in \Lambda$$

This ensures that $G_i[x]$ satisfies the homogeneous boundary conditions.

$$\partial_y^\lambda G_i[x] = 0, \; \lambda \in \Lambda, \text{ on } \Gamma$$

In other words, a Green's function $G_i[x]$ is the displacement of a structure if a concentrated conjugated load (conjugated with respect to the unknown quantity) acts at x and all the boundary conditions are zero.

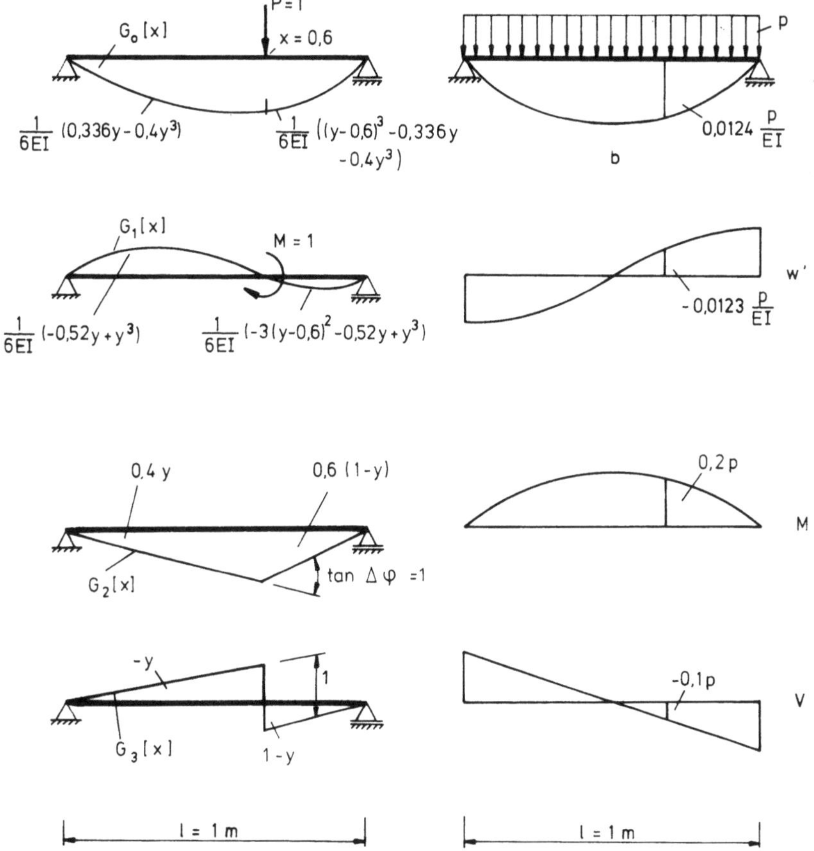

Figure 6.6

Well known is the application of Green's function in the analysis of frames and beams as in the case of the beam in Fig. 6.6b which is loaded with an evenly distributed force p. Assume we are interested in the values of w, w', M and V at the point $x = 0.6$. To find these values we load the same beam with conjugated quantities, these cause the deflections $G_i[x]$, and then calculate the reciprocal external work of the two states $G_i[x]$ and w, i.e. we formulate the second identity

$$B(G_i[x], w) = \partial^i w(x) + \left[V_i[x] w - M_i[x] w' \right]_0^1$$

$$- \left[G_i[x] V - G_i'[x] M \right]_0^1 - \int_0^1 G_i[x] p \, dy = 0$$

We easily verify that due to the properties of $G_i[x]$ and w the reciprocal work done on the boundary is zero and, hence the second identity simplifies in each case to

$$B(G_0[x], w) = w(x) - \int G_0[x] p \, dy = w(0.6) - 0.0124 \frac{p}{EI} = 0$$

$$B(G_1[x], w) = w'(x) - \int G_1[x]\, p\, dy = w'(0.6) + 0.0123\,\frac{p}{EI} = 0$$

$$B(G_2[x], w) = M(x) - \int G_2[x]\, p\, dy = M(0.6) - 0.12\, p = 0$$

$$B(G_3[x], w) = V(x) - \int G_3[x]\, p\, dy = V(0.6) + 0.10\, p = 0$$

The unknowns are simply the L_2-scalar product of $G_i[x]$ and the applied load p.

In practical applications the calculation of the deflection $w(x)$ often is not done by forming the L_2-scalar product of $G_0[x]$ and p but by forming the L_2-scalar product of the two bending moments $M_0[x]$ and M, see Fig. 6.7.

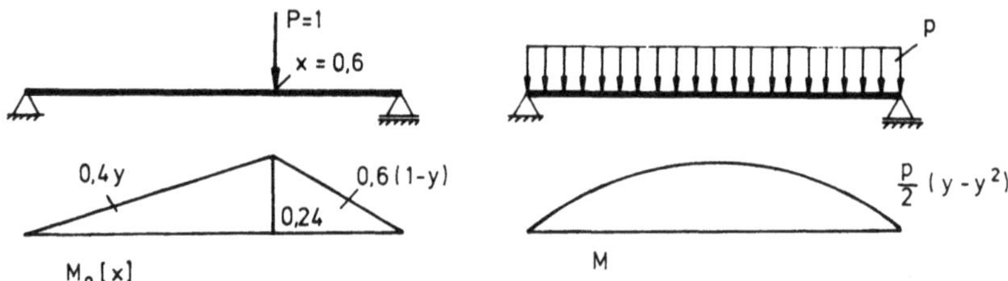

Figure 6.7

That is, the deflection $w(x)$ is calculated by applying the first identity (the principle of virtual displacements)

$$G(G_0[x], w) = w(x) + \left[V_0[x]\, w - M_0[x]\, w'\right]_0^1 - \int_0^1 \frac{M_0[x]\, M}{EI}\, dy$$

$$= w(x) - \int_0^1 \frac{M_0[x]\, M}{EI}\, dy = w(0.6) - 0.0124\,\frac{p}{EI} = 0$$

This is one of the most widely used formulas in structural mechanics.

In higher dimensions the equations of Green's functions are, in general, unknown and they must, therefore, be approximated by series expansions.

Fig. 6.8 shows the influence function, thus obtained, for the bending moment M_x in the middle of a simply supported quadratic plate, see [G1] p. 266.

The bending moment under the action of an evenly distributed load p is found if we multiply the volume V under the surface with the load intensity p

$$M_x = p\, V$$

This rule is based on the equation

$$B(G_2[x], w) = M_x(x) + \int_\Gamma \left(V_{v_2}[x]\, w - M_{v_2}[x]\, \frac{\partial w}{\partial v} + \frac{\partial}{\partial v}\, G_2[x]\, M_v \right.$$

$$\left. - G_2[x]\, V_v \right) ds_y + \left[\left[M_{v\tau}[x]\, w - G_2[x]\, M_{v\tau}\right]\right]$$

$$- \int_\Omega G_2[x]\, K\Delta\Delta w\, d\Omega_y = M_x(x) - \int_\Omega G_2[x]\, p\, d\Omega_y = 0$$

because the so-called influence surface is the function $G_2[x]$, the deflection of the quadratic plate when the slope jumps in the middle of the plate by an amount of

$$\lim_{\varepsilon \to 0} \int_{\Gamma_{N_\varepsilon}} \frac{\partial w}{\partial v}\, ds_y = 1$$

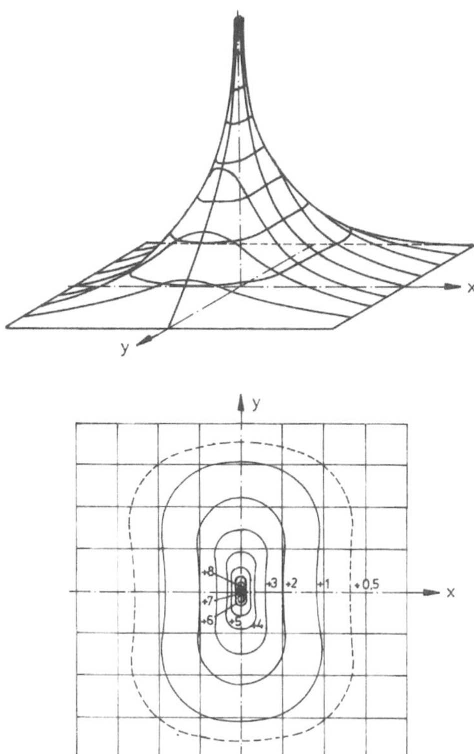

Figure 6.8

6.3 Compatibility on the Boundary

Consider the circular simply supported plate of radius R in Fig. 6.9.
Its deflection

$$w(r) = \frac{p}{64\,K}\left(\frac{5+v}{1+v}\,R^4 - 2\,\frac{3+v}{1+v}\,R^2\,r^2 + r^4\right)$$

satisfies the differential equation

$$K\,\Delta\Delta\,w = p \quad \text{in } \Omega$$

and has the boundary values

$$w = 0, \qquad \frac{\partial w}{\partial n} = \frac{\partial w}{\partial r} = -\frac{pR^3}{8(1+v)K}, \qquad M_n = 0, \qquad V_n = -p\frac{R}{2}$$

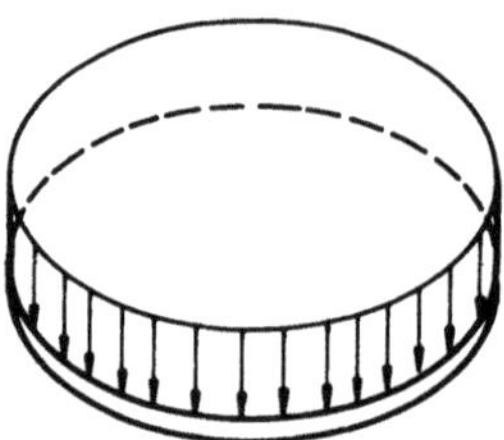

Figure 6.9

Because w belongs to $C^4(\bar{\Omega})$ we may formulate integral representations for $w = \partial^0 w$, $\partial w/\partial n = \partial^1 w$, $M_n = \partial^2 w$ and $V_n = \partial^3 w$,

$$c^i(x)\,\partial^i w(x) = \int_\Gamma \left(-V_{v_i}[x]\,0 + M_{v_i}[x]\frac{\partial w}{\partial v} - \frac{\partial}{\partial v}g_i[x]\,0 + g_i[x]\,V_v\right) ds_y$$

$$+ \int_\Omega g_i[x]\,p\,d\Omega_y \quad i = 0, 1, 2, 3 \tag{6.23}$$

in terms of the boundary functions $w = 0$, $\partial w/\partial v$, $M_v = 0$ and V_v and the load p in the domain.

The value of the characteristic function $c^{(i)}(x)$ on the boundary is 1/2. Hence, we have at all boundary points

$$\frac{1}{2}\left\{\begin{array}{c} 0 \\ \dfrac{\partial w}{\partial n} \\ 0 \\ V_n \end{array}\right\}(x) = \int_\Gamma \left(-V_{v_i}[x]\,0 + M_{v_i}[x]\frac{\partial w}{\partial v} - \frac{\partial}{\partial v}g_i[x]\,0 + g_i[x]\,V_v\right) ds_y$$

$$+ \int_\Omega g_i[x]\,p\,d\Omega_y \quad \begin{array}{c} i = 0 \\ i = 1 \\ i = 2 \\ i = 3 \end{array} \tag{6.24}$$

The functions on the left-hand side are boundary functions and they are the same functions as on the right-hand side, under the integral sign. Obviously, so we may conclude, the four boundary functions 0, $\partial w/\partial n$, 0 and V_n satisfy four integral equations on the boundary.

That this is not trivial is easily understood if we, for once, assume that we replace the functions 0, $\partial w/\partial n$, 0 and V_n on the right-hand side by arbitrary boundary functions f_0, f_1, f_2 and f_3. Then we may not expect the left-hand side to be the vector $1/2$ $\{f_0, f_1, f_2, f_3\}$ as in the case of the function w.

Hence we say, the four functions 0, $\partial w/\partial n$, 0 and V_n satisfy four compatibility conditions, four integral equations, on the boundary.

This result applies to the boundary values of all structural elements because the equation

$$B(g_i[x], u) = 0, \quad x \in \Gamma, \quad i = 0, 1, \ldots 2m - 1$$

which is the basis of Eq. (6.24) can be formulated for all structural elements.

Theorem 6.4 Compatibility on the boundary

$$p : 0 \leqslant i \leqslant 2m - 1, u \in C^{2m}(\bar{\Omega}), x \in \Gamma$$

$$q : c^{(i)}(x)\,\partial^i u(x) = \sum_{k=1}^{2m} (-1)^k \left[\partial^{2m-k} g_i[x], \partial^{k-1} u\right] + (g_i[x], Du) \quad (6.25)$$

This theorem asserts that the boundary values (with respect to an operator D) of a smooth function, $u \in C^{2m}$ satisfy $2m$ integral equations on the boundary.

In the case of one-dimensional structural elements these equations degenerate to systems of linear equations. The compatibility conditions are then, as we shall see in section 6.6, equivalent with the condition that the end displacements δ and end forces f satisfy the equation $K\delta = f + p$ where K is the stiffness matrix and p the negative support reactions of the clamped beam (bar).

Two examples will illustrate the usefullness of this theorem.

Let a bar with cross section $A = 5\,\text{cm}$, a Young's modulus of $E = 21\,000\,\text{kN/cm}^2$ and of length $l = 1000\,\text{cm}$ be loaded with a single force $P = 55\,\text{kN}$, see Fig. 6.10.

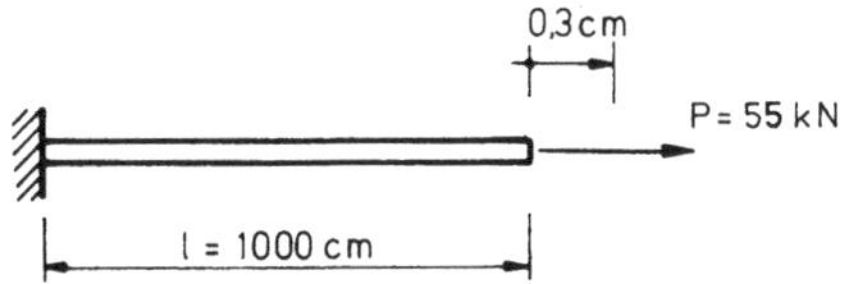

Figure 6.10

At the free end we observe a displacement of $0.3\,\text{cm}$. Is this measurement correct?

To test our data we check them with the compatibility conditions which require that the end forces $f_1 = -55\,\text{kN}$, $f_2 = 55\,\text{kN}$ and the end displacements $\delta_1 = 0$, $\delta_2 = 0.3\,\text{cm}$ satisfy the equation $K\delta = f$ or

$$\begin{bmatrix} 105 & -105 \\ -105 & 105 \end{bmatrix} \begin{bmatrix} 0 \\ 0.3 \end{bmatrix} = \begin{bmatrix} -55 \\ 55 \end{bmatrix}$$

where K is just the stiffness matrix of the bar, see Eq. (3.16).

Because $105 \cdot 0.3$ is not 55 our measurement must be wrong.

The concept of compatibility on the boundary is not new though it is never mentioned explicitly as such in the literature. The engineer knows, instinctively, that he cannot prescribe a force X and a displacement Y at one end of a bar at the same time. This is only admissible if the data match, i.e. if they satisfy the compatibility condition.

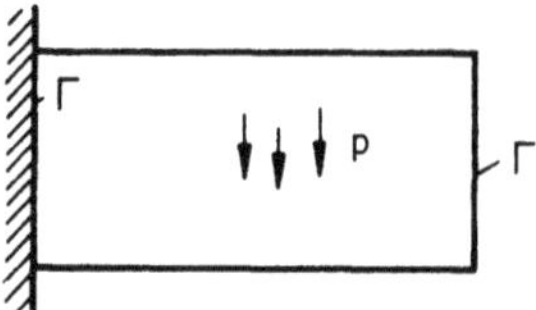

Figure 6.11

Our second example is the elastic plate in Fig. 6.11 which displaces under its own weight.

Assume we have measured the support reactions and the displacements on the free boundary. An interpolation of these discrete values with polynomials would render two vector-valued functions which represent, approximately, the boundary data of the field u. If those functions were the exact boundary values then they should, at every point x of the boundary, satisfy the first compatibility condition of an elastic plate, namely the integral equation

$$\dot{C}(x)\, u(x) = \int_{\Gamma} (U(y, x)\, t(y) - T(y, x)\, u(y))\, ds_y + \int_{\Omega} U(y, x)\, p(y)\, d\Omega_y, \quad x \in \Gamma$$

Discrepancies would hint to errors in the measurement or the interpolation.

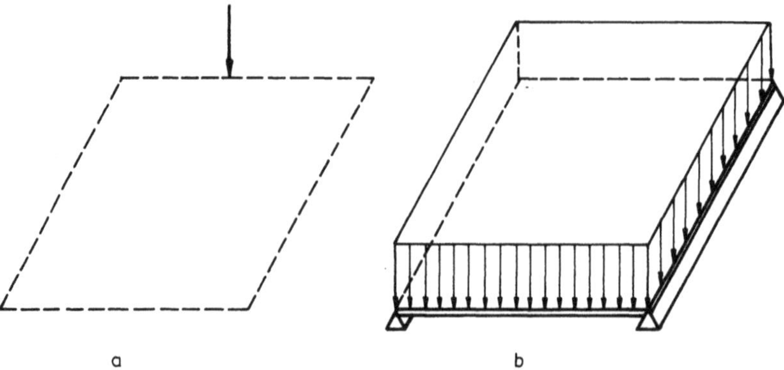

Figure 6.12

A mechanical interpretation of the compatibility condition provides Fig. 6.12. The deflection of the infinite plate on the left is $g_0[x] = g_0(y, x)$ and of the plate on the right $w(y)$. According to Betti's principle the reciprocal work of the two elastic states must be equal, i.e.

$$B(g_0[x], w) = 0 \qquad x \in \Gamma$$

If we repeat this equation at every point x on the boundary, that is if we let the concentrated force wander along the dashed line, then we have just formulated the first compatibility condition.

The second, third and fourth compatibility condition are obtained if we replace the concentrated force by a couple, a discontinuous rotation or a jump in deflection.

6.3.1 The Order of the Integral Operators

In a symbolic notation the four compatibility conditions of a Kirchhoff plate can be written as

$$\frac{1}{2} \begin{bmatrix} w \\ \dfrac{\partial w}{\partial n} \\ M_n \\ V_n \end{bmatrix} = \int\limits_{\Gamma} \begin{bmatrix} a_{00} & a_{01} & a_{02} & a_{03} \\ .. & a_{11} & .. & .. \\ .. & .. & .. & .. \\ .. & .. & a_{32} & a_{33} \end{bmatrix} \begin{bmatrix} w \\ \dfrac{\partial w}{\partial v} \\ M_v \\ V_v \end{bmatrix} ds_y + \begin{bmatrix} d_0 \\ d_1 \\ d_2 \\ d_3 \end{bmatrix}$$

where

$$d_i = \int\limits_{\Omega} g_i[x] \, K \Delta \Delta w \, d\Omega_y$$

Following Wendland, [W 3], we say an integral operator

$$\int\limits_{\Gamma} a_{ij}(y, x) f(y) \, ds_y = g(x)$$

has order 2α if it transforms a boundary function $f \in H^{r+2\alpha}(\Gamma)$ into a boundary function $g \in H^r(\Gamma)$ (r is any generic number for which this is true).

If $2\alpha > 0$ then the integral operator differentiates the function f and if $2\alpha < 0$ then the integral operator integrates the function f, the result, the function g, has a higher regularity than f.

With this definition in mind let us have a look at the first compatibility condition

$$\frac{1}{2} w = \int\limits_{\Gamma} a_{00} \, w \, ds_y + \int\limits_{\Gamma} a_{01} \frac{\partial w}{\partial v} \, ds_y + \int\limits_{\Gamma} a_{02} \, M_v \, ds_y + \int\limits_{\Gamma} a_{03} \, V_v \, ds_y + d_0$$

The first integral operator on the right-hand side has the order $2\alpha = 0$. It transforms w into itself (the function on the left-hand side). Nothing is gained or lost in regularity. The second operator has the order $2\alpha = -1$, it transforms the normal derivative ($\simeq$ 1st derivative) into w, i.e. it integrates once. The third operator has the order $2\alpha = -2$, it integrates the second derivative ($\simeq M_v$) twice, and the fourth operator has the order $2\alpha = -3$, it integrates the third derivative ($\simeq V_v$) thrice.

If we do this with all four compatibility conditions then we obtain the following picture

$$\begin{bmatrix} 0 \\ 1 \\ 2 \\ 3 \end{bmatrix} = \begin{bmatrix} 0 & -1 & -2 & -3 \\ 1 & 0 & -1 & -2 \\ 2 & 1 & 0 & -1 \\ 3 & 2 & 1 & 0 \end{bmatrix} \begin{bmatrix} 0 \\ 1 \\ 2 \\ 3 \end{bmatrix}$$

The "worst" compatibility condition is the last one because nearly all integral operators in this equation differentiate. The first operator even three times. Such an operator is hyper-singular. It requires very smooth splines (boundary elements) and very accurate integration schemes.

The order 2α plays an important role because it determines the lowest degree of splines admissible in any numerical scheme and because it is also responsible for the speed of the asymptotic convergence of the approximate solution. If h denotes the mesh-width of a boundary element net and $d-1$ the order of the splines then the error of the collocation solution f_h of the equation

$$\int_\Gamma a_{ij}(y, x) f_h(y)\, ds_y = g(x)$$

measured in the norm of the Sobolev space $H^{2\alpha}(\Gamma)$ behaves as $0(h^{d+1-2\alpha})$, see [A & W]. This means: an operator which differentiates, $2\alpha > 0$, has a slower convergence than an operator which integrates, $2\alpha < 0$.

6.3.2 The Essential Compatibility Conditions

Let us return to our model problem, the circular plate in Fig. 6.9. The boundary data of this plate satisfy the four integral equations (6.24).

From the beginning two of the four boundary functions, namely w and M_n, had to be zero. Hence, the two unknown functions, $\partial w/\partial n$ and V_n, the functions which do not appear in the formulation of the *bvp*, must satisfy the four integral equations

$$\frac{1}{2}\begin{bmatrix} 0 \\ \dfrac{\partial w}{\partial n} \\ 0 \\ V_n \end{bmatrix} = \int_\Gamma \begin{bmatrix} a_{00} & a_{01} & a_{02} & a_{03} \\ .. & a_{11} & .. & .. \\ .. & .. & .. & .. \\ .. & .. & a_{32} & a_{33} \end{bmatrix} \begin{bmatrix} 0 \\ \dfrac{\partial w}{\partial v} \\ 0 \\ V_v \end{bmatrix} ds_y + \begin{bmatrix} d_0 \\ d_1 \\ d_2 \\ d_3 \end{bmatrix}$$

But if two functions, $\partial w/\partial n$ and V_n, sastisfy four integral equations then not all four equations can be linearly independent. We may assume that only 2 of those 4 equations are. We formulate this as a theorem.

Theorem 6.5

Let the boundary Γ be smooth and assume that the functions $p \in C(\Omega), f_i \in C^{2m-i}(\Gamma)$, $0 \leqslant i \leqslant 2m - 1$, satisfy the first m integral equations $i = 0, 1 \ldots m - 1$

$$c^{(i)}(x) f_i(x) = \sum_{k=1}^{2m} (-1)^k \left[\partial^{2m-k} g_i[x], f_{k-1}\right] + (g_i[x], p), \quad x \in \Gamma \quad (6.26)$$

then these functions satisfy also the remaining integral equations (6.26) $i = m, \ldots 2m - 1$.

The proof of this theorem is simple in one dimension but no more so in higher dimensions. This is the reason why a proof in the latter case is, as far as we know, still missing.

Theorem 6.5 states that of the $2m$ integral equations which a smooth function $u \in C^{2m}(\bar{\Omega})$ must satisfy only the first m equations are linearly independent. The remaining m equations are satisfied automatically if the first m equations are.

Hence, the boundary data of a beam, $2m = 4$,

$$w(a), w(b) \quad w'(a), w'(b) \quad M(a), M(b) \quad V(a), V(b)$$

and of a Kirchhoff plate, $2m = 4$,

$$w \quad \partial w/\partial n \quad M_n \quad V_n \quad (F_k)$$

must satisfy two integral equations and the boundary data of a bar, $2m = 2$,

$$u(a), u(b) \quad N(a), N(b)$$

and an elastic plate or body, $2m = 2$,

$$u \quad \tau(u)$$

one integral equation.

We remarked at the beginning of this chapter that it is not permissible to substitute arbitrary boundary functions into the integral representations of a displacement function and to believe that the function, so constructed, has these functions as boundary data. For this to be true it is necessary and sufficient that the data satisfy the essential compatibility conditions as the next theorem explains.

Theorem 6.6

Let the boundary Γ be smooth and assume the functions $p \in C(\Omega), f_i \in C^{2m-i}(\Gamma)$, $i = 0, 1, \ldots 2m - 1$, satisfy the $2m$ integral equations (6.26) $i = 0, 1, \ldots 2m - 1$ (substitute $Du = p$, $\partial^i u = f_i$) then the function

$$u(x) = \sum_{k=1}^{2m} (-1)^k \left[\partial^{2m-k} g_0[x], f_{k-1}\right] + (g_0[x], p)$$

satisfies the differential equation

$$Du = p \quad \text{in } \Omega$$

and it is

$$\lim_{x \to \Gamma} \partial^i u = f_i \qquad 0 \leqslant i \leqslant 2m - 1$$

The proof of this theorem, again, is simple in one dimension, see the example in section 6.5, but no more so in higher dimensions. A proof for elastic plates and bodies is given in [H 1] but still missing, as far as we know, in the case of the Kirchhoff plate.

Theorem 6.6 is the converse of the integral representations in section 6.1. There we stated:

If a function $u \in C^{2m}$ exists and has the data Du in the interior and $\partial^0 u, \partial^1 u, \ldots \partial^{2m-1} u$ on the boundary then

$$u(x) = \sum_{k=1}^{2m} (-1)^k \left[\partial^{2m-k} g_0[x], \partial^{k-1} u \right] + (g_0[x], Du$$

is an integral representation of u.

Now the statement is: if there is a set of $2m + 1$ functions, $p, f_i, i = 0, 1, \ldots 2m - 1$ which satisfy the $2m$ integral equations (of which only m are linearly independent) then there exists a function

$$u(x) = \sum_{k=1}^{2m} (-1)^k \left[\partial^{2m-k} g_0[x], f_{k-1} \right] + (g_0[x], p) \tag{6.27}$$

whose data are the functions $p, f_i, i = 0, 1, \ldots 2m - 1$. In other words, we can construct a function which solves the equation $Du = p$ and assumes the boundary values f_i.

This theorem, in connection with the compatibility conditions, forms the basis of the boundary element method. Its usefullness stems from the fact that Eq. (6.27) allows to construct the displacement field explicitly once the boundary data (and the force in the field) are determined.

As an example consider the cantilever beam in Fig. 6.13.

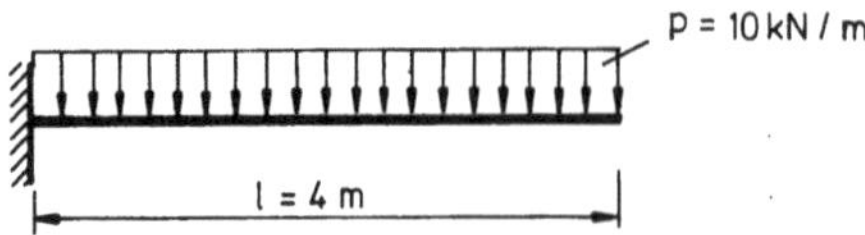

Figure 6.13

The boundary conditions require

$$w(0) = 0, \quad w'(0) = 0, \quad M(l) = 0, \quad V(l) = 0$$

This leaves $w(l)$, $w'(l)$, $M(0)$ and $V(0)$ as the unknown boundary data. Solving the compatibility conditions, see section 6.6, for these unknowns renders

$$w(l) = \frac{32p}{EI}, \quad w'(l) = \frac{10.67}{EI} p, \quad M(0) = -8p, \quad V(0) = 4p$$

Hence, all displacement- and force-terms of the deflection w are determined and on substituting these compatible data into Eq. (6.31) we obtain an integral representation of $w(x)$

$$w(x) = -\left[V_0\,[x]\,w - M_0\,[x]\,w' + g_0'\,[x]\,M - g_0\,[x]\,V\right]_0^4 + \int_0^4 g_0\,[x]\,p\,dy$$

$$= \frac{p}{6\,EI}\left(24\,x^2 - 4\,x^3 + \frac{x^4}{4}\right)$$

6.4 Summary

All these results which we often stated only in the abstract notation of section 2.10 will, in this section, be repeated, separately, for the single structural elements.

6.4.1 Bars

a) *Compatibility*: If $u(x)$ belongs to $C^2\,[a,b]$ then it satisfies the two equations $i = 0, 1$

$$c(x)\,\partial^i u(x) = -\left[N_i\,[x]\,u - g_i\,[x]\,N\right]_a^b + \int_a^b g_i\,[x]\,(-EA\,u'')\,dy \qquad (6.28)$$

on the boundary, at $x = a$ and $x = b$.

Formulating these equations consecutively for $x = a$ and for $x = b\;(i = 0)$ and then $x = a$ and $x = b\;(i = 1)$ results in the four equations

$$\begin{bmatrix} a & -a \\ -(1-b) & (1-b) \\ EA & -EA \\ EA & -EA \end{bmatrix}\begin{bmatrix} u(a) \\ u(b) \end{bmatrix} = \frac{1}{EA}\begin{bmatrix} -a(1-a) & a(1-b) \\ -a(1-b) & (1-b)\,b \\ -(1-a)\,EA & (1-b)\,EA \\ a\,EA & -b\,EA \end{bmatrix}\begin{bmatrix} N(a) \\ N(b) \end{bmatrix} + \begin{bmatrix} d_0(a) \\ d_0(b) \\ d_1(a) \\ d_1(b) \end{bmatrix}$$

where

$$d_i(x) = \int_a^b g_i\,[x]\,p\,dy$$

b) *Construction*: If four numbers $u(a)$, $u(b)$, $N(a)$, $N(b)$ and the function p satisfy Eq. (6.28) $i = 0$ at $x = a$ and $x = b$ then the function

$$u(x) = -\left[N_0\,[x]\,u - g_0\,[x]\,N\right]_a^b + \int_a^b g_0\,[x]\,p\,dy$$

$$= x\,u(b) + (1-x)\,u(a) + \Big\{(1-b)\,x\,N(b) - (1-x)\,a\,N(a)$$

$$+ (1-x)\int_a^x y\,p(y)\,dy + x\int_x^b (1-y)\,p(y)\,dy\Big\}\frac{1}{EA} \qquad (6.29)$$

solves the equation $-EA\,u''(x)=p(x)$, $a<x<b$, and then the four numbers

$$u(a),\ u(b),\ N(a),\ N(b)$$

are its boundary values.

6.4.2 Beams

a) *Compatibility*: If the function w belongs to $C^4[a,b]$ then it satisfies the four "integral equations" $i=0,1,2,3$

$$c(x)\,\partial^i w(x) = -\big[V_i[x]\,w - M_i[x]\,w' + g_i'[x]\,M - g_i[x]\,V\big]_a^b$$

$$+\int_a^b g_i[x]\,EIw^{IV}\,dy \tag{6.30}$$

at $x=a$ and $x=b$.

b) *Construction*: If eight numbers $w(a)$, $w(b)$; $w'(a)$, $w'(b)$; $M(a)$, $M(b)$; $V(a)$, $V(b)$ and the function p satisfy the Eqs. (6.30) $i=0,1$ at $x=a$ and $x=b$ then the function

$$w(x) = -\big[V_0[x]\,w - M_0[x]\,w' + g_0'[x]\,M - g_0[x]\,V\big]_a^b$$

$$+\int_a^b g_0[x]\,p\,dy = xw(b)+(1-x)\,w(a)$$

$$+x(1-b)\,w'(b)-a(1-x)\,w'(a)$$

$$+\frac{1}{6EI}\Big\{[-3(b-x)^2-\alpha(x)+3b^2(1-x)]\,M(b)$$

$$+[\alpha(x)-3a^2(1-x)]\,M(a)+[(b-x)^3+\alpha(x)\,b$$

$$-(1-x)\,b^3]\,V(b)-[\alpha(x)\,a-(1-x)\,a^3]\,V(a)$$

$$+\int_a^x [\alpha(x)\,y-(1-x)\,y^3]\,p(y)\,dy + \int_x^b [(y-x)^3$$

$$+\alpha(x)\,y-(1-x)\,y^3]\,p(y)\,dy\Big\} \tag{6.31}$$

$$\alpha(x)=x(1-x)(2-x)$$

solves the equation $EI\,w^{IV}=p$, $a<x<b$ and then the eight numbers are its boundary values.

6.4.3 Kirchhoff Plates

a) *Compatibility*: If w belongs to $C^4(\bar\Omega)$ then its boundary values $\partial^i w$, $i=0,1,2,3$ satisfy the four integral equations

$$\dot{c}^{(i)}(x)\,\partial^i\,w(x) = \int\limits_\Gamma \left[-V_{v_i}[x]\,w + M_{v_i}[x]\frac{\partial w}{\partial v} - \frac{\partial}{\partial v}\,g_i[x]\,M_v \right.$$

$$\left. + g_i[x]\,V_v \right] ds_y - \left[[M_{v\tau_i}[x]\,w - g_i[x]\,M_{v\tau}]\right]$$

$$+ \int\limits_\Omega g_i[x]\,K\Delta\Delta w\,d\Omega_y, \quad x \in \Gamma \tag{6.32}$$

b) *Construction*: If four boundary functions f_i, $i=0,1,2,3$ (substitute $\partial^i w = f_i$) and the domain function p (substitute $K\Delta\Delta w = p$) satisfy the integral equations (6.32) $i=0,1$ and if the boundary is smooth then the function

$$w(x) = \int\limits_\Gamma \left[-V_{v_0}[x]\,w + M_{v_0}[x]\frac{\partial w}{\partial v} - \frac{\partial}{\partial v}\,g_0[x]\,M_v + g_0[x]\,V_v \right] ds_y$$

$$+ \int\limits_\Omega g_0[x]\,K\Delta\Delta w\,d\Omega_y \tag{6.33}$$

solves the equation $K\Delta\Delta w = p$ in the interior and has the boundary values $\partial^i w = f_i$.

6.4.4 Elastic Plates and Bodies

a) *Compatibility*; If u belongs to $C^2(\bar{\Omega})$ then its boundary functions u and $\tau(u)$ satisfy the two integral equations

$$\dot{C}^{(0)}(x)\,u(x) = \int\limits_\Gamma [U(y,x)\,\tau(u)(y) - T(y,x)\,u(y)]\,ds_y$$

$$+ \int\limits_\Omega U(y,x)\,(-Lu(y))\,d\Omega_y \tag{6.34}$$

$$\dot{C}^{(1)}(x)\,\tau(u)(x) = \int\limits_\Gamma [T^*(y,x)\,\tau(u)(y) - Z(y,x)\,u(y)]\,ds_y$$

$$+ \int\limits_\Omega T^*(y,x)\,(-Lu(y))\,d\Omega_y \tag{6.35}$$

Here, $Z(y,x) = \tau_x\,T(y,x)$ is the matrix obtained when the operator $\tau(\ )$ acts on the columns of T.

b) *Construction*: If the boundary functions u and t (substitute $\tau(u)=t$) and the domain function p (substitute $-Lu=p$) satisfy the integral equation (6.34) and if the boundary is smooth then the function

$$u(x) = \int\limits_\Gamma [U(y,x)\,t(y) - T(y,x)\,u(y)]\,ds_y + \int\limits_\Omega U(y,x)\,p(y)\,d\Omega_y \tag{6.36}$$

solves the equation $-Lu = p$ and assumes the boundary values u and t.

6.5 An Example

Let us illustrate the preceding theorems with a simple example.

Consider the bar in Fig. 6.14a, which connects (e. g. in a truss) the points $x = 2$ and $x = 6$ and let $u(x)$ be the displacement of the bar. No distributed forces act on the bar, hence, u satisfies the equation

$$-EA\,u'' = 0$$

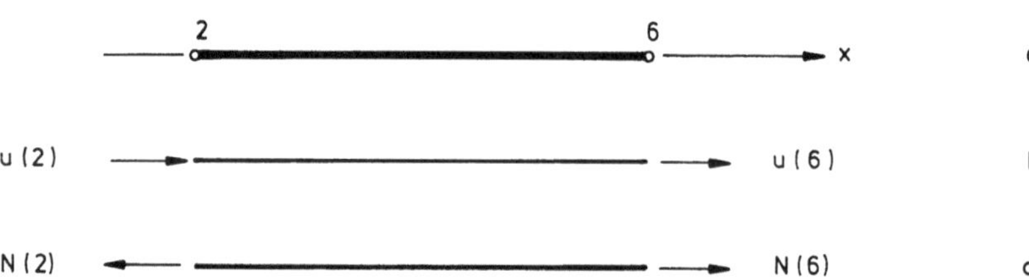

Figure 6.14

Because of this the displacement must be a polynominal of the first kind, hence it is in $C^\infty[2,6]$ and, a fortiori, in $C^2[2,6]$. As a consequence Eq. (6.29) applies and we obtain the following "integral representation" of u

$$u(x) = -\left[N_0[x]\,u - g_0[x]\,N\right]_2^6$$

$$= x\,u(6) + (1-x)\,u(2) + \frac{1}{EA}\,x\,(1-6)\,N(6) - \frac{1}{EA}\,(1-x)\,2\,N(2)$$

and with Eq. (6.28) of N as well

$$N(x) = -\left[N_1[x]\,u - g_1[x]\,N\right]_2^6 = \left(u(6) - u(2)\right)EA - 5N(6) + 2N(2)$$

The functions $g_i[x]$ $i = 0,1$ are the fundamental solutions from Eqs. (6.5a) and (6.5b) and

$$N_0[x] = \begin{cases} 1-x \\ -x, \end{cases} \qquad N_1[x] = \begin{cases} -EA & y \leqslant x \\ -EA & x \leqslant y \end{cases}$$

is the corresponding distribution of normal forces and $u(2)$, $u(6)$ and $N(2)$, $N(6)$ are the boundary values respectively.

Conversely, assume that

$$u(6),\,u(2) \qquad N(6),\,N(2)$$

are four numbers which satisfy the first compatibility condition, Eq. (6.28) $i = 0$, on the boundary, i. e. at $x = 6$ and at $x = 2$.

$$0 = 5\,u(6) - 5\,u(2) + \frac{(-30)}{EA}\,N(6) - \frac{(-10)}{EA}\,N(2)$$

$$0 = 2\,u(6) - 2\,u(2) + \frac{(-10)}{EA}\,N(6) - \frac{(-2)}{EA}\,N(2)$$

It then holds:
a) the four numbers satisfy also the second compatibility condition, Eq. (6.28)
   $i = 1$, on the boundary, i.e. at $x = 6$ and at $x = 2$, and
b) the function constructed with these four numbers

$$u(x) = -\left[N_0[x]\,u - g_0[x]\,N\right]_2^6$$

$$= x\,u(6) + (1 - x)\,u(2) + \frac{x(1 - 6)}{EA}\,N(6) - \frac{(1 - x)2}{EA}\,N(2)$$

solves the equation $-EA\,u'' = 0$ and has the boundary values $u(2)$, $\overset{\scriptscriptstyle\prime}{u}(6)$ and
$N(2)$, $N(6)$.
   The proof is easy: Eq. (6.28) $i = 0$ is identical with

$$\begin{bmatrix} 2 & -2 \\ 5 & -5 \end{bmatrix} \begin{bmatrix} u(2) \\ u(6) \end{bmatrix} = \begin{bmatrix} 2/EA & -10/EA \\ 10/EA & -30/EA \end{bmatrix} \begin{bmatrix} N(2) \\ N(6) \end{bmatrix} \tag{6.37}$$

and Eq. (6.28) $i = 1$ with

$$\begin{bmatrix} EA & -EA \\ EA & -EA \end{bmatrix} \begin{bmatrix} u(2) \\ u(6) \end{bmatrix} = \begin{bmatrix} 1 & -5 \\ 2 & -6 \end{bmatrix} \begin{bmatrix} N(2) \\ N(6) \end{bmatrix}$$

If we multiply the first row with $2/EA$ and the second row with $5/EA$ then we
obtain the equations of the first system. Hence, if the four numbers $u(2)$, $u(6)$, $N(2)$,
$N(6)$ satisfy the first system then also the second.
   That u is a solution of $-EA\,u'' = 0$ is true because u is a polynomial of the first kind
and the statement that the four numbers are the boundary values of $u(x)$ is true
because the four numbers satisfy Eq. (6.28) $i = 0$.

6.6 Stiffness Matrices and Compatibility Conditions

Every homogeneous displacement, $-EA\,u'' = 0$, of the truss element $[2, 6]$ satisfies the first compatibility condition, Eq. (6.28) $i = 0$,

$$c(x)\,u(x) = -\left[N_0[x]\,u\right]_2^6 + \left[g_0[x]\,N\right]_2^6$$

at $x = 2$ and at $x = 6$.
   These two conditions render an equation between the end displacements
$\{u(2), u(6)\}$ and end forces $\{N(2), N(6)\}$ of every such homogeneous displacement, see
Eq. (6.37).

Multiplying Eq. (6.37) from the left with the inverse of the right-hand matrix results in

$$EA \begin{bmatrix} -0.25 & 0.25 \\ -0.25 & 0.25 \end{bmatrix} \begin{bmatrix} u(2) \\ u(6) \end{bmatrix} = \begin{bmatrix} N(2) \\ N(6) \end{bmatrix}$$

or, with the notation of Fig. 3.11,

$$u(2) = \delta_1, \quad u(6) = \delta_2, \quad N(2) = -f_1, \quad N(6) = f_2$$

in

$$EA \begin{bmatrix} 0.25 & -0.25 \\ -0.25 & 0.25 \end{bmatrix} \begin{bmatrix} \delta_1 \\ \delta_2 \end{bmatrix} = \begin{bmatrix} f_1 \\ f_2 \end{bmatrix}$$

This matrix is, as a comparison with Eq. (3.16) shows, the stiffness matrix of a bar of length $l = 4$. Hence, the boundary data δ and f of a homogeneous displacement are compatible if they satisfy the equation $K\delta = f$.

Note that the left-hand matrix in Eq. (6.37) is singular but the right-hand matrix regular. In accordance with this the first bvp (only displacements are prescribed) is uniquely solvable but the second bvp (only end forces are prescribed) is not.

If u is a solution of the second bvp then $u + c$ where c is an arbitrary number (a rigid-body movement) is also a solution.

Next, consider the two essential compatibility conditions, see Eq. (6.30) $i = 0, 1$

$$c(x) \, \partial^i w(x) = - \left[V_i[x] \, w - M_i[x] \, w' + g_i'[x] \, M - g_i[x] \, V \right]_a^b$$

$$+ \int_a^b g_i[x] \, EI \, w^{IV} \, dy$$

for beams and assume that all the deflections w we test correspond to evenly distributed forces

$$EI \, w^{IV} = p \quad 2 < x < 6, \quad p = \text{constant}$$

At full length and in the notation of Fig. 3.15 these compatibility conditions read

$$\begin{bmatrix} 2 & 2 & -2 & -10 \\ 5 & 10 & -5 & -30 \\ 1 & 1 & -1 & -5 \\ 1 & 2 & -1 & -6 \end{bmatrix} \begin{bmatrix} \delta_1 \\ \delta_2 \\ \delta_3 \\ \delta_4 \end{bmatrix} = \frac{1}{6EI} \begin{bmatrix} 8 & -12 & 280 & -156 \\ 280 & -180 & 1\,800 & -660 \\ 12 & -14 & 180 & -86 \\ 156 & -86 & 660 & -182 \end{bmatrix} \begin{bmatrix} f_1 \\ f_2 \\ f_3 \\ f_4 \end{bmatrix} + \frac{p}{EI} \begin{bmatrix} 384 \\ 3\,520 \\ 288 \\ 1\,504 \end{bmatrix} \tag{6.38}$$

The single equations formulate, consecutively, Betti's principle with one of the elastic states on the left and the elastic state on the right in Fig. 6.15.

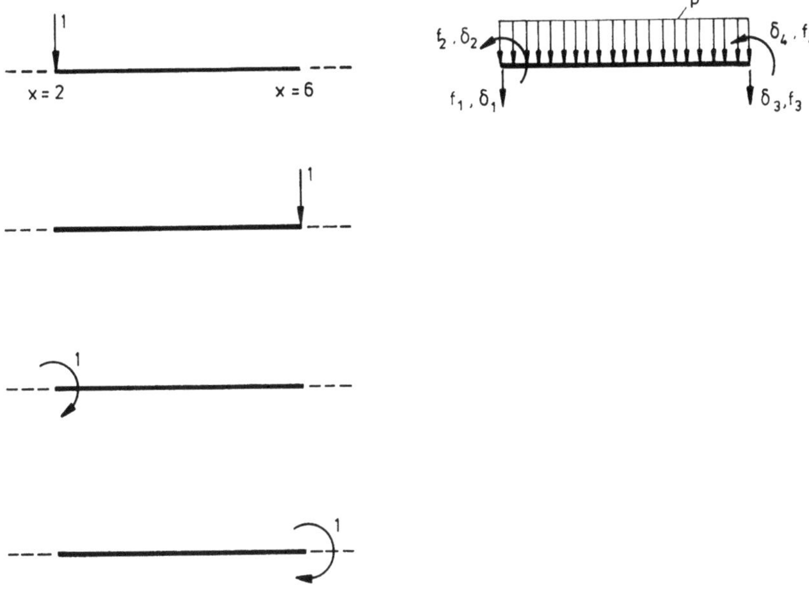

Figure 6.15

The matrix on the left-hand side in Eq. (6.38) is singular and the matrix on the right-hand side regular. Multiplying the equation from the left with the inverse of this matrix we obtain

$$EI\begin{bmatrix} 0.1875 & -0.375 & -0.1875 & -0.375 \\ & 1 & 0.375 & 0.5 \\ & & 0.1875 & 0.375 \\ \text{sym.} & & & 1 \end{bmatrix}\begin{bmatrix} \delta_1 \\ \delta_2 \\ \delta_3 \\ \delta_4 \end{bmatrix} = \begin{bmatrix} f_1 \\ f_2 \\ f_3 \\ f_4 \end{bmatrix} + p\begin{bmatrix} 2 \\ -1.33 \\ 2 \\ +1.33 \end{bmatrix} \tag{6.39}$$

The matrix in this equation is the stiffness matrix of a beam of length $l = 4$, see Eq. (3.18).

Let us check this result when the beam is clamped at both ends, see Fig. 6.16. In which case the displacement-terms are zero at both ends, $\delta_i = 0$. Hence the evaluation of the four equations renders consecutively

$$0 = f_1 + 2p, \quad 0 = f_2 - 1.33p, \quad 0 = f_3 + 2p, \quad 0 = f_4 + 1.33p$$

and a simple calculation confirms that the end forces f_i, indeed, satisfy these equations.

This check also gives us a clue toward finding the vector of the distributed forces associated with a stiffness matrix (usually this vector is, because it depends on the shape of the distributed load, neglected in the literature).

The components of this vector are the negative support reactions of the clamped beam under the action of the distributed forces.

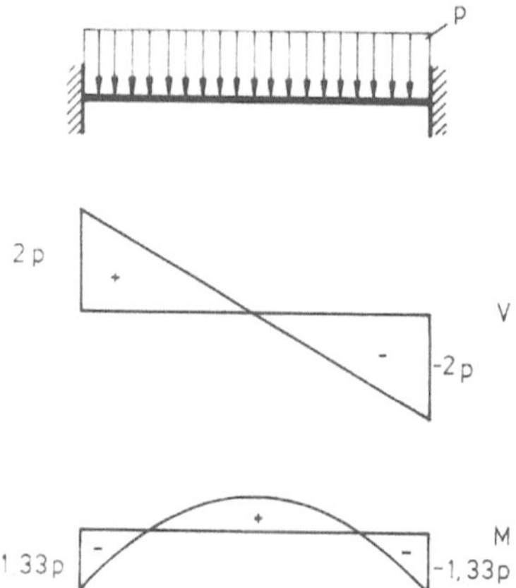

Figure 6.16

Experiencing this close connection between compatibility conditions and stiffness matrices why should we not try to formulate stiffness matrices this way also for elastic plates or bodies?

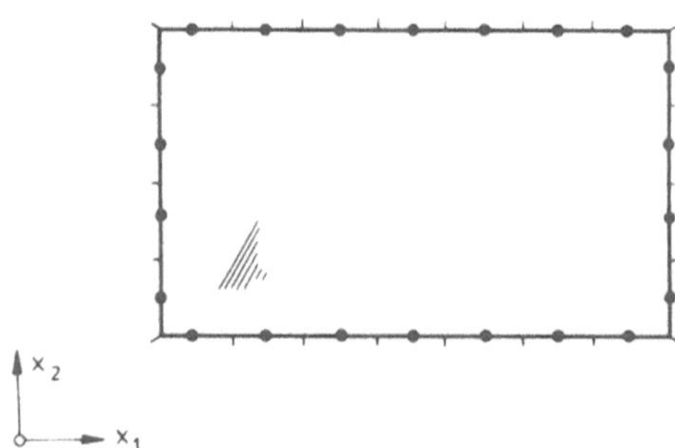

Figure 6.17

Consider the elastic plate in Fig. 6.17 whose boundary we partitioned into n boundary elements $R_j, j = 1, 2 \ldots n$, with mid-nodes x^j. We are searching for an expression that relates the nodal displacements $u_{ij} = u_i(x^j)$, $i = 1, 2; j = 1, 2 \ldots n$ of a homogeneous displacement field $(-Lu = 0)$ to the "equivalent nodal forces" $f_{ij} = f_i(x^j)$, $i = 1, 2; j = 1, 2 \ldots n$, that is for an equation as

$$\sum_{l=1}^{n} \sum_{k=1}^{2} K_{ijkl} u_{kl} = f_{ij} \quad i = 1, 2; j = 1, 2 \ldots n$$

How can we calculate this stiffness matrix?

To start, we observe, if $u(x)$ is a regular, homogeneous solution of the equation $-Lu = 0$ in $\bar{\Omega}$ then u must satisfy at every point of the boundary the compatibility condition, see Eq. (6.34),

$$\dot{C}(x)\,u(x) = \int_{\Gamma} [U(y,x)\,\tau(u)\,(y) - T(y,x)\,u(y)]\,ds_y \tag{6.40}$$

Let $\{\varphi_k(x)\}$ be a set of scalar-valued boundary functions with the property

$$\varphi_k(x^j) = \delta_{jk}$$

and let

$$u_i(x) = \sum_{k=1}^{n} u_{ik}\,\varphi_k(x), \quad t_i(x) = \sum_{k=1}^{n} t_{ik}\,\varphi_k(x), \quad i=1,2$$

be two vector-valued boundary functions whose components are these functions φ_k.

If we determine the coefficients u_{ik} and t_{ik} (substitute $(\tau(u))_i = t_i$) by a point collocation process at the nodes x^j we obtain a system of $2 \times n$ linear equations

$$\sum_{k=1}^{n} \sum_{l=1}^{2} H_{ijkl}\,u_{lk} = \sum_{k=1}^{n} \sum_{l=1}^{2} G_{ijkl}\,t_{lk} \quad i=1,2; j=1,2\ldots n \tag{6.41}$$

whose single coefficients have the following meaning

$$H_{ijkl} = \frac{1}{2}\,\varphi_k(x^j)\,\delta_{il} + \int_{\Gamma} T_{il}(y,x^j)\,\varphi_k(y)\,ds_y$$

$$G_{ijkl} = \int_{\Gamma} U_{il}(y,x^j)\,\varphi_k(y)\,ds_y$$

Multiplication of Eq. (6.41) with G^{-1} results in

$$\sum_{j,k=1}^{n} \sum_{i,l=1}^{2} G_{mrij}^{(-1)}\,H_{ijkl}\,u_{lk} = t_{mr} \quad m=1,2; r=1,2\ldots n \tag{6.42}$$

This equation is, contrary to bars and beams, not yet the final one. We need yet the concept of an "equivalent nodal force".

Let $u = \{u_1, u_2\}$ with

$$u_i(x) = \sum_{k=1}^{n} u_{ik}\,\varphi_k(x) \quad i=1,2$$

a virtual displacement of the boundary Γ. The corresponding movements of the nodes are just the coefficients of the expansion itself

$$u_{ik} = u_i(x^k) \quad i=1,2$$

We say the n nodal forces

$$f^k = \{f_{1k}, f_{2k}\}^T \quad k = 1, 2 \ldots n$$

are equivalent with the traction vector $t = \{t_1, t_2\}$

$$t_i(x) = \sum_{k=1}^{n} t_{ik} \, \varphi_k(x) \quad i = 1, 2$$

if for all virtual displacements $u = \{u_1, u_2\}$ the equivalence

$$\sum_{k=1}^{n} (f_{1k} u_{1k} + f_{2k} u_{2k}) = \int_{\Gamma} (t_1 u_1 + t_2 u_2) \, ds$$

holds.

The integral on the right-hand side is equal to

$$\sum_{i=1}^{2} \int_{\Gamma} u_i \, t_i \, ds = \sum_{i=1}^{2} \sum_{l,k=1}^{n} u_{ik} \, \Phi_{kl} \, t_{il}, \quad \Phi_{kl} = \int_{\Gamma} \varphi_k \, \varphi_l \, ds$$

Hence, given a traction vector t the equivalent nodal forces are uniquely determined by

$$f_{ik} = \sum_{l=1}^{n} \Phi_{kl} \, t_{il} = \sum_{l=1}^{n} \sum_{m=1}^{2} F_{iklm} \, t_{ml}, \quad F_{iklm} = \Phi_{kl} \, \delta_{im}$$

If we now multiply Eq. (6.42) from the left with the matrix F then we obtain an equation between the nodal displacements u_{kl} and the equivalent nodal forces f_{ik}.

$$\sum_{k,r,t,l=1}^{n} \sum_{i,m,j,s=1}^{2} F_{ikmr} \, G_{mrst}^{(-1)} \, H_{stjl} \, u_{jl} = \sum_{r=1}^{n} \sum_{m=1}^{2} F_{ikmr} \, t_{mr} = f_{ik}$$

or in simpler terms

$$FG^{-1} Hu = Ft = f$$

This is the relation we have been looking for. The matrix

$$K = FG^{-1} H$$

is the stiffness matrix of the elastic plate.

Naturally, this matrix is not the exact stiffness matrix. The exact matrix would have infinitely many rows and columns because the number of homogeneous displacement fields of this elastic plate, as of every elastic plate, is infinite.

To discuss the properties of the matrix K let $u^0 = \{u_{11}^0, u_{12}^0, \ldots u_{n2}^0\}$ the vector of a rigid-body movement of the plate, that is

$$u_i^0(x) = \sum_{k=1}^{n} u_{ik}^0 \, \varphi_k(x) \quad i = 1, 2$$

are the boundary values of a function $u = a + b \times x$. If K were a true stiffness matrix we would expect that K has the following properties.

(P1) Ker. $K u^0 = 0$
(P2) Equ. $u^{0T} K u = 0$
(P3) Sym. $u^T K \hat{u} = \hat{u}^T K u$
(P4) Pos. Def. $u^T K u > 0 \qquad u \neq u^0$

But normally, the matrix K has none of these properties exactly; in particular is K not symmetric.

The reason for this is the following:

The compatibility condition, Eq. (6.40), constitutes a relation between pairs $\{u, t\}$ of vector-valued boundary functions. On the set of all boundary functions u we could try to explain an operator which associates with every u the traction t, i.e. the function t which satisfies together with u the compatibility condition on the boundary.

$$\mathcal{K} : u \rightarrow \mathcal{K} u = t$$

But given a function u we may not expect that a function $t = \mathcal{K} u$ always exists. For this to be true u must satisfy certain conditions: it must, e. g., be the trace ($=$ boundary value) of a homogeneous displacement field and this in turn implies that u is continuous on Γ. Hence, the operator $\mathcal{K}$ is explained only on a certain class, say, M of boundary functions.

The question then is: do the functions $u = \{\varphi_i, \varphi_k\}$ belong to this class M and if so, can their tractions t be expanded in terms of the very same functions $\{\varphi_i, \varphi_k\}$? If the answer to both questions would be yes then the stiffness matrix K should have all four properties.

Stiffness matrices for homogeneous plate solutions, $K \Delta \Delta w = 0$, are obtained analogously by employing the compatibility conditions formulated in Eqs. (6.32) $i = 0, 1$. They lead, in a symbolical notation, to expressions such as

$$\begin{bmatrix} H \end{bmatrix} \begin{bmatrix} w \\ \partial w/\partial n \end{bmatrix} = \begin{bmatrix} G \end{bmatrix} \begin{bmatrix} M_n \\ V_n \end{bmatrix}$$

6.7 The Boundary Element Method

We learnt in the foregoing sections that the displacements in the interior of a structural element can be calculated by boundary and domain integrals.

These integral representations require (if we use the simple fundamental solution $g_0[x]$ in connection with the second identity, $B(g_0[x], u)$) the knowledge of all the displacement- and force-terms on the boundary and of the load in the domain. If our knowledge of the set of boundary functions is still incomplete then we can solve the compatibility conditions for the unknown boundary functions.

It is now a simple step to combine these two results and formulate it as the boundary element method.

Firstly, we solve the compatibility conditions for the unknown boundary values.

Secondly, we substitute the complete and compatible set of boundary functions into the integral representations of section 6.4; the problem is solved.

Let us illustrate this procedure with the cantilever beam in Fig. 6.18. The boundary conditions require that

$$w(0) = 0, \quad w'(0) = 0, \quad M(l) = 0, \quad V(l) = 0$$

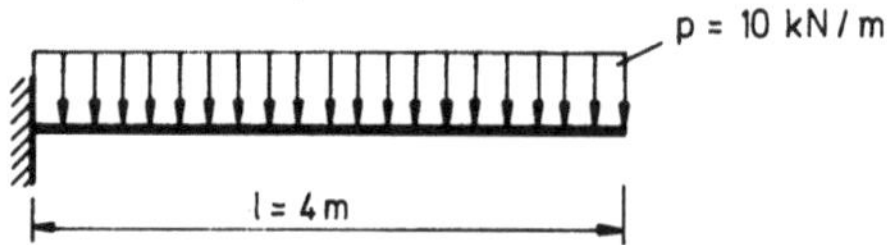

Figure 6.18

To determine the yet unknown boundary terms we turn to the compatibility conditions formulated in Eq. (6.39) and solve the third and fourth equation

$$EI \begin{bmatrix} 0.1875 & 0.375 \\ 0.375 & 1 \end{bmatrix} \begin{bmatrix} \delta_3 \\ \delta_4 \end{bmatrix} = 10 \begin{bmatrix} 2 \\ 1.33 \end{bmatrix}$$

for the displacements at the free end

$$\delta_3 = \frac{320}{EI}, \quad \delta_4 = -\frac{106.7}{EI}$$

and substitute these values into the first and second condition

$$EI \begin{bmatrix} -0.1875 & -0.375 \\ 0.375 & 0.5 \end{bmatrix} \begin{bmatrix} 320 \\ -106.7 \end{bmatrix} = \begin{bmatrix} f_1 \\ f_2 \end{bmatrix} + 10 \begin{bmatrix} 2 \\ 1.33 \end{bmatrix}$$

to calculate the support reactions,

$$f_1 = -40\,\text{kN}, \quad f_2 = 80\,\text{kNm}$$

Hence, all boundary terms are determined and it remains only to substitute these terms into Eq. (6.31)

$$w(x) = -\left[V_0[x]\, w - M_0[x]\, w' + g_0'[x]\, M - g_0[x]\, V \right]_0^4 + \int_0^4 g_0[x]\, p\, dy$$

$$= \frac{1.66}{EI}\left(24x^2 - 4x^3 + \frac{x^4}{4} \right)$$

The problem is solved.

In higher dimensions the compatibility conditions are integral equations ("the stiffness matrices have infinitely many rows and columns") and, therefore, to solve for the unknowns requires the solution of integral equations.

The boundary element method divides the boundary of the elastic body or plate into single elements and approximates the unknown functions on these elements by polynomial shape functions.

The coefficients of these polynominals are chosen so that the interpolates satisfy the compatibility condition at the element nodes. This renders a linear system of equations which can be solved for the unknown coefficients.

After the solution the complete and (nearly) compatible set of boundary functions is substituted into the integral representations of section 6.4.

This is, in essence, the boundary element method. For a detailed description we refer to the literature, see e. g. [B2], [B & B] and [B & T & W]. We only illustrate the method with a simple example.

Consider the plate in Fig. 6.19 which is stressed by tensile forces.

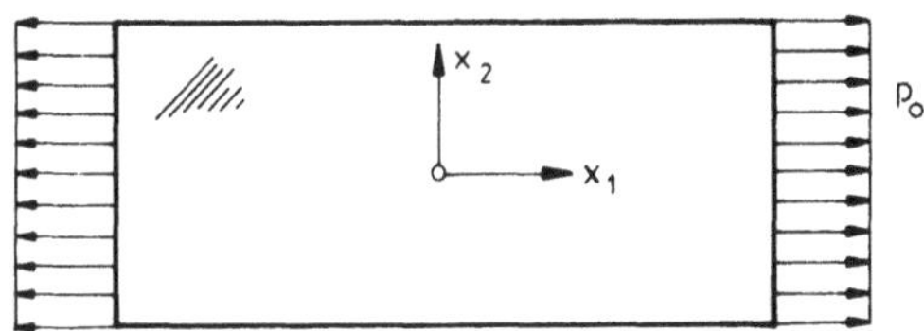

Figure 6.19

The displacement field $u = \{u_1, u_2\}$ satisfies the equations

$$-Lu = 0 \text{ in } \Omega, \qquad \tau(u) = \bar{t} \text{ on } \Gamma$$

where the components of $\bar{t} = \{\bar{t}_1, \bar{t}_2\}$ are

$$\bar{t}_1 = \begin{cases} 0 & \text{on the long sides} \\ \pm p_0 & \text{on the short sides} \end{cases}, \qquad \bar{t}_2 = 0 \text{ on all of } \Gamma$$

The solution of this bvp is the function u with the components

$$u_1(x) = \frac{p_0}{2\mu(1+v)}\, x_1, \qquad u_2(x) = -\frac{p_0\, v}{2\mu(1+v)}\, x_2$$

The displacement field, obviously, belongs to $C^2(\bar{\Omega})$ and, therefore, possesses the integral representation

$$C(x)\,u(x) = \int_{\Gamma} \left(U(y,x)\,\bar{t}(y) - T(y,x)\,u(y) \right) ds_y, \tag{6.43}$$

(we replaced $\tau(u)$ by $\bar{t}$).

As the boundary displacement u, the function u under the integral sign, does not appear in the formulation of the bvp we must first solve the compatibility condition

$$\dot{C}(x)\,u(x) + \int_{\Gamma} T(y,x)\,u(y)\,ds_y = \int_{\Gamma} U(y,x)\,\bar{t}(y)\,ds_y, \quad x \in \Gamma \tag{6.44}$$

for the unknown displacement $u(x)$ before we can employ this integral representation.

Note that the compatibility condition is simply obtained by placing in Eq. (6.43) the point x on the boundary Γ. The function $u(x)$ on the left-hand side is then the same function as the function under the integral sign, that is, the unknown boundary function and $C(x)$ becomes $\dot{C}(x) = 1/2\,I$ (if x is a smooth boundary point).

To calculate an approximation u_h^c we devide the edge Γ into n boundary elements $R_i, i = 1, 2 \ldots n$ and approximate u and $\bar{t}$ on each interval by piecewise constant vector-valued functions $u = \{u_i\}$, $\bar{t} = \{\bar{t}_i\}$

$$u_i(x) = \sum_{k=1}^{n} u_{ik}\,\varphi_k(x), \quad \bar{t}_i(x) = \sum_{k=1}^{n} t_{ik}\,\varphi_k(x)$$

where

$$\varphi_k(x) = \begin{cases} 1, & x \in R_k \\ 0, & \text{otherwise} \end{cases}$$

The condition that these piecewise constant functions satisfy the integral equation (6.44) at the mid-nodes renders a system of $2 \times n$ equations

$$\sum_{k=1}^{n} \sum_{l=1}^{2} H_{ijkl}\,u_{lk} = \sum_{k=1}^{n} \sum_{l=1}^{2} G_{ijkl}\,t_{lk} \quad i = 1, 2; j = 1, 2 \ldots n$$

where

$$H_{ijkl} = \frac{1}{2}\,\delta_{ij}\,\delta_{kl} + \int_{R_k} T_{il}(y, x^j)\,ds_y$$

$$G_{ijkl} = \int_{R_k} U_{il}(y, x^j)\,ds_y$$

For the particular choice

$$n = 38 \text{ boundary elements}, \quad v = 0.1, \quad p = 1000\,\text{kN/cm},$$
$$l = 13\,\text{cm}, \quad b = 6\,\text{cm}, \quad \mu = 9090.0\,\text{kN/cm}$$

the first component of the approximate solution, the step function u_1^c, and the exact solution are compared in Fig. 6.20.

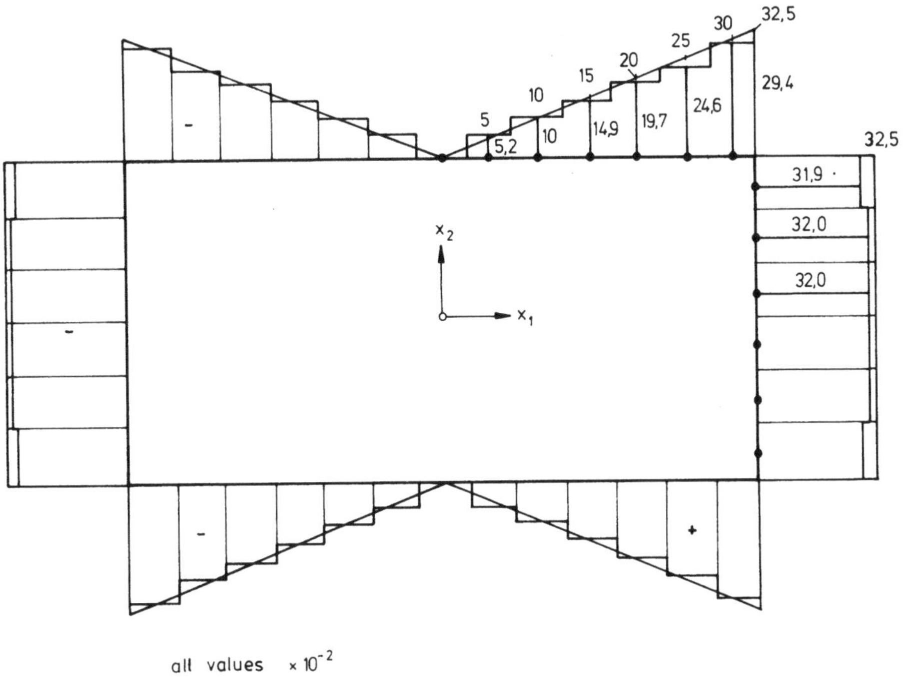

Figure 6.20

And the following table compares the approximate values with the values of the exact solution at some internal points.

	u_1	u_2	σ_x	σ_y	τ
$x = (0,3)$	0.00	0.0	995.0	-2.0	0.0
$(2,3)$	0.09	0.0	996.0	-4.0	0.0
$(4,3)$	0.20	0.0	999.0	-6.0	0.0
exact:					
$x = (0,3)$	0.00	0.0	1000.0	0.0	0.0
$(2,3)$	0.10	0.0	1000.0	0.0	0.0
$(4,3)$	0.20	0.0	1000.0	0.0	0.0

In chapter 11 we shall demonstrate that every finite element solution of a structure is the response of the structure to forces which approximate the original forces. The same is true with boundary element solutions. The plate solution

$$u_h(x) = \int_\Gamma [U(y, x)\,\bar{t}(y) - T(y, x)\,u_h^c(y)]\,ds_y \tag{6.45}$$

(u_h^c is the vector-valued step function which solves the collocation equations) is (mainly) the response of the plate to tensile forces at the short ends and to concentrated forces which act at the element interfaces.

These forces are responsible for the jump in displacements between the elements. They are infinite; the limit $\tau(u)(x)$, $x \to \Gamma$, does not exist at points where u_h^c jumps. Due to these infinite forces the energy is infinite too

$$E(u_h, u_h) = \infty$$

If instead of the piecewise continuous splines ("nonconforming boundary elements") we would use continuous splines ("conforming elements"), [H 11], the concentrated forces would disappear and $E(u_h, u_h)$ would be finite.

How did we find this out? With some Sobolev space theory as we will try to explain in the next section.

6.8 The Trace Theorem

The Sobolev space $H^k[a, b]$ is, see section 1.8, the completion of $C^k[a, b]$ in the norm

$$||u||_k = \left[\int_a^b (u^2 + u'^2 + u''^2 + \ldots u^{(k)2}) \, dx \right]^{1/2}$$

It consists of all those functions which either belong to $C^k[a, b]$ or which are arbitrary close to $C^k[a, b]$ in the sense of this norm.

Surprisingly it makes sense to define Sobolev spaces with fractional indices as $H^{0.5}$ or $H^{1.5}$ and even Sobolev spaces with negative indices as H^{-1}, $H^{-1.5}$ etc. Simply stated we can imagine to have a scale of Sobolev spaces which extends from $-\infty$ to $+\infty$

$$\ldots H^{-2} \subset H^{-1} \subset H^0 \subset H^1 \subset H^2 \ldots$$

with the fractional spaces in between. The worse a function in terms of smoothness the lower the index of the Sobolev space to which it belongs.

The "pivot-space" is the space H^0. It consists of all those functions whose integral

$$\int_a^b u^2 \, dx$$

is finite, whose square has finite area.

The $C_0^\infty[a, b]$ functions belong to every $H^k[a, b]$ whatever the value of k. But for other functions there is an upper limit to the index k, there is a "best" space.

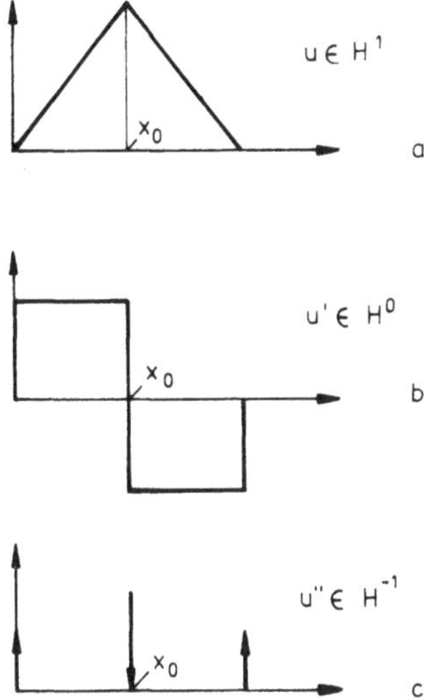

Figure 6.21

If the best space is a positive space, $u \in H^k[a, b]$ then this means that the $k - th$ derivative $u^{(k)}$ belongs to H^0,

$$u^{(k)} \in H^0$$

We may differentiate u k-times before an exotic function which has no H^0-measure ("no area"), such as a δ-function appears.

If the best space is a negative space, $H^{-k}[a, b]$, then this means that we must integrate u k-times before we are in H^0

$$\underbrace{\int \int \int \ldots \int}_{k\text{-times}} u \, dx \, dx \, dx \ldots dx \in H^0$$

If k is not a natural number, say $k = 1.5$, then the above remarks remain correct if k is replaced by $[k - 1] = 1$ (differentiation) or $[k] = 2$ (integration) where $[k] =$ the next integer which follows k.

The roof function in Fig. 6.21a belongs to H^1 (it is the best space). After one differentiation we are in H^0 and after a further differentiation we are in H^{-1}. The second derivative consists of three delta-functions (point loads).

Perhaps the following will acquaint the reader with these negative Sobolev spaces a little better:

If two functions $\{u, \hat{u}\}$ are in $H^0(\Omega)$ then their work integral—consider u to represent a displacement and $\hat{u}$ a distribution of forces —

$$\int_\Omega u \, \hat{u} \, d\Omega$$

is finite. Now given a function $\hat{u}$ (a load) in $H^{-k}(\Omega)$. From what space are the functions u (the displacements) which render the integral

$$\int_{\Omega} u\,\hat{u}\,d\Omega \tag{6.46}$$

finite?

It turns out that these are the functions in $H_0^k(\Omega)$, the space we obtain when we complete $C_0^k(\Omega)$ in the norm $\|u\|_k$.

That it is not the full space $H^k(\Omega)$ which is conjugated to $H^{-k}(\Omega)$ but only the subspace $H_0^k(\Omega)$ has something to do with the fact that the boundary values of functions in $H_0^k(\Omega)$ are zero and, hence, the boundary terms can be neglected when we do integration by parts.

The space $H^{-k}(\Omega)$ is called the dual of $H_0^k(\Omega)$; a function $\hat{u}$ belongs to $H^{-k}(\Omega)$ if and only if the integral (6.46) exists for all $u \in H_0^k(\Omega)$.

As an example consider the three functions u, u' and u'' in Fig. 6.21. Physically speaking it makes no sense to let the δ-functions u'', the point loads, act through the displacement u', because the step function u' is not uniquely defined at $x = 0$, x_0 and 1. In accordance with this the integral

$$\int_0^1 u'\,u''\,dx$$

does not exist in the distributional sense. But it would make sense to let the point loads $u'' \in H^{-1}$ act through the single-valued displacement $u \in H^1$. Not surprisingly the corresponding integral

$$\int_0^1 u\,u''\,dx = -\frac{1}{2}\,u(0) + 1\,u(x_0) - \frac{1}{2}\,u(l) = u(x_0)$$

exists.

The delta-functions we introduced in section 5.9 belong to negative Sobolev spaces

$$\delta_i \in H^{-s}(\Omega) \tag{6.47}$$

with index $s = \left[\dfrac{n}{2} + i\right]$, $n =$ dimension of the continuum, $i =$ degree of the singularity.

In the case of a beam ($n = 1$) or a plate ($n = 2$) this means that the delta-functions belong to the following Sobolev spaces

$n = 1$	H^{-1}	H^{-2}	H^{-3}	H^{-4}
$n = 2$	H^{-2}	H^{-3}	H^{-4}	H^{-5}

This is a simple corollary of Sobolev's Embedding Theorem: For the integral

$$\int_a^b w\,\delta_0(y - x_0)\,dy = w(x_0)\cdot 1$$

to make sense w must be single-valued at x_0. This is certainly the case if w is continuous in $[a, b]$. According to Sobolev's Embedding Theorem this is a property common to all the functions in $H^1[a, b]$. Hence δ_0 belongs to H^{-1}.

Or consider the function $\delta_3(y - x_0)$ which jumps at the source point. For the integral

$$\int_a^b w\,\delta_3(y - x_0)\,dy = V(x_0)\cdot 1$$

to make sense $V = -EIw'''$ must be single-valued at x_0. Sufficient for this is that w''' is in H^1, hence, w in $H^4[a, b]$. Consequently δ_3 belongs to $H^{-4}, s = \left[\dfrac{1}{2} + 3\right] = 4$.

Up to now we considered only domain functions, but the functions on the boundary Γ of a body or a plate form a scale of Sobolev spaces too.

$$\ldots H^{-2}(\Gamma) \subset H^{-1}(\Gamma) \subset H^0(\Gamma) \subset H^1(\Gamma) \subset H^2(\Gamma)\ldots$$

Hence, given a function $u \in H^k(\Omega)$ we may expect to find its trace ($=$ its boundary value) in some space $H^q(\Gamma)$. But which? Or stated otherwise: Given a function u, say in $H^1(\Omega)$, that is a function with the property that the measure of u (squared) and its first derivatives (squared) is finite with respect to Ω,

$$\|u\|_1 = \left[\int_\Omega (u^2 + u^2{,}_{x_1} + u^2{,}_{x_2})\,d\Omega\right]^{1/2} < \infty$$

Does this hold true for the trace too? That is, can we measure the trace u with respect to Γ in the, say, $H^1(\Gamma)$-norm

$$u \in H^1(\Omega) \longrightarrow \int_\Gamma (u^2 + u^2{,}_s)\,ds < \infty$$

whatever the function u? The answer is provided by the

Trace theorem

Let Ω be a smooth bounded domain and let $u \in H^k(\Omega)$ then its trace belongs to $H^{k-0.5}(\Gamma)$.

In our case k equals 1 and we, therefore, find the trace of u in $H^{0.5}(\Gamma)$. This implies that the measure of u^2 is finite on Γ.

$$\int_\Gamma u^2\,ds < \infty$$

but the same cannot be said, in general, about the traces of the derivatives.

Fig. 6.22 gives an idea how functions from fractional boundary spaces look like.

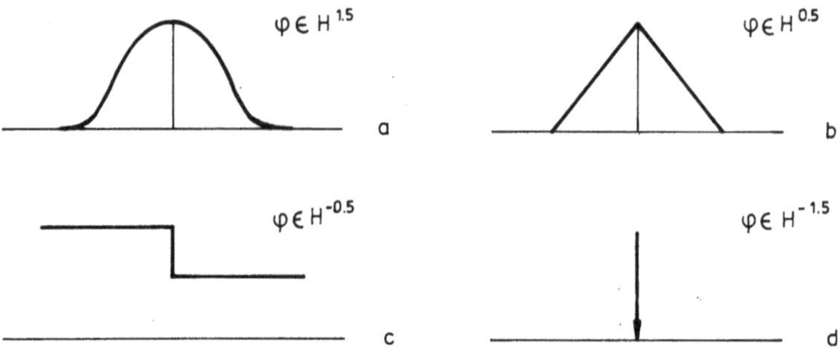

Figure 6.22

The Trace Theorem and Sobolev's Embedding Theorem are the basic tools to classify the functions with finite energy, the functions in $H^m(\Omega)$.

If u belongs to $H^m(\Omega)$ then, so we may state now, $\partial^i u$ belongs in the interior to $H^{m-i}(\Omega)$ and on the boundary to $H^{m-i-0.5}(\Gamma)$.

In terms of the single structural elements this means:

Bars $u \in H^1[a,b] \to u' \in H^0[a,b]$

Beams $w \in H^2[a,b] \to w' \in H^1, w'' \in H^0, w''' \in H^{-1}[a,b]$

Membranes $w \in H^1(\Omega) \quad \to \partial w/\partial n \in H^0(\Omega),$

$$w \in H^{0.5}(\Gamma), \quad \partial w/\partial n \in H^{-0.5}(\Gamma)$$

Elastic plates and bodies

$$\boldsymbol{u} \in H^1(\Omega) \quad \to \tau(\boldsymbol{u}) \in H^0(\Omega),$$

$$\boldsymbol{u} \in H^{0.5}(\Gamma), \tau(\boldsymbol{u}) \in H^{-0.5}(\Gamma)$$

Kirchhoff plates

$$w \in H^2(\Omega) \quad \to \partial w/\partial n \in H^1(\Omega), M_n(w) \in H^0(\Omega), V_n(w) \in H^{-1}(\Omega)$$

$$w \in H^{1.5}(\Gamma), \quad \partial w/\partial n \in H^{0.5}(\Gamma),$$

$$M_n \in H^{-0.5}(\Gamma), \quad V_n \in H^{-1.5}(\Gamma)$$

According to Sobolev's Embedding Theorem a function u in $H^k(\Omega)$ is also in $C(\bar{\Omega})$ if $k > n/2$ (n = dimension of the continuum). This means that a bar ($n = 1$) with finite energy has a continuous displacement

$$u \in H^1[a,b] \to u \in C[a,b] \quad \text{(because } 1 > 1/2 = n/2)$$

but not necessarily continuous first derivatives (normal forces)

$$u' \in H^0[a,b] \not\to u' \in C[a,b] \quad \text{(because } 0 < 1/2 = n/2)$$

Or consider a beam. A beam with finite energy has continuous deflection and slope, w and w', because w and w' both belong to H^1 if w is in H^2.

$$w, w' \in H^1[a, b] \rightarrow w, w' \in C[a, b] \quad (1 > 1/2 = n/2)$$

but not necessarily continuous second or third derivatives, w'' and w''', because these functions must not be better than H^0 and H^{-1}, resp. That is in beams with finite energy the bending moment and the shear force are allowed to jump, to be discontinuous.

And we also learn (to our surprise) that an elastic plate or body or a prestressed membrane with finite energy must not necessarily have a continuous displacement field or deflection. If u is in $H^1(\Omega)$ then u must not be in $C(\bar{\Omega})$ if the dimension of the continuum is $n = 2$ or $n = 3$.

Such a strange function is, e. g., $w = \ln(\ln r^{-1})$, see [S & F] p. 73. It is infinite and, therefore, discontinuous at $r = 0$ but it has finite energy in the circle $r < 1/2$,

$$\frac{1}{2} \int_\Omega (w^2{}_{,1} + w^2{}_{,2})\, d\Omega = \int_0^{2\pi} \int_0^{1/2} \left(w^2{}_{,r} + \left(\frac{w_{,\varphi}}{r}\right)^2 \right) r\, dr\, d\varphi$$

$$= \frac{1}{2} \int_0^{2\pi} \int_0^{1/2} (\ln r)^{-2}\, d(\ln r)\, d\varphi = \frac{1}{2} \int_0^{2\pi} \int_{-\infty}^{-\ln 2} x^{-2}\, dx\, d\varphi = \frac{\pi}{\ln 2}$$

In other words the deflection of such a prestressed membrane ($N = 1$) loaded with the distributed force $\Delta \ln(\ln r^{-1}) = (r^2 \ln r)^{-1}$

$$\Delta w = \frac{1}{r^2 \ln^2 r} \quad \text{in } \Omega = \left\{ 0 \leqslant r \leqslant \frac{1}{2}, 0 < \varphi \leqslant 2\pi \right\}, \quad w = 0 \text{ on } \Gamma$$

is infinite at $r = 0$ but, still, the internal energy is finite. The deflection $w = \ln(\ln r^{-1}) - \ln(\ln 2)$ belongs to $H^1(\Omega)$.

Let us check now the boundary element solution of the elastic plate in Eq. (6.45). For u_h to have finite energy it is, according to the trace theorem, necessary that

$$u_h \in H^{0.5}(\Gamma) \quad \tau(u_h) \in H^{-0.5}(\Gamma)$$

The boundary value, the trace, of u_h is the limit of u in Eq. (6.45) if x tends to Γ. But this limit is, up to small deviations, just the piecewise continuous collocation solution

$$\lim_{x \to \Gamma} u_h(x) = u_h^c(x) \hspace{4cm} \text{(at the collocation points)}$$

A piecewise continuous spline is not in $H^{0.5}(\Gamma)$ and, consequently, the necessary condition is violated, the displacement field $u_h(x)$ cannot have finite energy.

If we would use continuous splines on Γ to model the displacement then u_h^c would be in $H^{0.5}(\Gamma)$ and because the traction t is in $H^{-0.5}(\Gamma)$ it would then result that u_h has finite energy.

6.9 Elastic Potentials

Consider a massive ring with axis Γ and density $f(y)$. Given a particle of unit mass m at a point x, not on Γ, the attractive force P on m is the gradient

$$P = \operatorname{grad} u$$

of the gravitational field

$$u(x) = \frac{1}{4\pi} \int_\Gamma \frac{1}{r} f(y)\, ds_y, \quad r = |y - x| \tag{6.48}$$

This field is a potential field, that is when we move the particle m on a closed path the total work is zero.

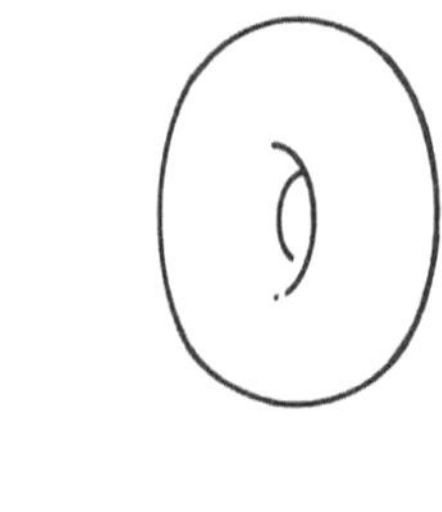

Figure 6.23

Potential fields satisfy the equation

$$\Delta u = 0$$

and also the potential in Eq. (6.48) complies with this condition. Its kernel is the $3 - D$ fundamental solution, $(4\pi r)^{-1}$, of the Laplace operator and, therefore,

$$\Delta u(x) = \frac{1}{4\pi} \int_\Gamma \Delta_x \frac{1}{r} f(y)\, ds_y = \int_\Gamma 0 f(y)\, ds_y = 0 \quad x \notin \Gamma$$

It is this property which renders $u(x)$, independently of its physical meaning, an ideal candidate to solve *bvps* of the Laplace operator as

$$\Delta u = 0 \quad \text{in } \Omega, \quad u = \bar{u} \quad \text{on } \Gamma \tag{6.49}$$

by the Trefftz method.

The idea is to select a function which satisfies the differential equation exactly but the boundary conditions only approximately.

To solve (6.49) we let

$$u(x) = \int_\Gamma \frac{1}{r} f(y)\, ds_y$$

(this function satisfies $\Delta u = 0$, whatever the value of $f(y)$) and, in order that the boundary condition

$$\lim_{x \to x_0 \in \Gamma} u(x) = \bar{u}(x_0) \qquad \forall\, x_0 \in \Gamma$$

is satisfied, we require that the layer satisfies the integral equation

$$\int_\Gamma \frac{1}{r} f(y)\, ds_y = \bar{u}(x_0), \qquad r = |y - x_0| \qquad \forall\, x_0 \in \Gamma$$

Such potentials ($=$ functions) exist now also in structural mechanics.

Let Γ denote the boundary (or a portion thereof) of an n-dimensional domain and $g_0[x]$ the fundamental solution of an operator D then the function

$$u(x) = \int_\Gamma \partial_y^i g_0(y, x) f(y)\, ds_y \qquad i \in \{0, 1 \ldots 2m - 1\}$$

where $f(y)$ is an arbitrary "boundary layer" is called an elastic potential. Such an elastic potential is a homogeneous solution

$$Du = \int_\Gamma D_x\, \partial_y^i g_0(y, x) f(y)\, ds_y = \int_\Gamma \partial_y^i D_x\, g_0(y, x) f(y)\, ds_y$$

$$= \int_\Gamma \partial_y^i 0 f(y)\, ds_y = 0$$

of the governing equation at all points x not on Γ.

The layers $f(y)$ which in other branches of physics are mass distributions or electrical charges are in structural mechanics displacements $\partial^0, \partial^1 \ldots \partial^{m-1}$ and forces $\partial^m, \partial^{m+1} \ldots \partial^{2m-1}$.

Consider, e.g., an infinite prestressed ($N = 1$) membrane which is loaded with forces $f(y)$ concentrated on a curve Γ, see Fig. 6.24.

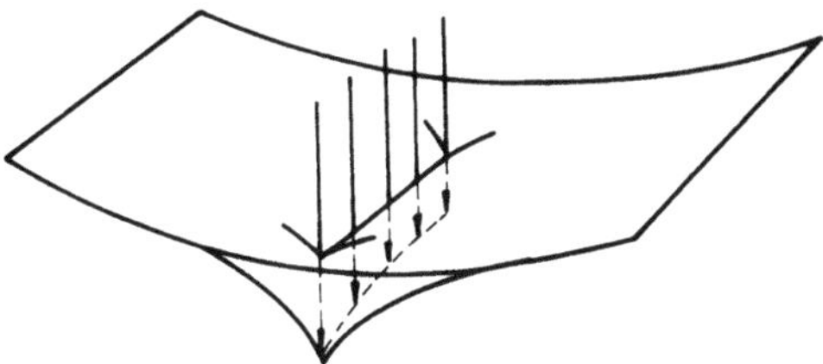

Figure 6.24

The deflection is the function

$$w(x) = \frac{1}{2\pi} \int_\Gamma \ln \frac{1}{r} f(y)\, ds_y$$

As all other regions are load free $w(x)$ must (and indeed does) satisfy the homogeneous equation

$$\Delta w = 0$$

at all points not on Γ.

Or assume the curve Γ is the locus of dislocations $d(y)$. Such dislocations cause at some distant point x not on Γ the deflection

$$w(x) = \int_\Gamma \partial_y^1 g_0(y, x)\, d(y)\, ds_y$$

The kernel of this influence function, this potential,

$$\partial_y^1 g_0(y, x) = \frac{\partial}{\partial v} g_0(y, x) = \frac{\partial}{\partial v} \frac{1}{2\pi} \ln \frac{1}{r} = \frac{-1}{2\pi r} \{ r_{,y_1} v_1(y) + r_{,y_2} v_2(y) \}$$

is the normal derivative of the fundamental solution $g_0(y, x)$ with respect to the normal $v(y)$ at the integration point y.

We expect now when we pass through Γ that the displacement jumps by the amount of the dislocation. That is we expect

$$\lim_{\varepsilon \to 0} \{ w(x_0 + \varepsilon n) - w(x_0 - \varepsilon n) \} = d(x_0)$$

when n is the normal at x_0 on Γ.

Analogously we expect a jump in $\partial w / \partial n$ when Γ is loaded with line forces $f(y)$

$$\lim_{\varepsilon \to 0} \left\{ \frac{\partial w}{\partial n}(x_0 + \varepsilon n) - \frac{\partial w}{\partial n}(x_0 - \varepsilon n) \right\} = f(x_0)$$

This property is common to all elastic potentials. That is, whenever we pass through Γ then the potential u itself or one of its "derivatives", $\partial^i u$, jumps.

We have the following simple rule:

Whenever $i + j = 2m - 1$ then we observe a jump of magnitude $f(x_0)$ in the function

$$\partial_x^j \int_\Gamma \partial_y^i g_0(y, x) f(y)\, ds_y$$

when passing through Γ at x. Otherwise not.

Besides boundary potentials there exist volume potentials

$$v(x) = \int_\Omega \partial_y^i g_0(y, x) p(y) \, d\Omega_y$$

These volume potentials are not homogeneous solutions, $Dv \neq 0$, instead they reproduce the layer p. That is D applied to v at an internal point $x \in \Omega$ renders

$$D v(x) = D \int_\Omega \partial_y^i g_0(y, x) p(y) \, d\Omega_y = \partial^i p(x)$$

The operator D reproduces the volume distribution p (or higher terms, $\partial^i p$, theoref if $i > 0$).

Consider an infinite prestressed ($N = 1$) membrane loaded in a region Ω with distributed forces $p(y)$. The deflection is

$$w(x) = \frac{1}{2\pi} \int_\Omega \ln \frac{1}{r} p(y) \, d\Omega_y$$

and because the equilibrium condition requires that

$$\Delta w = p$$

at all points in Ω we conclude that the fundamental solution $g_0(y, x) = -(2\pi)^{-1} \ln r$ must be such a reproducing kernel.

Elastic potentials are closely connected with influence functions because influence functions are sums of elastic potentials.

Potential theory is the branch of mathematics which studies the properties of these single functions. This section could only be a short introduction. For a thorough discussion of elastic potentials and their properties we must refer the reader to the literature [J & S], [K 2], [H 4], [H 5].

7 The Operators A

Up to now we were concerned with the differential equations which govern the displacement of the structural elements, as e.g. with the equation $-Lu=p$ which governs the displacement of an elastic body. Now we focus on the systems of three equations which, originally, preceded the displacement equations. In the case of an elastic body this was the system

$$
\begin{aligned}
E(u) - E &= 0 \\
C[E] - S &= 0 \\
- \operatorname{div} S &= p
\end{aligned}
\tag{7.1}
$$

The first equation expresses the strain-displacement relation, the second the stress-strain relation and the third formulates the equilibrium condition.

Consider now the equation

$$
\begin{bmatrix} 1 & -1 & 0 \\ 0 & 1 & -1 \\ 0 & 0 & 1 \end{bmatrix}
\begin{bmatrix} x_1 \\ x_2 \\ x_3 \end{bmatrix}
=
\begin{bmatrix} x_1 - x_2 \\ x_2 - x_3 \\ x_3 \end{bmatrix}
\tag{7.2}
$$

The matrix on the left-hand side can be termed an operator, A, which acts on the vector $x = \{x_1, x_2, x_3\}^T$. The right-hand side of Eq. (7.2) is the image, Ax, of the vector x.

To solve the system

$$
Ax = b
\tag{7.3}
$$

for x means to find the vector x whose image is the vector b. The scalar product of the vector Ax and the vector $\hat{x}$

$$
\hat{x}^T A x =: \langle A x, \hat{x} \rangle
$$

the number $\langle Ax, \hat{x} \rangle$, could be termed "virtual strain energy".

If $x = x_s$ is the solution of Eq. (7.3) then multiplication of Eq. (7.3) from the left with $\hat{x}$ results in

$$
\hat{x}^T A x = \hat{x}^T b
$$

which is the statement that "the virtual strain energy, $\hat{x}^T A x$, is equal to the external virtual work".

In the same sense the expression on the left-hand side of Eq. (7.1) can be termed the image of the "vector" $\Sigma = \{u, E, S\}$ under the action of an operator A. Gurtin, see [G 2] p. 95, calls Σ an "elastic state". The displacement field u, the strain tensor E and the stress tensor S that satisfy these three equations and the boundary conditions constitute the elastic state of the isotropic body in question.

Our decision to consider the left-hand side of Eq. (7.1) the image $A(\Sigma)$ of Σ under the action of an operator A allows us to apply all the formalism we developed for the operators D to the operators A as well.

The operators A are linear and they possess a first identity with a symmetric energy, $E(\Sigma, \hat{\Sigma}) = E(\hat{\Sigma}, \Sigma)$. Hence, all the principles of work and energy formulated for the operators D apply to the operators A as well.

The only difference with respect to the operators D is that the operators A act on three variables, these are of the kind u, ε, σ, and consequently their inner structure, their inner texture, is richer, more plentiful.

7.1 The Systems

Before we start we perform a small modification with respect to the triple $\{u, \varepsilon, \sigma\}$.

Let u_{el} denote the elastic displacement, u_T the temperature displacement and $u = u_{el} + u_T$ the total displacement. Then we consider from now on the function u in the triple $\{u, \varepsilon, \sigma\}$ to be the total displacement.

As a consequence we must modify the right-hand sides of the original equations as we explain in the following.

7.1.1 Bars

The differential equations

$$-EA\,u''_{el} = p, \qquad\qquad u'_T = \alpha_T\,T$$

which govern the elastic and the temperature displacement respectively are if we let

$$u = u_{el} + u_T$$

equivalent with the system

$$u' - \varepsilon = \alpha_T\,T$$
$$EA\varepsilon - N = 0$$
$$-N' = p$$

$$\Sigma = \{u, \varepsilon, N\}$$

7.1.2 Elastic Plates and Bodies

The differential equations

$$-L u_{el} = p, \qquad\qquad E(u_T) = \alpha_T\, T\, I$$

are if we let

$$u = u_{el} + u_T$$

equivalent with the system

$$
\begin{aligned}
E(u) - E &= \alpha_T\, T\, I & \qquad\qquad \tfrac{1}{2}\,(u_{i,j} + u_{j,i}) - \varepsilon_{ij} &= \alpha_T\, T\, \delta_{ij} \\
C[E] - S &= 0 & \widehat{=} \qquad\qquad C^{ijkl}\,\varepsilon_{kl} - \sigma_{ij} &= 0 \qquad\qquad (7.4) \\
-\operatorname{div} S &= p & -\sigma_{ij,j} &= p_i
\end{aligned}
$$

If there exist self-equilibrated initial stresses S^i, $\operatorname{div} S^i = 0$, in the elastic plate or body before the external load is applied then the second equation must be replaced by

$$C[E] - S = S^i$$

7.1.3 Beams

The differential equations

$$(E I w''_{el})'' = p, \qquad\qquad w''_T = -\alpha_T\,\frac{\Delta T}{h}$$

are if we let

$$w = w_{el} + w_T$$

equivalent with the system

$$
\begin{aligned}
\varepsilon - w'' &= \alpha_T\,\frac{\Delta T}{h} \\
E I \varepsilon + M &= 0 \\
-M'' &= p \\
\Sigma &= \{w, \varepsilon, M\}
\end{aligned}
$$

7.1.4 Kirchhoff Plates

The differential equations

$$K \Delta \Delta w_{el} = p, \qquad\qquad E(w_T) = -\alpha_T \theta\, I$$

are if we let

$$w = w_{el} + w_T$$

equivalent with the system

$$\begin{aligned}
E - E(w) &= \alpha_T \theta\, I & \varepsilon_{ij} - w_{,ij} &= \alpha_T \theta\, \delta_{ij} \\
C[E] + M &= 0 & \qquad \hat{=} \qquad C^{ijkl}\varepsilon_{kl} + M_{ij} &= 0 \\
-\operatorname{div}^2 M &= p & -M_{ij,ji} &= p
\end{aligned} \qquad (7.5)$$

$$\Sigma = \{w, E, M\}$$

As an example consider the clamped beam in Fig. 7.1 with its non-uniform temperature distribution.

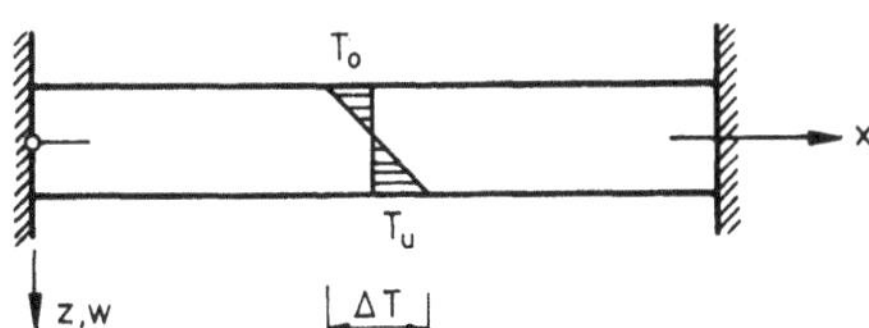

Figure 7.1

The elastic state $\Sigma = \{w, \varepsilon, M\}$ of the beam satisfies in $0 < x < l$ the three equations

$$\varepsilon - w'' = \alpha_T \frac{\Delta T}{h} \qquad (7.6\,\text{a})$$

$$EI\varepsilon + M = 0 \qquad (7.6\,\text{b})$$

$$-M'' = 0 \qquad (7.6\,\text{c})$$

and at $x = 0$ and at $x = l$ the boundary conditions

$$w(0) = w'(0) = w(l) = w'(l) = 0 \qquad (7.7)$$

To solve the problem, to find the elastic state, we start with Eq. (7.6c). According to this equation the bending moment M is linear, $M = ax + b$, and, hence, also the strain, $\varepsilon = -(EI)^{-1}(ax+b)$. Substituting ε into Eq. (7.6a) we obtain an expression for the second derivative

$$w'' = -\alpha_T \frac{\Delta T}{h} - \frac{1}{EI}(ax+b)$$

Consequently w itself must be of the form,

$$w(x) = -\alpha_T \frac{\Delta T x^2}{2h} - \frac{1}{EI}\left(\frac{ax^3}{6} + \frac{bx^2}{2}\right) + cx + d$$

and for it to satisfy the boundary conditions the integration constants must have the values

$$b = -\alpha_T \frac{\Delta T}{h} EI \qquad a = c = d = 0$$

Hence, the elastic state of the beam is determined

$$w(x) = -\alpha_T \frac{\Delta T}{h}\frac{x^2}{2} + \alpha_t \frac{\Delta T}{h}\frac{x^2}{2} = 0, \quad \varepsilon = \alpha_T \frac{\Delta T}{h}, \quad M = -\alpha_T \frac{\Delta T}{h} EI$$

and we recognize that the beam does not deflect. It stays in the neutral position because the sum of the two deflections, w_T and w_{el}, is zero at every point.

7.2 Identities

What really matters in structural mechanics are the identities of the operators. They are the foundation of all the work and energy principles and so too in the case of the operators A. Hence, the derivation of these identities has to be done first.

7.2.1 Elastic Plates and Bodies

Our basic tool, integration by parts, requires the single components of an elastic state to satisfy certain smoothness conditions. We, therefore, need some classifications. Let

$$\mathscr{S} = \{\Sigma = \{u, E, S\} | E \text{ and } S \text{ are symmetric}\}$$

denote the set of all Σ whose strain and stress tensors are symmetric. All the elastic states we shall work with will come from one of the following two subsets of $\mathscr{S}$.

$$\mathscr{S}^a = \{\Sigma \in \mathscr{S} | u \in C^1(\bar{\Omega}), E \in C(\bar{\Omega}), S \in C^1(\bar{\Omega})\}$$

$$\mathscr{S}^b = \{\Sigma \in \mathscr{S} | u \in C^1(\bar{\Omega}), E \in C(\bar{\Omega}), S \in C(\bar{\Omega})\}$$

The image $A(\Sigma)$ of an elastic state $\Sigma = \{u, E, S\}$ under the action of the operator A is the "vector"

$$A(\Sigma) = \begin{bmatrix} E(u) - E \\ C[E] - S \\ -\operatorname{div} S \end{bmatrix}$$

whose first two components are matrices and whose third component is a vector.

The derivation of the first identity for an operator D started with the integral

$$\int_\Omega Du \cdot \hat{u}\, d\Omega$$

Analogously, we multiply the three equations $A(\Sigma)$ with the conjugated quantities (scalar product of vectors or matrices, see section 1.1) integrate the result over the domain Ω and, thus, obtain a sum of three domain integrals which we denote by

$$\langle A(\Sigma), \hat{\Sigma} \rangle = \int_\Omega (E(u) - E) \cdot \hat{S}\, d\Omega + \int_\Omega (C[E] - S) \cdot \hat{E}\, d\Omega$$
$$+ \int_\Omega -\operatorname{div} S \cdot \hat{u}\, d\Omega$$

Remember that a dot denotes the scalar product of vectors and matrices,

$$E \cdot S = \varepsilon_{ij}\sigma_{ij} = \varepsilon_{11}\sigma_{11} + \varepsilon_{12}\sigma_{12} + \dots \quad \dots + \varepsilon_{33}\sigma_{33}$$

Due to the auxiliary lemma, (for a proof see Eq. (1.13)),

$$p: \ S \in C^1(\bar{\Omega}) \ \text{ and symmetric, } \hat{u} \in C^1(\bar{\Omega})$$

$$q: \ \int_\Omega -\operatorname{div} S \cdot \hat{u}\, d\Omega = -\int_\Gamma Sn \cdot \hat{u}\, ds + \int_\Omega S \cdot E(\hat{u})\, d\Omega \tag{7.8}$$

the last integral in $\langle A(\Sigma), \hat{\Sigma} \rangle$ can be replaced by the right-hand side of Eq. (7.8). This renders

$$\langle A(\Sigma), \hat{\Sigma} \rangle = \int_\Omega (E(u) - E) \cdot \hat{S}\, d\Omega + \int_\Omega (C[E] - S) \cdot \hat{E}\, d\Omega$$
$$+ \int_\Omega S \cdot E(\hat{u})\, d\Omega - \int_\Gamma Sn \cdot \hat{u}\, ds \tag{7.9}$$

The three domain integrals in this equation constitute on account of the symmetry of the elasticity tensor, $C[E] \cdot \hat{E} = E \cdot C[\hat{E}]$, a symmetric bilinear form

$$E(\Sigma, \hat{\Sigma}) = \int_\Omega (E(u) - E) \cdot \hat{S}\, d\Omega + \int_\Omega C[E] \cdot \hat{E}\, d\Omega + \int_\Omega S \cdot (E(\hat{u}) - \hat{E})\, d\Omega$$

which, for obvious reasons, we term strain energy.

Our result

$$\langle A(\Sigma), \hat{\Sigma} \rangle = E(\Sigma, \hat{\Sigma}) - \int_{\Gamma} \boldsymbol{S}\boldsymbol{n} \cdot \hat{\boldsymbol{u}}\, ds$$

is already the first identity of the operator A.

First identity

$$p:\ \Sigma, \hat{\Sigma} \in \mathscr{S}^a \times \mathscr{S}^b$$

$$p:\ G(\Sigma, \hat{\Sigma}) = \langle A(\Sigma), \hat{\Sigma} \rangle + \int_{\Gamma} \boldsymbol{S}\boldsymbol{n} \cdot \hat{\boldsymbol{u}}\, ds - E(\Sigma, \hat{\Sigma}) = 0$$

Due to the symmetry of the energy a second identity exists as well.

Second identity

$$p:\ \Sigma, \hat{\Sigma} \in \mathscr{S}^a$$

$$q:\ B(\Sigma, \hat{\Sigma}) = \langle A(\Sigma), \hat{\Sigma} \rangle + \int_{\Gamma} \boldsymbol{S}\boldsymbol{n} \cdot \hat{\boldsymbol{u}}\, ds - \int_{\Gamma} \boldsymbol{u} \cdot \hat{\boldsymbol{S}}\boldsymbol{n}\, ds - \langle \Sigma, A(\hat{\Sigma}) \rangle = 0$$

7.2.2 Kirchhoff Plates

The operator A of a Kirchhoff plate applied to $\Sigma = \{w, E, M\}$ renders

$$A(\Sigma) = \begin{bmatrix} E - E(w) \\ C[E] + M \\ -\operatorname{div}^2 M \end{bmatrix}$$

We introduce the following classes

$$\mathscr{S} = \{\Sigma = \{w, E, M\} \,|\, E \text{ and } M \text{ are symmetric}\}$$

$$\mathscr{S}^a = \{\Sigma \in \mathscr{S} \,|\, w \in C^2(\bar{\Omega}),\ E \in C(\bar{\Omega}),\ M \in C^2(\bar{\Omega})\}$$

$$\mathscr{S}^b = \{\Sigma \in \mathscr{S} \,|\, w \in C^2(\bar{\Omega}),\ E \in C(\bar{\Omega}),\ M \in C^1(\bar{\Omega})\}$$

and, again, start with an auxiliary lemma, see [H 7] A 8.4.1.

$$p:\ M \in C^2(\bar{\Omega}) \text{ and symmetric, } \hat{w} \in C^2(\bar{\Omega})$$

$$q:\ \int_{\Omega} -\operatorname{div}^2 M\, \hat{w}\, d\Omega = -\int_{\Gamma} \left[V_n \hat{w} - M_n \frac{\partial \hat{w}}{\partial n} \right] ds - [[M_{nt}\hat{w}]]$$

$$-\int_{\Omega} M \cdot E(\hat{w})\, d\Omega \tag{7.10}$$

where

$$V_n = \frac{d}{ds} M_{nt} + Q_n, \quad M_{nt} = M_{ij} n_i t_j, \quad M_n = M_{ij} n_i n_j, \quad Q_n = M_{ij,i} n_j$$

and n and t are the normal and tangent on the boundary.

We next multiply the three equations of system A with the conjugated quantities (scalar product!), integrate the result over the domain Ω and, thus, obtain

$$\langle A(\Sigma), \hat{\Sigma} \rangle = \int_\Omega (E - E(w)) \cdot \hat{M} \, d\Omega + \int_\Omega (C[E] + M) \cdot \hat{E} \, d\Omega$$
$$+ \int_\Omega - \operatorname{div}^2 M \, \hat{w} \, d\Omega$$

With Eq. (7.10) follows

$$\langle A(\Sigma), \hat{\Sigma} \rangle = \int_\Omega (E - E(w)) \cdot \hat{M} \, d\Omega + \int_\Omega (C[E] + M) \cdot \hat{E} \, d\Omega$$
$$- \int_\Omega M \cdot E(\hat{w}) \, d\Omega$$
$$- \int_\Gamma \left[V_n \hat{w} - M_n \frac{\partial \hat{w}}{\partial n} \right] ds - [[M_{nt} \hat{w}]]$$

Finally we term the three domain integrals in this equation the energy

$$E(\Sigma, \hat{\Sigma}) = \int_\Omega (E - E(w)) \cdot \hat{M} \, d\Omega + \int_\Omega C[E] \cdot \hat{E} \, d\Omega + \int_\Omega M \cdot (\hat{E} - E(\hat{w})) \, d\Omega$$

and call our result the first identity of the operator A.

First identity

$$p: \ \Sigma, \hat{\Sigma} \in \mathscr{S}^a \times \mathscr{S}^b$$

$$q: \ G(\Sigma, \hat{\Sigma}) = \langle A(\Sigma), \hat{\Sigma} \rangle + \int_\Gamma \left[V_n \hat{w} - M_n \frac{\partial \hat{w}}{\partial n} \right] ds$$
$$+ [[M_{nt} \hat{w}]] - E(\Sigma, \hat{\Sigma}) = 0$$

On account of the symmetry of $E(\Sigma, \hat{\Sigma})$ a second identity exists as well.

Second identity

$$p: \ \Sigma, \hat{\Sigma} \in \mathscr{S}^a$$

$$q: \ B(\Sigma, \hat{\Sigma}) = \langle A(\Sigma), \hat{\Sigma} \rangle + \int_\Gamma \left[V_n \hat{w} - M_n \frac{\partial \hat{w}}{\partial n} \right] ds + [[M_{nt} \hat{w}]]$$
$$- [[w \hat{M}_{nt}]] - \int_\Gamma \left[w \hat{V}_n - \frac{\partial w}{\partial n} \hat{M}_n \right] ds - \langle \Sigma, A(\hat{\Sigma}) \rangle = 0$$

7.2.3 Bars

In the case of bars and beams we simply list the equations.
The product:

$$\langle A(\Sigma), \hat{\Sigma} \rangle = \int_a^b (u' - \varepsilon) \, \hat{N} \, dx + \int_a^b (EA\varepsilon - N) \, \hat{\varepsilon} \, dx + \int_a^b - N' \hat{u} \, dx \qquad (7.11)$$

The auxiliary lemma:

$$p: \; N, \hat{u} \in C^1[a, b]$$

$$q: \; \int_a^b - N' \, \hat{u} \, dx = - [N\hat{u}]_a^b + \int_a^b N \, \hat{u}' \, dx \qquad (7.12)$$

On substituting Eq. (7.12) into Eq. (7.11) we obtain

$$\langle A(\Sigma), \hat{\Sigma} \rangle = \int_a^b (u' - \varepsilon) \, \hat{N} \, dx + \int_a^b (EA\varepsilon - N) \, \hat{\varepsilon} \, dx + \int_a^b N \hat{u}' \, dx - [N\hat{u}]_a^b$$

The energy is the expression

$$E(\Sigma, \hat{\Sigma}) = \int_a^b (u' - \varepsilon) \, \hat{N} \, dx + EA \int_a^b \varepsilon \, \hat{\varepsilon} \, dx + \int_a^b N(\hat{u}' - \hat{\varepsilon}) \, dx$$

First identity

$$p: \; \Sigma \in \mathscr{S}^a = \{\Sigma = \{u, \varepsilon N\} | u, N \in C^1, \, \varepsilon \in C^0\}$$

$$\Sigma \in \mathscr{S}^b = \{\Sigma = \{u, \varepsilon, N\} | u \in C^1, \varepsilon, N \in C^0\}$$

$$q: \; G(\Sigma, \hat{\Sigma}) = \langle A(\Sigma), \hat{\Sigma} \rangle + [N\hat{u}]_a^b - E(\Sigma, \hat{\Sigma}) = 0$$

7.2.4 Beams

The product

$$\langle A(\Sigma), \hat{\Sigma} \rangle = \int_a^b (\varepsilon - w'') \, \hat{M} \, dx + \int_a^b (EI\varepsilon + M) \, \hat{\varepsilon} \, dx + \int_a^b - M'' \hat{w} \, dx \qquad (7.13)$$

The auxiliary lemma

$$p: \; M, \hat{w} \in C^2[a, b]$$

$$q: \; \int_a^b - M'' \, \hat{w} \, dx = - [M' \hat{w} - M \hat{w}']_a^b - \int_a^b M \hat{w}'' \, dx \qquad (7.14)$$

On substituting Eq. (7.14) into Eq. (7.13) we obtain

$$\langle A(\Sigma), \hat{\Sigma}\rangle = \int_a^b (\varepsilon - w'')\, \hat{M}\, dx + \int_a^b (EI\varepsilon + M)\, \hat{\varepsilon}\, dx$$

$$- \int_a^b M\hat{w}''\, dx - [M'\hat{w} - M\hat{w}']_a^b$$

The energy is the expression

$$E(\Sigma, \hat{\Sigma}) = \int_a^b (\varepsilon - w'')\, \hat{M}\, dx + EI \int_a^b \varepsilon\hat{\varepsilon}\, dx + \int_a^b M(\hat{\varepsilon} - \hat{w}'')\, dx$$

First identity

$$p: \ \Sigma \in \mathscr{S}^a = \{\Sigma = \{w, \varepsilon, M\} | w, M \in C^2, \varepsilon \in C^0\}$$

$$\hat{\Sigma} \in \mathscr{S}^b = \{\Sigma = \{w, \varepsilon, M\} | w \in C^2, \varepsilon, M \in C^0\}$$

$$q: \ G(\Sigma, \hat{\Sigma}) = \langle A(\Sigma), \hat{\Sigma}\rangle + [M'\hat{w} - M\hat{w}']_a^b - E(\Sigma, \hat{\Sigma}) = 0$$

All these identities are, as in chapter 2, the mathematical basis of the work and energy principles which we attribute to the single structural elements and their elastic states Σ.

Corollary 1, principle of virtual displacements

$$p: \ \Sigma \in \mathscr{S}^a$$

$$q: \ G(\Sigma, \hat{\Sigma}) = 0 \qquad \forall\, \hat{\Sigma} \in \mathscr{S}^b$$

Corollary 2, principle of virtual forces

$$p: \ \Sigma \in \mathscr{S}^b$$

$$q: \ G(\hat{\Sigma}, \Sigma) = 0 \qquad \forall\, \hat{\Sigma} \in \mathscr{S}^a$$

Corollary 3, "eigenwork = int. energy"

$$p: \ \Sigma \in \mathscr{S}^a$$

$$q: \ \frac{1}{2}\, G(\Sigma, \Sigma) = 0$$

Corollary 4, Betti's principle

$$p: \ \Sigma, \hat{\Sigma} \in \mathscr{S}^a$$

$$q: \ B(\Sigma, \hat{\Sigma}) = 0$$

Corollary 5, Equilibrium

$$p:\ \Sigma \in \mathcal{S}^a, \Sigma_r = elastic\ state\ of\ a\ rigid\text{-}body\ movement$$

$$q:\ G(\Sigma, \Sigma_r) = 0$$

7.3 Energy Principles

In this section we shall formulate the energy principles for the solutions of the boundary value problems governed by the operators *A*.

7.3.1 Elastic Plates and Bodies

Let $\{\Sigma, \hat{\Sigma}\}$ be two elastic states from $\mathcal{S}^a$. The second identity for this pair reads

$$(-1)\,B(\Sigma, \hat{\Sigma}) = -\langle A(\Sigma), \hat{\Sigma}\rangle - \int_\Gamma \boldsymbol{Sn} \cdot \hat{\boldsymbol{u}}\,ds + \int_\Gamma \boldsymbol{u} \cdot \hat{\boldsymbol{S}}\boldsymbol{n}\,ds$$

$$+ \langle \Sigma, A(\hat{\Sigma})\rangle = 0$$

With, see Eq. (7.9),

$$\langle \Sigma, A(\hat{\Sigma})\rangle = E(\Sigma, \hat{\Sigma}) - \int_\Gamma \boldsymbol{u} \cdot \hat{\boldsymbol{S}}\boldsymbol{n}\,ds$$

the identity transforms into

$$V(\Sigma, \hat{\Sigma}) = \{ -\langle A(\Sigma), \hat{\Sigma}\rangle - \int_\Gamma \boldsymbol{Sn} \cdot \hat{\boldsymbol{u}}\,ds + \int_\Gamma \boldsymbol{u} \cdot \hat{\boldsymbol{S}}\boldsymbol{n}\,ds\}$$

$$+ E(\Sigma, \hat{\Sigma}) - \int_\Gamma \boldsymbol{u} \cdot \hat{\boldsymbol{S}}\boldsymbol{n}\,ds = 0$$

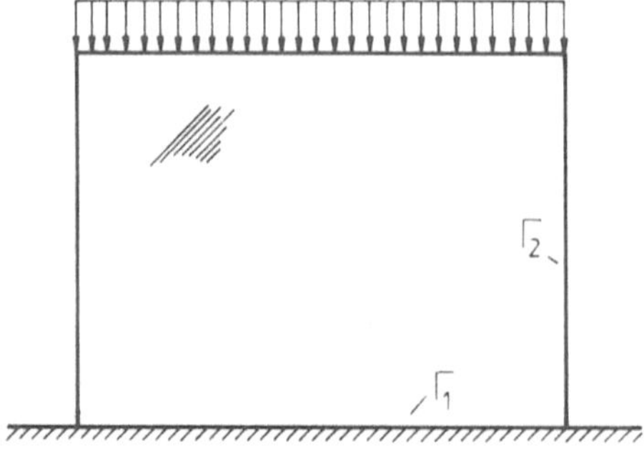

Figure 7.2

Energy principles are obtained if we substitute, as in chapter 4, the data of a *bvp* into this form V.

Consider, e. g., the plate in Fig. 7.2 which, besides being acted upon by external loads, undergoes some temperature change $T(x)$. The whole set of equations which must be satisfied by the elastic state Σ in such a general case is

$$
\begin{aligned}
E(u) - E &= \alpha_T T\,I & u &= \bar{u} \quad \text{on} \quad \Gamma_1 \\
C[E] - S &= 0 \qquad \text{in } \Omega & & \\
-\operatorname{div} S &= p & \tau(u) &= \bar{t} \quad \text{on} \quad \Gamma_2
\end{aligned}
\tag{7.15}
$$

Let us assume that the solution Σ is in $\mathscr{S}^a$. The requirements placed on Σ in the domain, i. e. the three equations on the left-hand side in (7.15), imply that Σ solves the equation

$$
A(\Sigma) = \begin{bmatrix} \alpha_T T I \\ 0 \\ p \end{bmatrix}
$$

Hence, the data of the *bvp* are

$$
\{\alpha_T T\,I, 0, p\} \text{ in } \Omega, \quad \bar{u} \text{ on } \Gamma_1, \quad \bar{t} \text{ on } \Gamma_2
$$

These data substituted into the form $V(\Sigma, \hat{\Sigma})$ render

$$
\begin{aligned}
\delta \Pi(\Sigma, \hat{\Sigma}) = &-\int_\Omega \alpha_T T\,I \cdot \hat{S}\,d\Omega - \int_\Omega p \cdot \hat{u}\,d\Omega - \int_{\Gamma_1} Sn \cdot \hat{u}\,ds - \int_{\Gamma_2} \bar{t} \cdot \hat{u}\,ds \\
&+ \int_{\Gamma_1} (\bar{u} - u) \cdot \hat{S}n\,ds + E(\Sigma, \hat{\Sigma}) = 0 \quad \forall\,\hat{\Sigma} \in \mathscr{S}^b
\end{aligned}
\tag{7.16}
$$

which is the first variation of **the basic functional**

$$
\begin{aligned}
\Pi(\Sigma) = &\frac{1}{2} E(\Sigma, \Sigma) - \int_\Omega \alpha_T T\,I \cdot S\,d\Omega \\
&- \int_\Omega p \cdot u\,d\Omega + \int_{\Gamma_1} (\bar{u} - u) \cdot Sn\,ds - \int_{\Gamma_2} \bar{t} \cdot u\,ds
\end{aligned}
$$

Because all pairs $\{\Sigma_s, \hat{\Sigma}\}, \Sigma_s = \text{solution of the } bvp, \hat{\Sigma} \in \mathscr{S}^b$, are zeros of $\delta \Pi(\Sigma_s, \hat{\Sigma})$ we conclude that Σ_s is a stationary point of the functional $\Pi(\Sigma)$, i. e.

$$
\delta \Pi(\Sigma_s, \hat{\Sigma}) = \frac{d}{d\varepsilon}(\Sigma_s + \varepsilon\,\hat{\Sigma})|_{\varepsilon = 0} = 0 \quad \forall\,\hat{\Sigma} \in \mathscr{S}^b
$$

Conversely, let Σ be an elastic state from $\mathscr{S}^a$ which satisfies Eq. (7.16). Integrating the energy by parts the expression $\delta \Pi(\Sigma, \hat{\Sigma})$ becomes

$$
\begin{aligned}
\langle A(\Sigma), \hat{\Sigma} \rangle &- \int_\Omega \alpha_T T\,I \cdot \hat{S}\,d\Omega - \int_\Omega p \cdot \hat{u}\,d\Omega \\
&- \int_{\Gamma_2} (\bar{t} - Sn) \cdot \hat{u}\,ds + \int_{\Gamma_1} (\bar{u} - u) \cdot \hat{S}n\,ds = 0 \quad \forall\,\hat{\Sigma} \in \mathscr{S}^b
\end{aligned}
$$

and we there conclude, with the help of lemma 1 and 2 (see chapter 4), that Σ solves the $b\,v\,p$.

We have thus found the following theorem.

Theorem 7.1 The basic principle (Hu-Washizu principle)

The elastic state $\Sigma \in \mathscr{S}^a$ solves the bvp in Eq. (7.15) if and only if $\Sigma \in \mathscr{S}^a$ is a stationary point of the functional $\Pi(\Sigma)$ with respect to all variations $\hat{\Sigma} \in \mathscr{S}^a$.

Other energy principles are obtained if we introduce subsets of $\mathscr{S}$.

$$\mathscr{S}_0 = \{\Sigma \in \mathscr{S}^a | C^{-1}[S] - E = 0\}$$

$$\mathscr{S}_1 = \{\Sigma \in \mathscr{S}^b | E(u) - E = \alpha_T TI,\ C[E] - S = 0,\ u = \bar{u} \text{ on } \Gamma_1\}$$

$$\mathscr{S}_2 = \{\Sigma \in \mathscr{S}^a | C^{-1}[S] - E = 0,\ -\operatorname{div} S = p,\ Sn = \bar{t} \text{ on } \Gamma_2\}$$

and study the basic functional on these subsets.

The elastic states which belong to $\mathscr{S}_0$ satisfy the stress-strain relation.

The states in $\mathscr{S}_1$ are the geometrically admissible states. They satisfy the first two equations in (7.15) and the geometric boundary conditions.

The states in $\mathscr{S}_2$ are the statically admissible states, they satisfy the last two equations in (7.15) and the static boundary conditions.

$$\mathscr{S}_1 = bc + \begin{cases} E(u) - E = \alpha_T TI \\ C[E] - S = 0 \\ \quad -\operatorname{div} S = p \end{cases} \Bigg\} + bc = \mathscr{S}_2$$

What characterises these subsets are the side conditions. Every elastic state in $\mathscr{S}_0$ satisfies the condition $E = C^{-1}[S]$ and, hence, every state Σ in $\mathscr{S}_0$ is already determined by its two components u and S.

It is only consequential if we replace in the basic functional $\Pi(\Sigma)$ the strain tensor E by $C^{-1}[S] = E$ if Σ is from $\mathscr{S}_0$. The result is the restriction of the basic functional $\Pi(\Sigma)$ to the class $\mathscr{S}_0$.

$$\Pi(\Sigma)|_{\mathscr{S}_0} = \Pi_0(u, S) = -\frac{1}{2} \int_\Omega C^{-1}[S] \cdot S \, d\Omega + \int_\Omega E(u) \cdot S \, d\Omega$$

$$- \int_\Omega \alpha_T TI \cdot S \, d\Omega - \int_\Omega p \cdot u \, d\Omega$$

$$- \int_{\Gamma_2} \bar{t} \cdot u \, ds + \int_{\Gamma_1} (\bar{u} - u) \cdot Sn \, ds$$

or, in more popular terms, the functional of the Hellinger-Reissner principle.

Theorem 7.2 Hellinger-Reissner principle

The elastic state $\Sigma \in \mathscr{S}_0$ solves the bvp in Eq. (7.15) if and only if $\Sigma \in \mathscr{S}_0$ is a stationary point of the functional $\Pi_0(u, S)$ with respect to all variations $\hat{\Sigma} \in \mathscr{S}_0$.

This theorem is a simple corollary of the basic principle, hence, there is no need for a proof.

The restriction of the basic functional to the class $\mathscr{S}_1$ is

$$\Pi(\Sigma)|_{\mathscr{S}_1} = \Pi_1(u) = \frac{1}{2} \int_\Omega C[E(u)] \cdot E(u)\, d\Omega - \int_\Omega C[E(u)] \cdot \alpha_T T I\, d\Omega$$

$$+ \frac{1}{2} \int_\Omega C[\alpha_T T I] \cdot \alpha_T T I\, d\Omega - \int_\Omega p \cdot u\, d\Omega - \int_{\Gamma_2} \bar{t} \cdot u\, ds \qquad (7.17)$$

This expression is the functional $\Pi_1(u)$ of the principle of minimum potential energy, see Eq. (4.32).

Theorem 7.3 Principle of minimum potential energy

The elastic state $\Sigma \in \mathscr{S}_1$ solves the bvp in Eq. (7.15) if and only if $\Sigma \in \mathscr{S}_1$ satisfies the inequality

$$\Pi_1(u + \hat{u}) - \Pi_1(u) = \frac{1}{2} E(\hat{u}, \hat{u}) \geq 0 \qquad \forall\, \hat{u} \in \mathscr{S}_{1,0}$$

Here, $\mathscr{S}_{1,0}$ is the space

$$\mathscr{S}_{1,0} = \{\Sigma \in \mathscr{S}^b | E(u) - E = 0,\ C[E] - S = 0,\ u = 0 \text{ on } \Gamma_1\}$$

The constant term in the functional, see Eq. (7.17), naturally, can be deleted because the thus modified functional attains its minimum at the same function u.

A proof of this theorem was given for the case $T = 0$ in section 4.3. If $T \neq 0$ then the proof is done as in the case of Theorem 7.4.

The restriction of the basic functional to the class $\mathscr{S}_2$ is

$$\Pi(\Sigma)|_{\mathscr{S}_2} = \Pi_2(S) = -\frac{1}{2} \int_\Omega C^{-1}[S] \cdot S\, d\Omega - \int_\Omega \alpha_T T I \cdot S\, d\Omega + \int_{\Gamma_1} \bar{u} \cdot Sn\, ds$$

This expression is, up to the factor (-1), the functional of the principle of minimum complementary energy.

Theorem 7.4 Principle of minimum complementary energy

The elastic state $\Sigma \in \mathscr{S}^a$ solves the bvp in Eq. (7.15) if and only if $\Sigma \in \mathscr{S}^a$ satisfies the inequality

$$\Pi_2(S + \hat{S}) - \Pi_2(S) = -\int_\Omega C^{-1}[\hat{S}] \cdot \hat{S}\, d\Omega \leq 0 \qquad \forall\, \hat{\Sigma} \in \mathscr{S}_{2,0} \qquad (7.18)$$

Here, $\mathscr{S}_{2,0}$ is the space

$$\mathscr{S}_{2,0} = \{\Sigma \in \mathscr{S}^a | C^{-1}[S] - E = 0,\ -\operatorname{div} S = 0,\ Sn = 0 \text{ on } \Gamma_2\}$$

The functional $\Pi_2(\Sigma)$ attains at Σ_s its maximum and the functional $(-1)\,\Pi_2(\Sigma)$, therefore, at Σ_s its minimum.

To prove this theorem we first demonstrate that the solution Σ_s of the bvp satisfies inequality (7.18).

The left-hand side of the inequality can be expressed as

$$\Pi_2(S+\hat{S})-\Pi_2(S)=-\int_\Omega C^{-1}[S]\cdot\hat{S}\,d\Omega-\frac{1}{2}\int_\Omega C^{-1}[\hat{S}]\cdot\hat{S}\,d\Omega$$

$$-\int_\Omega \alpha_T TI\cdot\hat{S}\,d\Omega+\int_{\Gamma_1} \bar{u}\cdot\hat{S}n\,ds \qquad (7.18\,a)$$

Due to Eq. (7.8) and $\hat{S}n=0$ on Γ_2 and div $\hat{S}=0$ in Ω the last integral can be written as

$$\int_{\Gamma_1} \bar{u}\cdot\hat{S}n\,ds=\int_\Gamma u\cdot\hat{S}n\,ds=\int_\Omega u\cdot\text{div}\,\hat{S}\,d\Omega+\int_\Omega E(u)\cdot\hat{S}\,d\Omega$$

$$=\int_\Omega (E+\alpha_T TI)\cdot\hat{S}\,d\Omega=\int_\Omega C^{-1}[S]\cdot\hat{S}\,d\Omega+\int_\Omega \alpha_T TI\cdot\hat{S}\,d\Omega$$

If we substitute this expression into Eq. (7.18 a) then we obtain just the statement

$$\Pi_2(S+\hat{S})-\Pi_2(S)=-\frac{1}{2}\int_\Omega C^{-1}[\hat{S}]\cdot\hat{S}\,d\Omega=-\frac{1}{2}\int_\Omega \hat{E}\cdot C[\hat{E}]\,d\Omega \quad \leqslant 0$$

and therewith the proof in one direction is done.

Conversely, let Σ an elastic state from $\mathscr{S}_2$ which satisfies (7.18). Then we must have as well

$$\Pi_2(S+\varepsilon\hat{S})-\Pi_2(S)=\varepsilon\{-\int_\Omega C^{-1}[S]\cdot\hat{S}\,d\Omega-\int_\Omega \alpha_T TI\cdot\hat{S}\,d\Omega$$

$$+\int_{\Gamma_1} \bar{u}\cdot\hat{S}n\,ds\}-\frac{1}{2}\varepsilon^2\int_\Omega C^{-1}[\hat{S}]\cdot\hat{S}\,d\Omega \quad \leqslant 0$$

where $\hat{\Sigma}$ is in $\mathscr{S}_{2,0}$ and ε in $(-\infty, +\infty)$. This expression is a polynomial which attains its maximum at $\varepsilon=0$. Hence, the first derivative at this point is zero, i.e.

$$-\int_\Omega C^{-1}[S]\cdot\hat{S}\,d\Omega-\int_\Omega \alpha_T TI\cdot\hat{S}\,d\Omega+\int_{\Gamma_1} \bar{u}\cdot\hat{S}n\,ds=0 \quad \forall\,\hat{\Sigma}\in\mathscr{S}_2,0$$

$$(7.19)$$

With the help of Eq. (7.9) the first integral in this equation can be expressed as

$$-\int_\Omega C^{-1}[S]\cdot\hat{S}\,d\Omega=-\int_\Omega E\cdot\hat{S}\,d\Omega=-\int_\Omega E(u)\cdot\hat{S}\,d\Omega-\int_\Omega S\cdot E(\hat{u})\,d\Omega$$

$$+\langle A(\Sigma),\hat{\Sigma}\rangle+\int_\Gamma Sn\cdot\hat{u}\,ds \qquad (7.20)$$

We now apply the auxiliary lemma, see Eq. (7.8), to the first two integrals in the last equation of (7.20) and substitute the full expression into Eq. (7.19). The result is

$$\int_{\Gamma_1} (\bar{u} - u) \cdot \hat{S} n \, ds + \int_{\Omega} (E(u) - E - \alpha_T T I) \cdot \hat{S} \, d\Omega = 0 \quad \forall \, \hat{\Sigma} \in \mathscr{S}_{2,0} \quad (7.21)$$

If this Eq. (7.21) would be valid for all $\hat{\Sigma} \in \mathscr{S}^a$ then we could with the help of lemma 1 and 2 conclude immediately that $\Sigma = \Sigma_s$, i.e. that Σ solves the bvp. But as it is, Eq. (7.21) is only valid for all $\hat{\Sigma}$ in $\mathscr{S}_{2,0}$. Hence, we need a further lemma, see [G 2] p. 119, to draw our conclusion.

Lemma 3

Assume (i) Ω *is simply connected,* (ii) *the boundary* Γ *is convex* (iii) u *is continuous on* Γ_1, (iv) $T \in C(\bar{\Omega}) \cap C^3(\Omega)$, (v) $E \in C(\bar{\Omega}) \cap C^3(\Omega)$ *and that*

$$- \int_{\Omega} (E + \alpha_T T I) \cdot \hat{S} \, d\Omega + \int_{\Gamma_1} \bar{u} \cdot \hat{S} n \, ds = 0$$

is valid for all $\hat{S} \in C^\infty(\bar{\Omega})$ *which vanish on* $\Gamma_2 = \Gamma - \Gamma_1$ *and which satisfy the equation div* $\hat{S} = 0$, *then there exists a displacement field* u *with the property*

$$E + \alpha_T T I = E(u), \quad u = \bar{u} \quad \text{on } \Gamma_1$$

With this lemma the result that Σ solves the bvp is now readily reached and the proof, thus, complete.

Note that we did not mention the assumption of lemma 3 (as, e. g., the assumption that Ω is simply connected, etc.) in the statement of Theorem 7.4 but they, naturally, are an integral part of the theorem.

By splitting $\mathscr{S}$ into further subsets we could formulate further energy principles, see e. g. [O 1] and [W 5], but we leave this as an exercise to the reader.

The connection between the principle of minimum potential energy and the principle of maximum (or minimum) complementary energy is the same as in chapter 4.

Because Σ_s belongs to $\mathscr{S}_1$ as to $\mathscr{S}_2$ and because the two functionals $\Pi_1(\Sigma)$ and $\Pi_2(\Sigma)$ are the restrictions of the same basic functional $\Pi(\Sigma)$ to the classes $\mathscr{S}_1$ and $\mathscr{S}_2$ the two functionals Π_1 and Π_2 have at Σ_s the same value

$$\Pi_2(\Sigma_s) = \Pi(\Sigma_s) = \Pi_1(\Sigma_s)$$

and because Σ_s is the solution of the two complementary energy principles

$$\Pi_1(\Sigma') \to \text{Min}, \quad \Sigma' \in \mathscr{S}_1, \quad \Pi_2(\Sigma'') \to \text{Max}, \quad \Sigma'' \in \mathscr{S}_2$$

see Theorems 7.3 and 7.4, we have for arbitrary states Σ'', $\Sigma' \in \mathscr{S}_2$, $\mathscr{S}_1$

$$\Pi_2(\Sigma'') \leqslant \Pi_2(\Sigma_s) = \Pi(\Sigma_s) = \Pi_1(\Sigma_s) \leqslant \Pi_1(\Sigma') \quad (7.22)$$

That is, the elastic states in $\mathscr{S}_1$ and $\mathscr{S}_2$ render upper and lower bounds for the total energy of the equilibrium state Σ_s.

These inequalities (7.22) are not the only bounds available for the total energy $\Pi(\Sigma_s)$. There exists a second estimate which is of some relevance because it is the basis of the hyper-circle method.

To formulate this second estimate we need some preparatory results.

a) As in section 4.4 we consider the linear manifolds $\mathscr{S}_1$ and $\mathscr{S}_2$ to be the direct sum of two arbitrary elements, $\Sigma^{(1)}$ or $\Sigma^{(2)}$, and the two subspaces $\mathscr{S}_{1,0}$ or $\mathscr{S}_{2,0}$ respectively.

$$\mathscr{S}_1 = \Sigma^{(1)} \oplus \mathscr{S}_{1,0} \qquad \mathscr{S}_2 = \Sigma^{(2)} \oplus \mathscr{S}_{2,0}$$

Fig. 7.3 illustrates the meaning of the two equations.

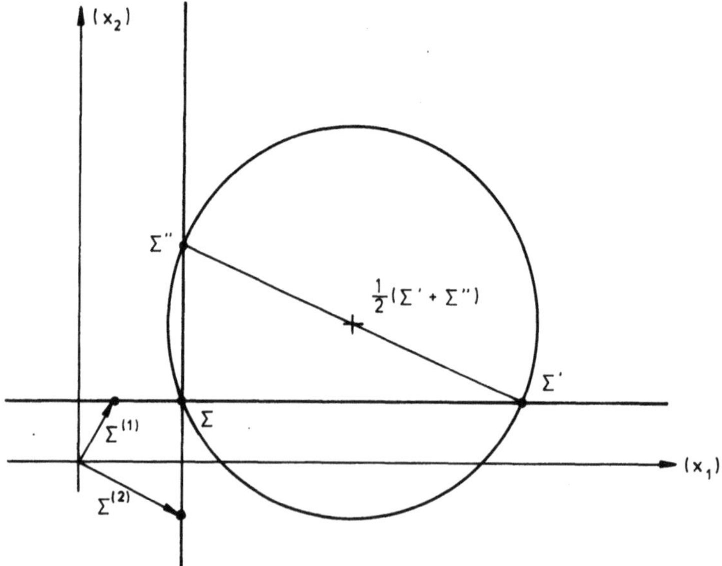

Figure 7.3

b) On the set of all strain tensors with square integrable coefficients we define the scalar product

$$((E,\hat{E})):=\int_\Omega C[E]\cdot\hat{E}\,d\Omega$$

which is a measure of the internal virtual work done when the stresses $C[E]$ act through the virtual strains $\hat{E}$. This scalar product has the properties:

(i) $\qquad ((E,\hat{E}))=((\hat{E},E))$

(ii) $\qquad ((aE_1+bE_2,\hat{E}))=a((E_1,\hat{E}))+b((E_2,\hat{E}))$

(iii) $((E, E)) \geqslant 0$

(iv) $((E, E)) = 0 \leftrightarrow E = 0$ (see $[G\,2]$ p. 85)

(v) $((E, \hat{E}))^2 \leqslant ((E, E))\,((\hat{E}, \hat{E}))$ (Schwarz' inequality)

The norm of a strain tensor is the number

$$\|E\| = ((E, E))^{1/2}$$

It has the properties

(i) $\|E\| = 0 \leftrightarrow E = 0$

(ii) $\|a E\| = |a|\,\|E\|, \quad a \in \mathbb{R}^1$ (7.23)

(iii) $\|E + \hat{E}\| \leqslant \|E\| + \|\hat{E}\|$

The last equation, the so-called triangular inequality, is easily verified with the help of the equation

$$\|E + \hat{E}\|^2 = \|E\|^2 + 2\left((E, \hat{E})\right) + \|\hat{E}\|^2$$

and Schwarz' inequality.

c) Let $\Sigma' = \{u', E', S'\} \in \mathscr{S}_{1,0}$ and $\Sigma'' = \{u'', E'', S''\} \in \mathscr{S}_{2,0}$. These two states substituted into Eq. (7.8),

$$\int\limits_{\Omega} - \operatorname{div} S'' \cdot u'\, d\Omega = - \int\limits_{\Gamma} S'' n \cdot u'\, ds + \int\limits_{\Omega} S'' \cdot E(u')\, d\Omega$$

render, because of $\operatorname{div} S'' = 0$ in Ω, $S'' n = 0$ on Γ_2, $u' = 0$ on Γ_1 and $\Gamma = \Gamma_1 \cup \Gamma_2$,

$$0 = \int\limits_{\Omega} S'' \cdot E(u')\, d\Omega = \int\limits_{\Omega} C[E''] \cdot E'\, d\Omega = ((E'', E'))$$ (7.24)

Hence, two strain tensors E' and E'' from $\mathscr{S}_{1,0}$ and $\mathscr{S}_{2,0}$ are orthogonal with respect to the scalar product introduced above; the interior virtual work of two such states is zero.

We are now in a position to formulate the second energy estimate.

Theorem 7.5

$$p: \ \Sigma' = \{u', E', S'\} \in \mathscr{S}_1, \ \Sigma'' = \{u'', E'', S''\} \in \mathscr{S}_2$$

$$q: \ \frac{1}{8}\left\|\frac{E' + E''}{2} - E\right\|^2 = \frac{1}{2}\|E' - E''\|^2 = \Pi_1(u') - \Pi_2(S'') \ \geqslant 0$$ (7.25)

This theorem states: the average $1/2\,(E' + E'')$ of two tensors E', E'' from $\mathscr{S}_1$ and $\mathscr{S}_2$ approximates the tensor E of the true state the better, the smaller the difference $\Pi_1(u') - \Pi_2(S'')$ or, what is equivalent, the smaller the difference $E' - E''$.

This equation is the basis of the hyper-circle method of Prager and Synge, see Fig. 7.3. Given two arbitrary points E', E'' and the point E a circle can be drawn which passes through these three points and has the radius

$$\left\| \frac{E' + E''}{2} - E \right\|$$

The smaller the radius, the smaller the distance of the center, $1/2\,(E' + E'')$, of the circle from the tensor E.

To prove Theorem 7.5 we start with the observation that for arbitrary tensors E' and E'' holds

$$\frac{1}{2}\,\|E' - E''\| = \frac{1}{2}\,\|E'\|^2 - ((E', E'')) + \frac{1}{2}\,\|E''\|^2 \tag{7.26}$$

Substituting $\Sigma'' = \{u'', E'', S''\}$ and $\Sigma' = \{u', E', S'\}$ into Eq. (7.9), i.e. formulating

$$E(\Sigma'', \Sigma') = \langle A(\Sigma''), \Sigma' \rangle + \int_\Gamma S'' n \cdot u'\, ds$$

we obtain

$$\int_\Omega (E(u'') - E'') \cdot S'\, d\Omega + \int_\Omega C[E''] \cdot E'\, d\Omega + \int_\Omega S'' \cdot \alpha_T TI\, d\Omega$$

$$= \int_\Omega (E(u'') - E'') \cdot S'\, d\Omega + \int_\Omega p \cdot u'\, d\Omega + \int_{\Gamma_1} S'' n \cdot \bar{u}\, ds + \int_{\Gamma_2} \bar{t} \cdot u'\, ds$$

or

$$\int_\Omega C[E''] \cdot E'\, d\Omega = ((E', E'')) = \int_\Omega p \cdot u'\, d\Omega + \int_{\Gamma_1} S'' n \cdot \bar{u}\, ds$$

$$+ \int_{\Gamma_2} \bar{t} \cdot u'\, ds - \int_\Omega S'' \cdot \alpha_T TI\, d\Omega$$

This result substituted into Eq. (7.26) yields

$$\frac{1}{2}\,\|E' - E''\| = \frac{1}{2}\,\|E'\|^2 - \int_\Omega p \cdot u'\, d\Omega - \int_{\Gamma_1} S'' n \cdot \bar{u}\, ds$$

$$- \int_{\Gamma_2} \bar{t} \cdot u'\, ds + \int_\Omega S'' \cdot \alpha_T TI\, d\Omega + \frac{1}{2}\,\|E''\|^2$$

or with

$$\|E'\|^2 = \int_\Omega C[E'] \cdot E'\, d\Omega = \int_\Omega C[E(u')] \cdot E(u')\, d\Omega$$

$$- 2 \int_\Omega C[E(u')] \cdot \alpha_T TI\, d\Omega + \int_\Omega C[\alpha_T TI] \cdot \alpha_T TI\, d\Omega$$

and

$$\|E''\|^2 = \int_\Omega C[E''] \cdot E''\, d\Omega = \int_\Omega C^{-1}[S''] \cdot S''\, d\Omega$$

finally

$$\frac{1}{2}\|E' - E''\|^2 = \Pi_1(u') - \Pi_2(S'') \tag{7.27}$$

The left-hand side can be written as

$$\frac{1}{2}\|E' - E''\|^2 = \frac{1}{2}\|(E' - E) - (E'' - E)\|^2 = \frac{1}{2}\|E' - E\|^2$$
$$- ((E' - E, E'' - E)) + \frac{1}{2}\|E' - E\|^2 = \ldots$$

The term in the middle is zero, $((E' - E, E'' - E)) = 0$, because the strain energy of two tensors $(\Sigma' - \Sigma_s) \in \mathscr{S}_{1,0}$ and $(\Sigma'' - \Sigma_s) \in \mathscr{S}_{2,0}$ from $\mathscr{S}_{1,0}$ and $\mathscr{S}_{2,0}$ is zero, see Eq. (7.24).

Hence, we do not commit a mistake if we change the sign of the term in the middle and continue as follows

$$\ldots = \frac{1}{2}\|E' - E\|^2 + ((E' - E, E'' - E)) + \frac{1}{2}\|E'' - E\|^2$$
$$= \frac{1}{2}\|(E' - E) + (E'' - E)\|^2 = \frac{1}{2}\|E' + E'' - 2E\|^2 = \frac{1}{8}\left\|\frac{E' + E''}{2} - E\right\|^2$$

The proof of Eq. (7.25) is done.

The formulation of the same principles for other elements would be a mere repetition of the foregoing results. Hence, we stop here and leave the formulation of the same results for bars, beams, membranes and Kirchhoff and Reissner plates as an exercise to the reader.

We only want to comment on the abstract properties of the operators A because these operators A appear, also, in nonlinear structural mechanics.

7.4 Sufficient Conditions

The question is: which properties must an operator A have for the energy

$$E(\Sigma, \hat{\Sigma}) = \int_\Omega \left[(E(u) - E) \cdot \hat{S} + C[E] \cdot \hat{E} + S \cdot (E(\hat{u}) - \hat{E}) \right] d\Omega$$

in the first identity

$$G(\Sigma, \hat{\Sigma}) = \langle A(\Sigma), \hat{\Sigma} \rangle + \int_\Gamma r(S, u) \cdot \hat{u} \, ds - E(\Sigma, \hat{\Sigma}) = 0$$

to be the first variation of a functional $F(\Sigma)$? (Recall that this is the essential condition).

The answer is the following: let $E(\ )$, $C[\ ]$, $d(\ )$ and $r(\ )$ denote arbitrary operators then it is sufficient that

(i)
$$\int_\Omega C[E] \cdot \hat{E}\, d\Omega = \frac{d}{d\varepsilon}\, W(E + \varepsilon\hat{E})|_{\varepsilon=0} \qquad (7.28)$$

i.e. there exists a function $W(E)$ whose first variation is the integral on the left-hand side, and it must hold that

(ii)
$$\int_\Omega d(S,u) \cdot \hat{u}\, d\Omega = \int_\Gamma r(S,u) \cdot \hat{u}\, ds - \int_\Omega E_u(\hat{u}) \cdot S\, d\Omega \qquad (7.29)$$

i.e. the equation $d(S,u)$ must be the Euler-equation of the functional

$$P(u) = \int_\Omega E(u) \cdot S\, d\Omega$$

when variation is done with respect to u.

This condition has, in this chapter, been the subject of the auxiliary lemmas, as Eq. (7.8) etc.

7.5 Other Mixed Formulations

A mixed formulation is understood to be a formulation where higher derivatives are introduced as independent unknowns. In this sense the formulations in this chapter are mixed formulations. These formulations are in no way unique.

The beam equation, $EIw^{IV} = p$, can, e.g., be split up as

$$-w'' - M = 0$$
$$-M'' = \frac{p}{EI} \qquad \Sigma = \{w, M\}$$

The first identity of this system is the expression

$$G(\Sigma, \hat{\Sigma}) = \int_0^l [(-w'' - M)\, \hat{M} - M''\, \hat{w}]\, dx + [w'\, \hat{M} + M'\, \hat{w}]_0^l - E(\Sigma, \hat{\Sigma})$$

where

$$E(\Sigma, \hat{\Sigma}) = \int_0^l (w'\, \hat{M}' + M'\, \hat{w}' - M\, \hat{M})\, dx$$

is the first variation of the functional

$$F(\Sigma) = \frac{1}{2} \int_0^l (2w'\, M' - M^2)\, dx$$

Solutions of beam problems are stationary points of this functional.

The advantage this formulation has is that it requires less smooth shape functions. C^0-shape functions are sufficient. The same shape functions would be nonconforming in the case of the system A.

The disadvantage is the lack of mechanical meaning because M is introduced as $M = -w''$ and not as we would expect as $M = -EIw''$.

This is also seen if we split the plate equation $K\Delta\Delta w = p$ into

$$-\Delta w + v = 0$$

$$-\Delta v = \frac{p}{K} \qquad\qquad \Sigma = \{w, v\}$$

The first identity is the expression

$$G(\Sigma, \hat{\Sigma}) = \int_\Omega [(-\Delta w + v)\,\hat{v} - \Delta v\,\hat{w}]\,d\Omega$$

$$+ \int_\Gamma \left[\frac{\partial w}{\partial n}\,\hat{v} + \frac{\partial v}{\partial n}\,\hat{w}\right] ds - E(\Sigma, \hat{\Sigma}) = 0$$

where

$$E(\Sigma, \hat{\Sigma}) = \int_\Omega (\operatorname{grad} w \cdot \operatorname{grad} \hat{v} + \operatorname{grad} \hat{w} \cdot \operatorname{grad} v + v\hat{v})\,d\Omega$$

is the first variation of

$$F(\Sigma) = \frac{1}{2} \int_\Omega (2\operatorname{grad} w \cdot \operatorname{grad} v + v^2)\,d\Omega$$

Only in the case of a clamped plate

$$w = 0 \qquad \frac{\partial w}{\partial n} = 0 \quad \text{on } \Gamma$$

is the application of this identity trivial.

7.6 The Babuŝka-Brezzi Condition

Consider the *bvp* of a prestressed membrane

$$-N\,\Delta w = p \quad w = 0 \quad \text{on } \Gamma \quad (N = 1)$$

and assume we decompose it as follows

$$\nabla w - \boldsymbol{v} = \boldsymbol{0}_{(2)} \qquad\qquad w = 0 \quad \text{on } \Gamma$$

$$-\operatorname{div} \boldsymbol{v} = p_{(1)} \qquad\qquad \Sigma = \{w, \boldsymbol{v}\} \tag{7.30}$$

Recall that the derivatives $Nw,_1$ and $Nw,_2$ are the stresses in a membrane, see section 1.9.5. Hence, the vector $v = \{v_1, v_2\}$ can be considered the stress vector of the membrane.

To the system (7.30) belongs the identity

$$G(\Sigma, \hat{\Sigma}) = \int_\Omega [(\nabla w - v) \cdot \hat{v} - \operatorname{div} v \hat{w}] \, d\Omega - \int_\Gamma w \hat{v} \cdot n \, ds$$
$$+ \int_\Omega (\operatorname{div} v \hat{w} + \operatorname{div} \hat{v} w + v \cdot \hat{v}) \, d\Omega \tag{7.31}$$

and to the bvp, therefore, the potential energy functional

$$\Pi_1(\Sigma) = \frac{1}{2} \int_\Omega (2 \operatorname{div} v w + v \cdot v) \, d\Omega + \int_\Omega p w \, d\Omega$$

Assume we approximate the deflection and the stresses with piecewise polynomials φ_i, $\boldsymbol{\Psi}_j = \{\Psi_{1j}, \Psi_{2j}\}$ which are globally in $C^0(\bar{\Omega})$

$$w = \sum_{i=1}^{d} \delta_i \varphi_i \qquad \varphi_i = 0 \quad \text{on } \Gamma, \qquad v = \sum_{j=1}^{s} \sigma_j \boldsymbol{\Psi}_j$$

The conditions

$$\frac{\partial \Pi_1}{\partial \delta_i} = \int_\Omega \operatorname{div} v \varphi_i \, d\Omega + \int_\Omega p \varphi_i \, d\Omega = 0 \qquad i = 1, 2 \ldots d$$

$$\frac{\partial \Pi_1}{\partial \sigma_j} = \int_\Omega (\operatorname{div} \boldsymbol{\Psi}_j w + v \cdot \boldsymbol{\Psi}_j) \, d\Omega = 0 \qquad j = 1, 2 \ldots s$$

render the system

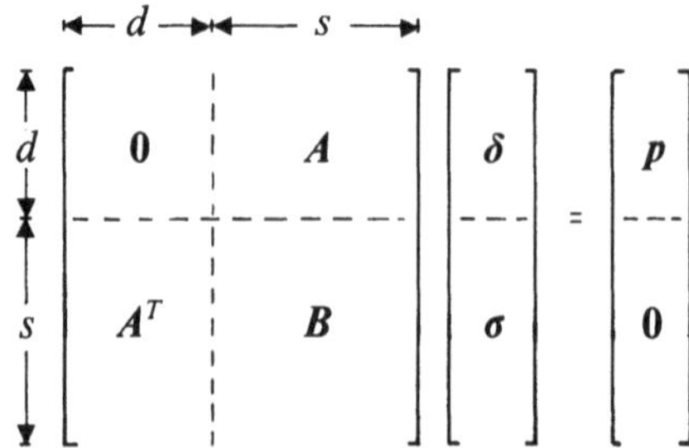

where

$$a_{ij} = \int_\Omega \operatorname{div} \boldsymbol{\Psi}_j \varphi_i \, d\Omega \qquad b_{ij} = \int_\Omega \boldsymbol{\Psi}_i \cdot \boldsymbol{\Psi}_j \, d\Omega \qquad p_i = \int_\Omega -p \varphi_i \, d\Omega$$

(Note that due to $\varphi_i = 0$ on Γ we have, see Eq. (1.12),

$$a_{ij} = \int_\Omega \operatorname{div} \boldsymbol{\Psi}_j \varphi_i \, d\Omega = - \int_\Omega \boldsymbol{\Psi}_j \cdot \nabla \varphi_i \, d\Omega$$

A formulation preferred by some authors).

With regard to d and s, the degrees of freedom of the deflection w and the stress vector v, we have three options

$$d < s \quad d = s \quad d > s$$

If we opt for $d > s$ then the system

$$\begin{bmatrix} & \\ & A & \\ & \end{bmatrix} \begin{bmatrix} \sigma \end{bmatrix} = \begin{bmatrix} p \end{bmatrix}$$

is overdetermined and if we let $d = s$ then A is quadratic

$$\begin{bmatrix} & A & \end{bmatrix} \begin{bmatrix} \sigma \end{bmatrix} = \begin{bmatrix} p \end{bmatrix}$$

which means that the stresses are uniquely determined (if A is regular) by the external forces alone. This, too, does not make sense.

Hence, the only reasonable choice is $d < s$.

$$\begin{bmatrix} & A & \end{bmatrix} \begin{bmatrix} \sigma \end{bmatrix} = \begin{bmatrix} p \end{bmatrix} \tag{7.32}$$

This agreed upon the next question is: under which conditions does this underdetermined system (7.32) admit a solution σ?

The answer is that p must be orthogonal to the nullspace of the adjoint operator A^T. That is we must have

$$p^T \delta = 0 \tag{7.33}$$

with respect to all solutions δ of

$$A^T \delta = 0 \tag{7.34}$$

The best what could happen is, that (7.34) has only the trivial solution, $\delta = 0$. Because then condition (7.33) is always satisfied, then a solution σ of (7.32) always exists.

Now, for $\delta = 0$ to be the only solution of (7.34) the columns of A^T must be linearly independent and this property is, as we will explain, equivalent with the existence of a constant $\beta > 0$ such that

$$0 < \beta|\delta| < \inf_{\delta \in R^d} \sup_{\sigma \in R^s} \frac{\sigma^T A^T \delta}{|\sigma|} \tag{7.35}$$

Assume, first, that though the columns of A^T are linearly independent there exists a vector δ with the property that

$$\sup_{\sigma \in R^s} \frac{\sigma^T A^T \delta}{|\sigma|} = 0$$

(in which case $\beta = 0$)

Choosing for σ consecutively the unit vectors $e_1, e_2, \ldots e_s$ we obtain

$$\sum_{j=1}^{d} a_{ji}\delta_j = 0 \quad i = 1, 2 \ldots s$$

which, because not all δ_j are zero, contradicts the claim that the columns of A^T are linearly independent. Hence, no such δ can exist. Instead this negative result implies (as could be demonstrated) that a constant $\beta > 0$ must exist.

Conversely assume that the inequality (7.35) holds. Then the problem:

$$A^T \delta = 0 \quad \delta \neq 0$$

has no solution, that is, the columns of A^T are linearly independent.

The inequality (7.35) is the discrete analogue of the Babuška-Brezzi condition which, for the mixed problem (7.30) to have a solution in $W \times V$, requires that there exists a constant $\beta > 0$ such that

$$0 < \beta\|w\|_W = \inf_{w \in W} \sup_{v \in V} \frac{\int w \operatorname{div} v \, d\Omega}{\|v\|_V}$$

(W and V are appropriate subspaces of Sobolev spaces, see [C 5] p. 415).

This Babuška-Brezzi conditions plays an important role in the theory of Lagrange multipliers and mixed methods.

In the case of pure displacement formulations the only condition for a solution to exist is the coerciveness

$$E(u, u) > c\|u\|_U^2 \quad U = \text{``solution space''}$$

This property entails that the stiffness matrix is positive definite and, therefore, also the finite element equation

$$K\delta = f$$

uniquely solvable.

In the case of the mixed system (7.30) the coerciveness

$$\int_\Omega \boldsymbol{v} \cdot \boldsymbol{v} \, d\Omega = \|\boldsymbol{v}\|_{H^0}^2$$

alone is not sufficient. In addition the Babuška-Brezzi condition must hold on $W \times V$, see [B 1] and [B 4].

The discrete analogues of these two conditions (coerciveness and B.-B. cond.) are the conditions

$$\boldsymbol{\delta}^T \boldsymbol{B} \boldsymbol{\delta} > 0 \quad \boldsymbol{\delta} \neq \boldsymbol{0} \quad \boldsymbol{A}^T \boldsymbol{\delta} = \boldsymbol{0} \rightarrow \boldsymbol{\delta} = \boldsymbol{0}$$

which, if satisfied, guarantee that the discrete problem has a solution, namely

$$\boldsymbol{\delta} = -(\boldsymbol{A} \boldsymbol{B}^{-1} \boldsymbol{A}^T)^{-1} \boldsymbol{p}, \quad \boldsymbol{\sigma} = -\boldsymbol{B}^{-1} \boldsymbol{A}^T \boldsymbol{\delta}$$

Consider a simple example, the bar in Fig. 7.4.

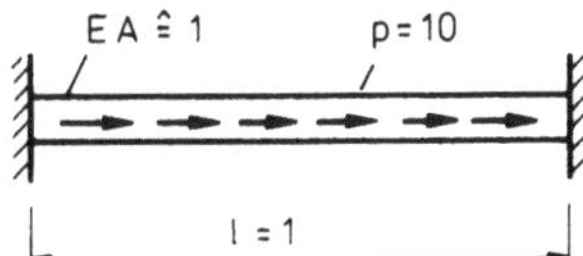

Figure 7.4

Here, the Hellinger-Reissner functional is the expression

$$\Pi_0(\boldsymbol{\Sigma}) = -\frac{1}{2} \int_0^1 \frac{N^2}{EA} \, dx + \int_0^1 u' N \, dx - \int_0^1 p u \, dx - [Nu]_0^1$$

(The last term can be dropped if all trial functions satisfy the geometric boundary conditions $u(0) = u(1) = 0$)

We first let, see Fig. 7.5a

$$u_h = \delta \varphi \quad N_h = \sigma \psi$$

This renders the system

$$\begin{bmatrix} 0 & 0 \\ \hline 0 & 1 \end{bmatrix} \begin{bmatrix} \delta \\ \sigma \end{bmatrix} = \begin{bmatrix} 5 \\ 0 \end{bmatrix}$$

which has no solution because the Babuška-Brezzi condition is violated, the matrix $\boldsymbol{A}^T$—which actually consists of just one element, $a_{11} = 0$—is the zero matrix.

If we keep the one d. o. f. approximation for u but allow two d. o. f. for N, see Fig. 7.5b

$$u_h = \delta\varphi \qquad N_h = \sigma_1\psi_1 + \sigma_2\psi_2$$

then this results in

$$\begin{bmatrix} 0 & 1 & -1 \\ 1 & -0.5 & 0 \\ -1 & 0 & -0.5 \end{bmatrix} \begin{bmatrix} \delta \\ \sigma_1 \\ \sigma_2 \end{bmatrix} = \begin{bmatrix} 5 \\ 0 \\ 0 \end{bmatrix}$$

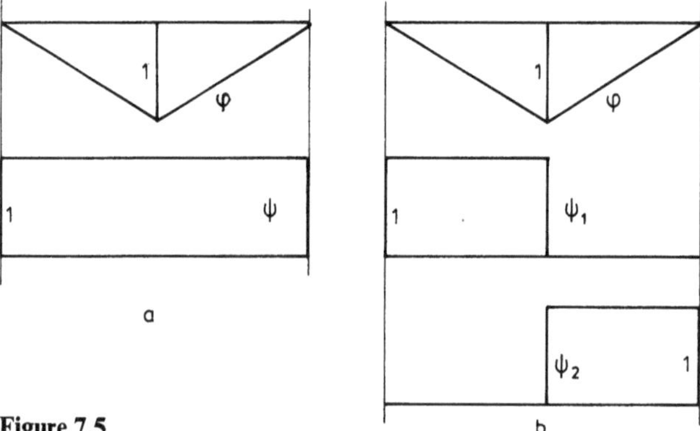

Figure 7.5

As A^T consists of only one column ($\neq \mathbf{0}$) the Babuška-Brezzi condition is satisfied

$$A^T\delta = \mathbf{0} \rightarrow \delta = 0$$

and because, in addition, $-\mathbf{B}$ is positive definite the discrete problem has a solution, the one in Fig. 7.6.

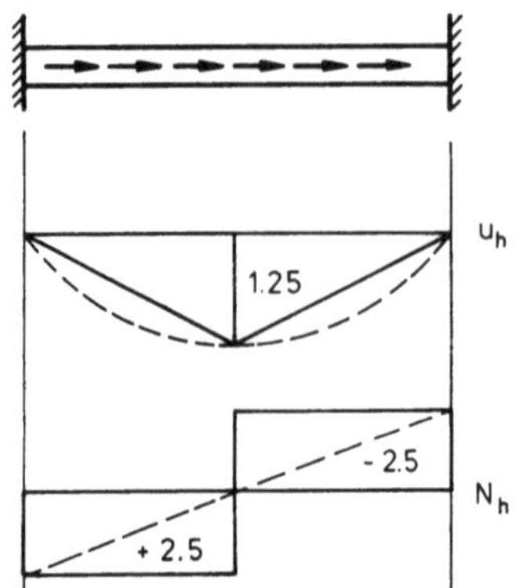

Figure 7.6

In the case of elastic plates the Hellinger-Reissner functional is the expression

$$\Pi_0(\Sigma) = -\frac{1}{2} \int_\Omega C^{-1}[S] \cdot S \, d\Omega + \int_\Omega E(u) \cdot S \, d\Omega - \int_\Omega p \cdot u \, d\Omega - \int_{\Gamma_2} \bar{t} \cdot u \, ds$$
$$+ \int_{\Gamma_1} (\bar{u} - u) \cdot Sn \, ds$$

If the trial functions satisfy $u_h = \bar{u}$ on Γ_1 then the last integral, the one over Γ_1, can be neglected and the Babuška-Brezzi condition reads

$$0 < \beta \|u\| < \inf_u \sup_S \frac{\int E(u) \cdot S \, d\Omega}{\|S\|}$$

This condition, if satisfied, guarantees that the class of all approximating strain tensors, $E(\varphi_i)$, and the class of all approximating stress tensors, S_j, match, that is the latter set contains all the strains derived from the displacement approximations.

"The mixed methods remain a delicate and intriguing class of finite element methods for linear elliptic problems", [C&O] p. 134. The numerical analyst working with mixed methods faces a number of strange effects such as "spurious modes", "zero energy modes", "locking" etc. Topics which cannot be covered in the context of this book. For a further discussion of mixed methods and their applications we must refer to the finite element literature, [C&O], [M&O], [T1], [T&W] and [W6].

8 Shells

In this chapter we will extend our approach to shells. As there are many, many different formulations for shells we had to decide for one particular model. We opted for Koiter's model because the mathematical properties of this model are fully worked out. But as anyone who is familiar with shells will recognize all what is said in the following applies to different models (nearly) as well. The reader will, certainly, also realize that the mathematics of shells closely fits into the general picture.

8.1 Shells as Surfaces

A shell—or to be more exact—its middle surface is a manifold embedded into Euclidian space. To do calculations on this manifold we construct a mapping between the points $\boldsymbol{\Theta} = \{\theta^1, \theta^2\}$ of a plane domain Ω and the points of the shell surface.

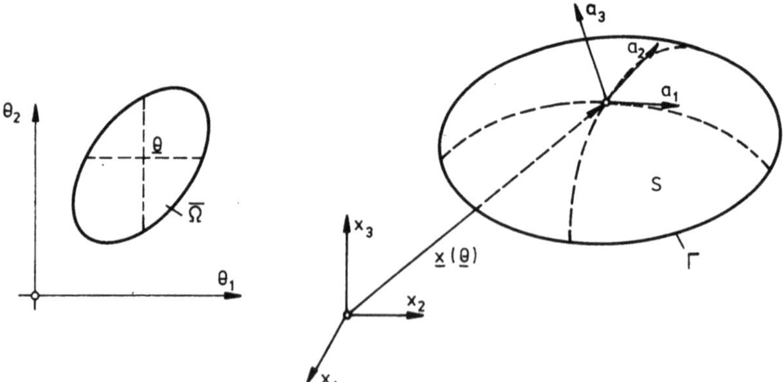

Figure 8.1

This mapping is the position vector, see Fig. 8.1,

$$x(\boldsymbol{\Theta}) = x(\theta^1, \theta^2) = \{x_1(\theta^1, \theta^2),\, x_2(\theta^1, \theta^2),\, x_3(\theta^1, \theta^2)\}^T$$

We assume Ω to be a regular domain, i.e. integration by parts is applicable to domain integrals and we assume, furthermore, that the position vector $x(\boldsymbol{\Theta})$ belongs to $C^3(\bar{\Omega})$ and that at every point $\boldsymbol{\Theta}$ of Ω the two vectors

$$a_1 = \frac{\partial x}{\partial \theta^1}, \qquad a_2 = \frac{\partial x}{\partial \theta^2}$$

are linearly independent. These vectors constitute together with the vector

$$a_3 = \frac{a_1 \times a_2}{|a_1 \times a_2|}$$

at every point $x(\Theta)$ of the surface the base vectors of a curvilinear system of coordinates.

If we employ in the following the summation convention

$$t^a g_\alpha = \sum_{\alpha=1}^{2} t^\alpha g_\alpha, \qquad t^i g_i = \sum_{i=1}^{3} t^i g_i$$

then the range of Greek indices is $\{1, 2\}$ and of Latin indices $\{1, 2, 3\}$.

The tensor

$$a_{ik} = a_{ki} = a_i \cdot a_k$$

is *the first fundamental form of the surface*. Its determinant

$$a = a_{11} a_{22} - (a_{12})^2 = |a_1 \times a_2|^2$$

has, on account of the linear independence of the base vectors a_1 and a_2, no zero in Ω.

With the covariant base vectors a_i we associate contravariant base vectors defined through

$$a_\alpha \cdot a^\beta = \delta_\alpha^\beta = \begin{cases} 1, \ \alpha = \beta \\ 0, \ \alpha \neq \beta \end{cases} \qquad a^3 = a_3 \tag{8.1}$$

Their metric tensor is

$$a^{ik} = a^{ki} = a^i \cdot a^k$$

Between the single base vectors exists the relation

$$a_\alpha = a_{\alpha\beta} a^\beta, \qquad a^\alpha = a^{\alpha\beta} a_\beta$$

so that their vector products express as

$$a_\alpha \times a_\beta = \varepsilon_{\alpha\beta} a_3, \qquad a^\alpha \times a^\beta = \varepsilon^{\alpha\beta} a_3,$$

$$a_3 \times a_\beta = \varepsilon_{\beta\lambda} a^\lambda, \qquad a_3 \times a^\beta = \varepsilon^{\beta\lambda} a_\lambda$$

where

$$\varepsilon_{\alpha\beta} = a^{1/2} e_{\alpha\beta}, \qquad \varepsilon^{\alpha\beta} = a^{-1/2} e^{\alpha\beta}, \qquad e^{\alpha\beta} = e_{\alpha\beta} \,\hat{=}\, \begin{bmatrix} 0 & 1 \\ -1 & 0 \end{bmatrix}$$

The second fundamental form of the surface is the tensor

$$b_{\alpha\beta} = b_{\beta\alpha} = -\boldsymbol{a}_\alpha \cdot \boldsymbol{a}_{3,\beta} = \boldsymbol{a}_3 \cdot \boldsymbol{a}_{\alpha,\beta} = \boldsymbol{a}_3 \cdot \boldsymbol{a}_{\beta,\alpha}$$

where

$$\boldsymbol{a}_{\alpha,\beta} = \frac{\partial \boldsymbol{a}_\alpha}{\partial \theta^\beta}, \qquad \boldsymbol{a}_{3,\beta} = \frac{\partial \boldsymbol{a}_3}{\partial \theta^\beta}$$

With the *Christoffel symbols*

$$\Gamma^\alpha_{\beta\gamma} = \Gamma^\alpha_{\gamma\beta} = \frac{1}{2} a^{\alpha\lambda}(a_{\lambda\beta,\gamma} + a_{\gamma\lambda,\beta} - a_{\beta\gamma,\lambda}) \qquad a_{\lambda\beta,\gamma} = \frac{\partial a_{\lambda\beta}}{\partial \theta^\gamma}$$

covariant derivatives of surface tensors T_α, $T_{\alpha\beta}$,

$$T_\alpha|_\gamma = T_{\alpha,\gamma} - \Gamma^\lambda_{\alpha\gamma} T_\lambda$$

$$T_{\alpha\beta}|_\gamma = T_{\alpha\beta,\gamma} - \Gamma^\lambda_{\alpha\gamma} T_{\lambda\beta} - \Gamma^\lambda_{\beta\gamma} T_{\lambda\alpha}$$

are defined.

Let $\boldsymbol{\Theta}(t) = \{\theta^1(t), \theta^2(t)\}$ be a curve in the parameter domain Ω then $\boldsymbol{x}(\boldsymbol{\Theta}(t))$ is the image of this curve on the surface S. Its line element is

$$ds = [a_{\alpha\beta}\, \dot\theta^\alpha\, \dot\theta^\beta]^{1/2}\, dt \qquad \dot\theta^\alpha = \frac{d\theta^\alpha}{dt}$$

If the parameter t is the arc-length s then the equation of the tangent-vector becomes

$$\boldsymbol{\tau} = \frac{d\boldsymbol{x}(\boldsymbol{\Theta}(s))}{ds} = \frac{d\theta^\alpha}{ds}\frac{\partial \boldsymbol{x}}{\partial \theta^\alpha} = \tau^\alpha \boldsymbol{a}_\alpha$$

The normal $\boldsymbol{v}$ at the curve is the vector, see Fig. 8.2,

$$\boldsymbol{v} = \boldsymbol{\tau} \times \boldsymbol{a}_3 = \varepsilon_{\alpha\beta}\, \tau^\beta\, \boldsymbol{a}^\alpha = v_\alpha\, \boldsymbol{a}^\alpha \tag{8.2}$$

and the surface element dS on S is

$$dS = |\boldsymbol{a}_1 \times \boldsymbol{a}_2|\, d\theta^1\, d\theta^2 = a^{1/2}\, d\theta^1\, d\theta^2$$

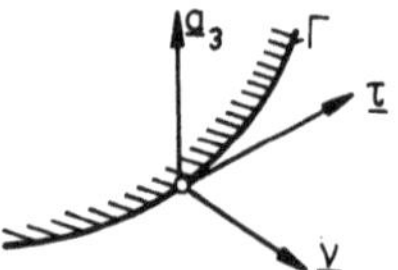

Figure 8.2

The integration by parts rule employed in the case of shells is the following:

Theorem 8.1

Let $u(\Theta) = \{u_1, u_2\} \in C^1(\bar{\Omega})$ a vector field in the surface S, $x(\Theta) \in C^3(\bar{\Omega})$, then holds

$$\int_S \text{div}_S u \, dS = \int_\Gamma u \cdot v \, ds \tag{8.3}$$

where

$$\text{div}_S u := u^\alpha|_\alpha = u^1|_1 + u^2|_2$$

and $v = \{v_1, v_2\}$ is the normal at the edge Γ.

8.2 Statics

Let $x(\Theta)$ the original and $\tilde{x}(\Theta)$ the displaced position of a point on the surface then

$$u(\Theta) = \tilde{x}(\Theta) - x(\Theta)$$

is the displacement vector of the point $x(\Theta)$. We measure its components with respect to the base vectors of the undeformed shell surface.

$$u = u_i \, a^i$$

The shell problem is solved if the displacement field $u(\Theta)$ of the shell surface is determined.

The position vector of the deformed surface is

$$\tilde{x}(\Theta) = x(\Theta) + u(\Theta)$$

Hence, the base vectors of the deformed surface are, see [B&C] (2.3−1)

$$\tilde{a}_\alpha = a_\alpha + \gamma_\alpha^\beta \, a_\beta + \varphi \times a_\alpha, \qquad \tilde{a}_3 = a_3 + \varphi \times a_3 \tag{8.4}$$

where

$$\gamma_\alpha^\beta = a^{\beta\lambda} \gamma_{\lambda\alpha}, \qquad \varphi = \varepsilon^{\lambda\beta} \, \Phi_\beta \, a_\lambda + \chi \, a_3,$$

$$\Phi_\beta = u_3|_\beta + b_\beta^\alpha \, u_\alpha, \qquad \chi = \frac{1}{2} \, \varepsilon^{\alpha\beta} \, u_\beta|_\alpha \tag{8.5}$$

The coordinate lines on the shell surface are easily extendable to a three-dimensional curvilinear system of coordinates θ^1, θ^2, θ^3 if we let the third coordinate, θ^3, measure the orthogonal distance to the surface.

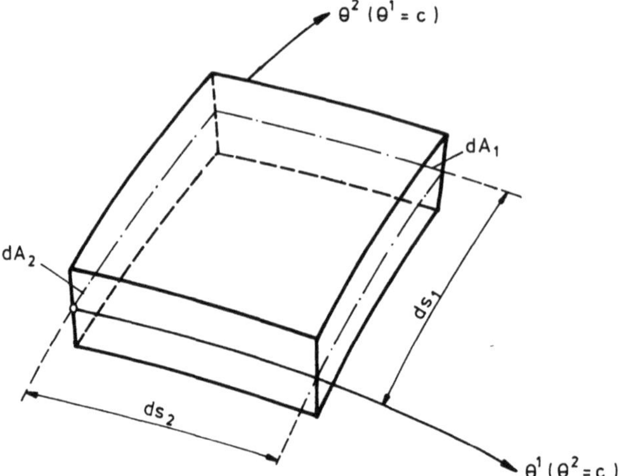

Figure 8.3

Assume that the stress tensor of the three-dimensional shell continuum refers to this curvilinear system, $\sigma_{ik} = \sigma_{ik}(\theta^1, \theta^2, \theta^3)$ and let us consider the infinitesimal parallelepiped in Fig. 8.3.

On the face dA_α, $(\theta^\alpha = c.)$ acts the traction vector

$$\boldsymbol{t}^\alpha = \sigma^{ak}(g^{(\alpha\alpha)})^{-1/2}\,\boldsymbol{g}_k$$

$((\ )$ indicates no sum on $\alpha)$
and hence, the force

$$\boldsymbol{t}^\alpha \, dA_\alpha = \boldsymbol{t}^\alpha (a^{(\alpha\alpha)}\,a)^{1/2}\,d\theta^\beta\,d\theta^3 =: \boldsymbol{T}^\alpha\,d\theta^\beta\,d\theta^3, \qquad \alpha \neq \beta$$

Integrating this force and its moment with respect to the shell thickness h and dividing by the line element

$$ds_\alpha = (a_{(\beta\beta)})^{1/2}\,d\theta^\beta = (aa^{(\alpha\alpha)})^{1/2}\,d\theta^\beta$$

of the coordinate line $\theta^\alpha = c$ we obtain stress-resultants

$$\boldsymbol{n}^\alpha = \int_{-h/2}^{h/2} \boldsymbol{T}^\alpha\,d\theta^3\,d\theta^\beta (ds_\alpha)^{-1} = n^{\alpha\lambda}\,\boldsymbol{a}_\lambda + q^\alpha\,\boldsymbol{a}_3\,(a^{(\alpha\alpha)})^{-1/2}$$

$$\boldsymbol{m}^\alpha = \int_{-h/2}^{h/2} (\theta^3\,\boldsymbol{a}_3 \times \boldsymbol{T}^\alpha)\,d\theta^3\,d\theta^\beta\,(ds_\alpha)^{-1} = m^{\alpha\beta}\,\varepsilon_{\beta\lambda}\,\boldsymbol{a}^\lambda\,(a^{(\alpha\alpha)})^{-1/2}$$

of dimension [force/length] or [moment/length], respectively.

The 10 functions $n^{\alpha\beta}$, $m^{\alpha\beta}$ and q^α depend on the components σ_{ik} of the stress tensor. We shall later make appropriate assumptions about these functions, see section 8.3.

Next, consider a curvilinear triangle on a surface, see Fig. 8.4, whose sides coincide with two parameter lines $\theta^1 = c_1$, $\theta^2 = c_2$ and a curve c. If the triangle shrinks till it is mere point P on c then the equilibrium conditions require that the internal actions satisfy at P the equations (see [N1] p. 494 (9.11) and p. 495 (9.19))

$$n = n^\alpha \, v_\alpha \, (a^{(\alpha\alpha)})^{1/2} = n^\lambda \, a_\lambda + n^3 \, a_3$$

$$m = m^\alpha \, v_\alpha \, (a^{(\alpha\alpha)})^{1/2} = \varepsilon_{\beta\lambda} \, m^\beta \, a_\lambda$$

where the components

$$n^\lambda = n^{\alpha\lambda} \, v_\alpha, \qquad n^3 = q^\alpha \, v_\alpha, \qquad m^\beta = m^{\alpha\beta} \, v_\alpha \tag{8.6}$$

depend on the normal $v = \{v_1, v_2\}$ on the curve c at P.

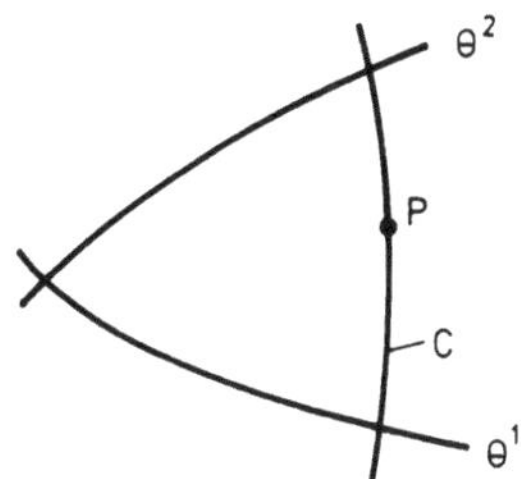

Figure 8.4

The couple vector m can be split into two vectors, one parallel to the normal v and one parallel to the tangent τ

$$m = m_v \, v + m_\tau \, \tau$$

The single components are

$$m_v = m \cdot v = \varepsilon_{\beta\lambda} \, m^\beta \, a^\lambda \, v_\gamma \, a^\gamma = \varepsilon_{\beta\lambda} \, m^\beta \, v^\lambda = - \varepsilon_{\lambda\beta} \, m^\beta \, v^\lambda$$

$$m_\tau = m \cdot \tau = \varepsilon_{\beta\lambda} \, m^\beta \, a^\lambda \, \tau_\gamma \, a^\gamma = \varepsilon_{\beta\lambda} \, m^\beta \, \tau^\lambda = v_\beta \, m^\beta$$

or with $m^\beta = m^{\beta\gamma} \, v_\gamma$

$$m_v = - \varepsilon_{\lambda\beta} \, m^{\beta\gamma} \, v_\gamma \, v^\lambda, \qquad m_\tau = v_\beta \, m^{\beta\gamma} \, v_\gamma \tag{8.7}$$

Assume the shell is loaded with continuous distributed surface forces

$$p = p^\alpha \, a_\alpha + p^3 \, a_3$$

then the force which acts on a surface element dS, see Fig. 8.5, is

$$p \, dS = (p^\alpha \, a_\alpha + p^3 \, a) \, a^{1/2} \, d\theta^1 \, d\theta^2 =: P \, d\theta^1 \, d\theta^2$$

Multiplying the stress-resultants $\boldsymbol{n}^\alpha$, $\boldsymbol{m}^\alpha$ acting on the coordinate lines $\theta^\alpha = c$ with ds_α we obtain resulting forces $\boldsymbol{N}^\alpha\, d\theta^\beta$ and $\boldsymbol{M}^\alpha\, d\theta^\beta$

$$\boldsymbol{n}^{(\alpha)}\, ds_{(\alpha)} = (n^{\alpha\lambda}\, \boldsymbol{a}_\lambda + q^\alpha\, \boldsymbol{a}_3)\, a^{1/2}\, d\theta^\beta = \boldsymbol{N}^\alpha\, d\theta^\beta, \qquad \alpha \neq \beta$$

$$\boldsymbol{m}^{(\alpha)}\, ds_{(\alpha)} = (m^{\alpha\beta}\, \varepsilon_{\beta\lambda}\, \boldsymbol{a}^\lambda)\, a^{1/2}\, d\theta^\beta = \boldsymbol{M}^\alpha\, d\theta^\beta, \qquad \alpha \neq \beta$$

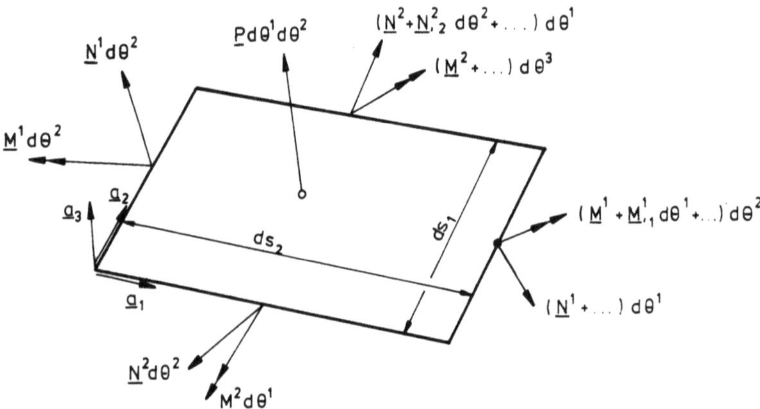

Figure 8.5

The equilibrium conditions require that, in the limit, $d\theta^1, d\theta^2 \to 0$, the two equations

$$\boldsymbol{N}^\alpha,_\alpha + \boldsymbol{P} = 0 \qquad \boldsymbol{M}^\alpha,_\alpha + \boldsymbol{a}_\alpha \times \boldsymbol{N}_\alpha = 0$$

or in components

$$n^{\alpha\beta}|_\alpha - b^\beta_\alpha\, q^\alpha + p^\beta = 0 \qquad \beta = 1, 2 \tag{$8.8_{1,2}$}$$

$$n^{\alpha\beta}\, b_{\alpha\beta} + q^\alpha|_\alpha + p^3 = 0 \tag{8.8_3}$$

$$m^{\alpha\beta}|_\alpha - q^\beta = 0 \qquad \beta = 1, 2 \tag{$8.8_{4,5}$}$$

$$\varepsilon_{\lambda\beta}(n^{\lambda\beta} - m^{\alpha\beta}\, b^\lambda_\alpha) = 0 \tag{8.8_6}$$

are satisfied.

If we eleminate from this system with the help of $(8.8_{4,5})$ the shear forces q^α then the equations reduce to a system of 4 equations for eight unknowns $n^{\alpha\beta}$, $m^{\alpha\beta}$.

8.3 Koiter's Model

To reduce the number of unknowns further we introduce new symmetric tensors

$$\bar{n}^{\alpha\beta} = n^{\alpha\beta} + b^{\beta}_{\gamma}\, m^{\gamma\alpha} \tag{8.9}$$

$$\bar{m}^{\alpha\beta} = \frac{1}{2}\,(m^{\alpha\beta} + m^{\beta\alpha}) \tag{8.10}$$

(the symmetry of $\bar{n}^{\alpha\beta}$ is a consequence of Eq. (8.8_6)).

These six functions satisfy because of Eq. (8.8) (for a proof see [N2] p. 47 and [K1] p. 175 (3.11/12)) the three equations

$$-(\bar{n}^{\alpha\beta} - b^{\beta}_{\lambda}\,\bar{m}^{\lambda\alpha})|_{\alpha} + b^{\beta}_{\alpha}\,\bar{m}^{\lambda\alpha}|_{\lambda} = p^{\beta}, \qquad \beta = 1,2 \tag{8.11_{1,2}}$$

$$-b_{\alpha\beta}(\bar{n}^{\alpha\beta} - b^{\beta}_{\lambda}\,\bar{m}^{\lambda\alpha}) - \bar{m}^{\alpha\beta}|_{\alpha\beta} = p^{3} \tag{8.11_3}$$

Koiter assumes the following relation to exist between the tensors $\bar{n}^{\alpha\beta}$, $\bar{m}^{\alpha\beta}$ and the displacement vector $\boldsymbol{u}$ (see [B&C] p. 21 (2.4–1/2))

$$\bar{n}^{\alpha\beta} = h\, C^{\alpha\beta\lambda\delta}\, \gamma_{\lambda\delta} \qquad \bar{m}^{\alpha\beta} = \frac{h^3}{12}\, C^{\alpha\beta\lambda\mu}\, \rho_{\lambda\mu}$$

where

$$C^{\alpha\beta\lambda\delta} = C^{\lambda\delta\alpha\beta} = \mu\left[a^{\alpha\lambda}\,a^{\beta\delta} + a^{\alpha\delta}\,a^{\beta\lambda} + \frac{2\nu}{1-\nu}\,a^{\alpha\beta}\,a^{\lambda\delta}\right] \tag{8.12}$$

is the elasticity tensor of the shell;

$$\gamma_{\alpha\beta} = \gamma_{\beta\alpha} = \frac{1}{2}\,(\tilde{a}_{\alpha\beta} - a_{\alpha\beta}) = \frac{1}{2}\,(u_{\alpha}|_{\beta} + u_{\beta}|_{\alpha}) - b_{\alpha\beta}\,u_3$$

is the strain tensor and

$$\rho_{\alpha\beta} = \rho_{\beta\alpha} = -(\tilde{b}_{\alpha\beta} - b_{\alpha\beta}) = -\{u_3|_{\alpha\beta} - b^{\lambda}_{\alpha}\,b_{\lambda\beta}\,u_3 + b^{\lambda}_{\alpha}\,u_{\lambda}|_{\beta}$$
$$+ b^{\lambda}_{\beta}\,u_{\lambda}|_{\alpha} + b^{\lambda}_{\beta}|_{\alpha}\,u_{\lambda}\}$$

the change of curvature tensor.

On substituting the so defined $\bar{n}^{\alpha\beta}$ and $\bar{m}^{\alpha\beta}$ into Eq. (8.11) we obtain a system of three partial differential equations for the three components of the displacement $\boldsymbol{u}(\boldsymbol{\Theta})$ which we denote by

$$-\boldsymbol{D}\boldsymbol{u} = \boldsymbol{p}$$

8.4 The first Identity

We multiply the system (8.11) of the three equations with a displacement field $\hat{u}$ (scalar product) and integrate the product over the shell surface and, thus, obtain the result

$$\int_S - \boldsymbol{D}\boldsymbol{u} \cdot \hat{\boldsymbol{u}}\, dS = \int_\Gamma \left[(-\{\bar{n}^{\alpha\beta} - b^\beta_\lambda \bar{m}^{\lambda\alpha}\}|_\alpha + b^\beta_\alpha \bar{m}^{\lambda\alpha}|_\lambda)\, \hat{u}_\beta \right.$$

$$+ \left. (-b_{\alpha\beta}(\bar{n}^{\alpha\beta} - b^\beta_\lambda \bar{m}^{\lambda\alpha}) - \bar{m}^{\alpha\beta}|_{\alpha\beta})\, \hat{u}_3 \right]\, dS$$

Integration by parts, see Theorem 8.1, applied to this integral renders the first identity.

First identity

$$p: \boldsymbol{u}, \hat{\boldsymbol{u}} \in C^4(\bar{\Omega}) \times C^2(\bar{\Omega}),\ \boldsymbol{x}(\boldsymbol{\Theta}) \in C^3(\bar{\Omega})$$

$$q: G(\boldsymbol{u}, \hat{\boldsymbol{u}}) = \int_S - \boldsymbol{D}\boldsymbol{u} \cdot \hat{\boldsymbol{u}}\, dS + \int_\Gamma [\bar{n}^{\alpha\beta} v_\beta - 2\bar{m}^{\times\beta} b^\alpha_\times v_\beta]\, \hat{u}_\alpha\, ds$$

$$+ \int_\Gamma [-\bar{m}^{\alpha\beta}\, \hat{u}_3|_\alpha\, v_\beta + \bar{m}^{\alpha\beta}|_\alpha\, \hat{u}_3\, v_\beta]\, ds - E(\boldsymbol{u}, \hat{\boldsymbol{u}}) = 0 \qquad (8.13)$$

Here $E(\boldsymbol{u}, \hat{\boldsymbol{u}})$ denotes the strain energy

$$E(\boldsymbol{u}, \hat{\boldsymbol{u}}) = \int_S [\bar{n}^{\alpha\beta}(\boldsymbol{u})\, \gamma_{\alpha\beta}(\hat{\boldsymbol{u}}) + \bar{m}^{\alpha\beta}(\boldsymbol{u})\, \rho_{\alpha\beta}(\hat{\boldsymbol{u}})]\, dS$$

$$= 2\mu h \int_S \left\{ \left[\gamma^\alpha_\beta \hat{\gamma}^\alpha_\beta + \frac{v}{1-v}\, \gamma^\alpha_\alpha \hat{\gamma}^\beta_\beta \right] + \frac{h^2}{12} \left[\rho^\alpha_\beta \hat{\rho}^\alpha_\beta + \frac{v}{1-v}\, \rho^\alpha_\alpha \hat{\rho}^\beta_\beta \right] \right\}\, dS$$

which is a symmetric bilinear form.

With the help of Eqs. (8.6_1) and (8.6_3) the first boundary integral, I_1, in Eq. (8.13) can be written

$$I_1 = \int_\Gamma (\bar{n}^\alpha - 2\bar{m}^\times b^\alpha_\times)\, \hat{u}_\alpha\, ds$$

Furthermore, if s is the arc-length on the edge then (see [N1] p. 552 (15.22))

$$\hat{u}_3|_\alpha = \hat{u}_{3,\alpha} = v_\alpha \frac{\partial \hat{u}_3}{\partial v} - \varepsilon_{\alpha\lambda}\, v^\lambda \frac{d\hat{u}_3}{ds}$$

Substituting this into I_2, the second boundary integral, we obtain with the help of Eqs. $(8.7_{1,2})$ and (8.6_2)

$$I_2 = \int_\Gamma \left[-\bar{m}^{\alpha\beta} \left(v_\alpha \frac{\partial \hat{u}_3}{\partial v} - \varepsilon_{\alpha\lambda} v^\lambda \frac{d\hat{u}_3}{ds} \right) + \bar{m}^{\alpha\beta}|_\alpha \hat{u}_3 \right] v_\beta \, ds$$

$$= \int_\Gamma \left[-\bar{m}^{\alpha\beta} v_\alpha v_\beta \frac{\partial \hat{u}_3}{\partial v} + (\bar{m}^{\alpha\beta} \varepsilon_{\alpha\lambda} v^\lambda v_\beta \hat{u}_3)_{,s} - (\bar{m}^{\alpha\beta} \varepsilon_{\alpha\lambda} v^\lambda v_\beta)_{,s} \hat{u}_3 \right.$$

$$\left. + \bar{m}^{\alpha\beta}|_\alpha v_\beta \hat{u}_3 \right] ds = \int_\Gamma \left[-\bar{m}^{\alpha\beta} v_\alpha v_\beta \frac{\partial \hat{u}_3}{\partial v} - (\bar{m}^{\alpha\beta} \varepsilon_{\alpha\lambda} v^\lambda v_\beta)_{,s} \hat{u}_3 \right.$$

$$\left. + \bar{m}^{\alpha\beta}|_\alpha v_\beta \hat{u}_3 \right] ds - \left[\left[\bar{m}^{\alpha\beta} \varepsilon_{\alpha\lambda} v^\lambda v_\beta \hat{u}_3 \right]\right]$$

$$= \int_\Gamma \left[-\bar{m}_\tau \frac{\partial \hat{u}_3}{\partial v} + \frac{d\bar{m}_v}{ds} u_3 + \bar{n}^3 \hat{u}_3 \right] ds + \left[\left[\bar{m}_v \hat{u}_3 \right]\right] \qquad (\ )_{,s} = d/ds$$

With the boundary integrals so modified the first identity becomes

$$G(u, \hat{u}) = \int_S -Du \cdot \hat{u} \, dS + \int_\Gamma \left[(\bar{n}^\alpha - 2\bar{m}^\varkappa b_\varkappa^\alpha) \hat{u}_\alpha - \bar{m}_\tau \frac{\partial \hat{u}_3}{\partial v} \right.$$

$$\left. + \left(\bar{n}^3 + \frac{d\bar{m}_v}{ds} \right) \hat{u}_3 \right] ds + \left[\left[\bar{m}_v \hat{u}_3 \right]\right] - E(u, \hat{u}) = 0$$

and it is now evident that the first identity is, if the curvature of the shell is zero, the sum of the identities of an elastic plate and a Kirchhoff plate, see Eq. (2.20) and (2.18), resp.

Next, we shall establish the connection between the first identity and the original problem, the four equations $(8.8_{1,2,3,6})$ for the eight unknowns $n^{\alpha\beta}$, $m^{\alpha\beta}$.

With Eqs. (8.9), (8.10), $(8.8_{4,5})$ and (8.6) the two boundary integrals in Eq. (8.13) can be written as

$$I_1 + I_2 = \int_\Gamma \left[n^{\alpha\beta} v_\beta \hat{u}_\alpha + \bar{m}^{\alpha\beta}|_\alpha v_\beta \hat{u}_3 - \bar{m}^{\varkappa\beta} b_\varkappa^\alpha v_\beta \hat{u}_\alpha - \bar{m}^{\alpha\beta} v_\beta \hat{u}_3|_\alpha \right] ds$$

$$= \int_\Gamma \left(\bar{n}^\alpha \hat{u}_\alpha + \bar{n}^3 \hat{u}_3 - \bar{m}^\varkappa b_\varkappa^\alpha \hat{u}_\alpha - \bar{m}^\alpha \hat{u}_3|_\alpha \right) ds$$

where

$$\bar{n}^3 = \bar{q}^\beta v_\beta = \bar{m}^{\alpha\beta}|_\alpha v_\beta, \qquad \bar{m}^\alpha = \bar{m}^{\alpha\beta} v_\beta$$

Let us assume that the tensor $m^{\alpha\beta}$ which solves Eq. (8.8) is symmetric then

$$\bar{m}^{\alpha\beta} = m^{\alpha\beta}$$

and, hence, also

$$\bar{n}^3 = n^3 \qquad \bar{m}^\alpha = m^\alpha$$

If we consider, additionally, Eqs. (8.4) and (8.5) then follows

$$I_1 + I_2 = \int_\Gamma (\boldsymbol{n} \cdot \hat{\boldsymbol{u}} + \boldsymbol{m} \cdot \hat{\boldsymbol{\varphi}})\, ds$$

where $\hat{\boldsymbol{\varphi}}$ is the vector from Eq. (8.4).

The first identity, thus, becomes

$$G(\boldsymbol{u}, \hat{\boldsymbol{u}}) = \int_S - \boldsymbol{D}\boldsymbol{u} \cdot \hat{\boldsymbol{u}}\, dS + \int_\Gamma (\boldsymbol{n} \cdot \hat{\boldsymbol{u}} + \boldsymbol{m} \cdot \hat{\boldsymbol{\varphi}})\, ds - E(\boldsymbol{u}, \hat{\boldsymbol{u}})$$

and the mechanical meaning of the single integrals is now evident.

With the formulation of the first identity the mathematics is done. The formulation of the work and energy principles is a mere repetition of the steps done already so often in the foregoing chapters.

We close with a theorem which assesses the properties of the energy, $E(\boldsymbol{u}, \hat{\boldsymbol{u}})$, see [B & C].

Theorem 8.2

$$p: \boldsymbol{x}(\boldsymbol{\Theta}) \in C^3(\bar{\Omega}),\ \boldsymbol{u} = \{u_1, u_2, u_3\}^T \in C^1 \times C^1 \times C^2(\bar{\Omega})$$

$$q: E(\boldsymbol{u}, \boldsymbol{u}) = \int_S [\bar{n}^{\alpha\beta}(\boldsymbol{u})\, \gamma_{\alpha\beta}(\boldsymbol{u}) + \bar{m}^{\alpha\beta}(\boldsymbol{u})\, \rho_{\alpha\beta}(\boldsymbol{u})]\, dS \ \geqslant 0$$

The energy is zero if and only if $\boldsymbol{u}$ is a rigid body-movement:

$$\boldsymbol{u} = \boldsymbol{a} + \boldsymbol{b} \times \boldsymbol{x}(\boldsymbol{\Theta}) \ \leftrightarrow\ E(\boldsymbol{u}, \boldsymbol{u}) = 0$$

A system A, naturally, exists for shells too; it can be found in [A1] or [M1].

9 Second-Order Analysis

If the equations of equilibrium are established using the geometry of the displaced structure then we speak of second-order analysis.

In this chapter we shall formulate the identities associated with the differential equations of second-order analysis for beams and plates.

9.1 Beams

Consider the beam on elastic supports (stiffness c) in Fig. 9.1 which is loaded with a longitudinal force λP at the free end, with longitudinal distributed forces λp_x which keep their horizontal direction when the beam deflects and with vertical distributed forces p_z.

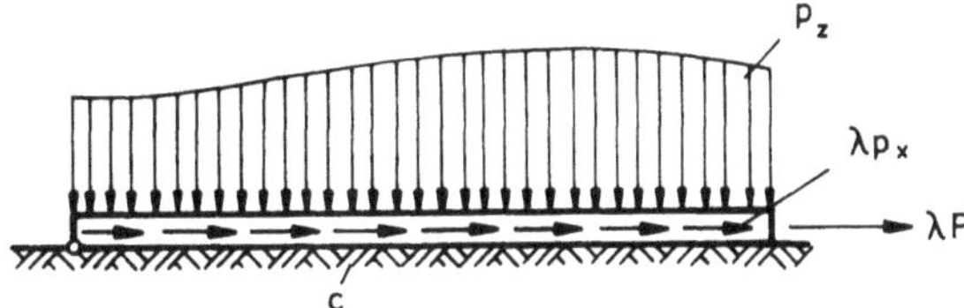

Figure 9.1

The deflection of the beam satisfies the differential equation

$$EI\, w^{IV} - \lambda(L\, w')' + c\, w = p_z \tag{9.1}$$

where

$$L(x) = P + \int_0^x p_x(y)\, dy$$

is the longitudinal force in the beam corresponding to a load multiple $\lambda = 1$.

Eq. (9.1) can be abbreviated as

$$D\, w = p_z$$

The integral

$$\int_0^l D w \hat{w}\, dx = \int_0^l \left(EI w^{IV} - \lambda (L w')' + c w \right) \hat{w}\, dx$$

is the virtual work of the load $D w$ acting through the virtual displacement $\hat{w}$.

Integration by parts applied to this integral renders the first identity of the operator D

$$p: w, \hat{w} \in C^4 \times C^2$$

$$q: G(w, \hat{w}) = \int_0^l D w\, \hat{w}\, dx + [(V + \lambda L w')\, \hat{w} - M \hat{w}']_0^l - E(w, \hat{w}) = 0 \qquad (9.2)$$

where

$$E(w, \hat{w}) = \int_0^l \left(\frac{M \hat{M}}{EI} + \lambda w' L \hat{w}' + c w\, \hat{w} \right) dx$$

is the symmetric energy.

The second identity is

$$p: w, \hat{w} \in C^4$$

$$q: B(w, \hat{w}) = \int_0^l D w\, \hat{w}\, dx + [(V + \lambda L w')\, \hat{w} - M \hat{w}']_0^l$$

$$\qquad\qquad - [w(\hat{V} + \lambda L \hat{w}') - w' \hat{M}]_0^l - \int_0^l w\, D \hat{w}\, dx = 0$$

Figure 9.2

Hence, the boundary operators associated with the operator D are

$$\partial^0 = \mathrm{id.}, \quad \partial^1 = \frac{d}{dx}, \quad \partial^2 = -EI \frac{d^2}{dx^2}, \quad \partial^3 = -EI \frac{d^3}{dx^3} + \lambda L \frac{d}{dx}$$

The last term

$$\partial^3 w = -EI w''' + \lambda L w' = V + \lambda L w' = T \qquad (9.3)$$

is the transverse force. Its direction is orthogonal to the original neutral position of the axis, see Fig. 9.2

The basic functional, the total energy, of the beam in Fig. 9.1 is

$$\Pi(w) = \frac{1}{2} E(w, w) - \int_0^l p_z \, w \, dx - [w(V + \lambda L w')]_0^l$$

and its restriction to the set

$$R_1 = \{w \in C^2[0, l] \,|\, w(0) = w(l) = 0\}$$

the set of all geometrically admissible functions is the potential energy functional

$$\Pi_1(w) = \frac{1}{2} E(w, w) - \int_0^l p_z \, w \, dx$$

Let $w = w_s$ be the solution of the bvp in Fig. 9.1 and $\hat{w} \in R_{1,0} = R_1$. Standard procedures reveal that the difference in potential energy between $w_s + \hat{w}$ and w_s is expressed as

$$\Pi_1(w_s + \hat{w}) - \Pi_1(w_s) = \delta \Pi_1(w_s, \hat{w}) + \frac{1}{2} E(\hat{w}, \hat{w}) = \frac{1}{2} E(\hat{w}, \hat{w})$$

Hence, the solution of the bvp is a minimum of the functional Π_1 if the energy $E(\hat{w}, \hat{w})$ is positive definite on $R_{1,0} - \{0\}$, that is if

$$E(\hat{w}, \hat{w}) = \int_0^l \left(\frac{\hat{M}^2}{EI} + \lambda L \hat{w}'^2 + c \hat{w}^2 \right) dx > 0 \quad \forall \hat{w} \in R_1 - \{0\}$$

This condition is equivalent with the condition

$$\lambda > -\frac{\int (\hat{M}^2/EI + c \hat{w}^2) \, dx}{\int L \hat{w}'^2 \, dx} = -\frac{A}{B} \quad \forall \hat{w} \in R_1 - \{0\}$$

That is, λ must be greater than the maximum the ratio $-A/B$ attains on $R_1 - \{0\}$.

The maximum is, as demonstrated in the literature, see e. g. [C6] p. 158, a negative number

$$\sup_{\hat{w} \in R_1 - \{0\}} -\frac{A(\hat{w})}{B(\hat{w})} = -\lambda_1$$

Hence, the energy $E(\hat{w}, \hat{w})$ is positive definite as long as the load multiple λ satisfies the inequality

$$\lambda > -\lambda_1$$

This is just the stability criterion, see Fig. 9.3.

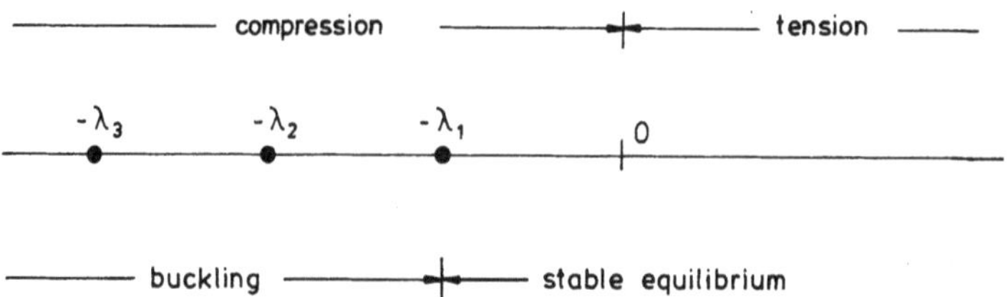

Figure 9.3

In the rest of this section we shall consider the differential equation (9.1) when the coefficients c and $p_x(x)$ are both zero, i.e. when (9.1) simplifies to

$$EIw^{IV} - \lambda P w'' = p_z \tag{9.4}$$

Dividing by EI and introducing the term

$$\varepsilon^2 = -\frac{\lambda P}{EI} l^2$$

the equation assumes the more convenient form

$$w^{IV} + \frac{\varepsilon^2}{l^2} w'' = \frac{p_z}{EI} \tag{9.5}$$

In case $\varepsilon^2 > 0$ (compression) the general homogeneous solution of the thus modified Eq. (9.5) is

$$w = c_1 \sin\left(\frac{\varepsilon x}{l}\right) + c_2 \cos\left(\frac{\varepsilon x}{l}\right) + c_3 x + c_4 \tag{9.6}$$

and in case $\varepsilon^2 < 0$ (tension)

$$w = c_1 \sinh\left(\frac{|\varepsilon| x}{l}\right) + c_2 \cosh\left(\frac{|\varepsilon| x}{l}\right) + c_3 x + c_4$$

The fundamental solution $g_0[x]$ of the compressed beam, $\varepsilon^2 > 0$ is, see [P2] p. 210,

$$g_0[x] = \frac{l^3}{\varepsilon^2 EI} \begin{cases} \dfrac{\sin\left(\varepsilon(1-\xi)\right)}{\varepsilon \sin \varepsilon} \cdot \sin\left(\dfrac{\varepsilon y}{l}\right) - (1-\xi)\dfrac{y}{l} & y \leqslant x \\[3ex] \dfrac{\sin \varepsilon \xi}{\varepsilon \sin \varepsilon} \cdot \sin\left(\varepsilon\left(1 - \dfrac{y}{l}\right)\right) - \xi\left(1 - \dfrac{y}{l}\right) & x \leqslant y \end{cases}$$

where $\xi = x/l$ and $l > 0$. This function is the deflection of the beam in Fig. 9.4.

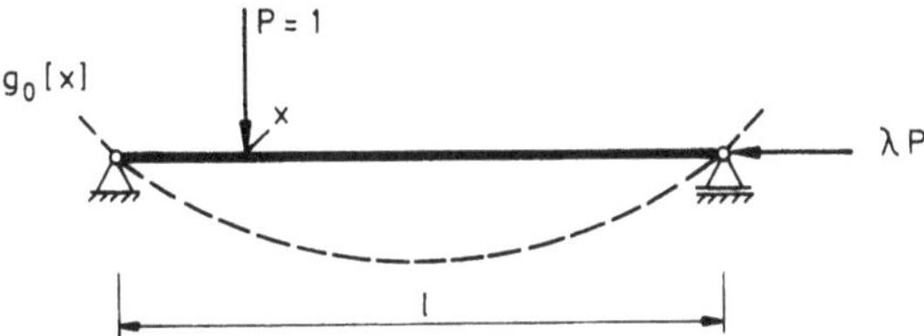

Figure 9.4

Next, we shall derive the second-order stiffness matrix of a beam and we define to this aim new sets of displacement and force terms δ_i and f_i on the boundary which replace w, w', M and T (see Fig. 9.5).

In these new terms the work done on the boundary is expressed as

$$[Tw - Mw']_0^l = T(l)\,w(l) - T(0)\,w(0) - M(l)\,w'(l) + M(0)\,w'(0)$$

$$= f_1\,\delta_1 + f_2\,\delta_2 + f_3\,\delta_3 + f_4\,\delta_4 = f^T \delta$$

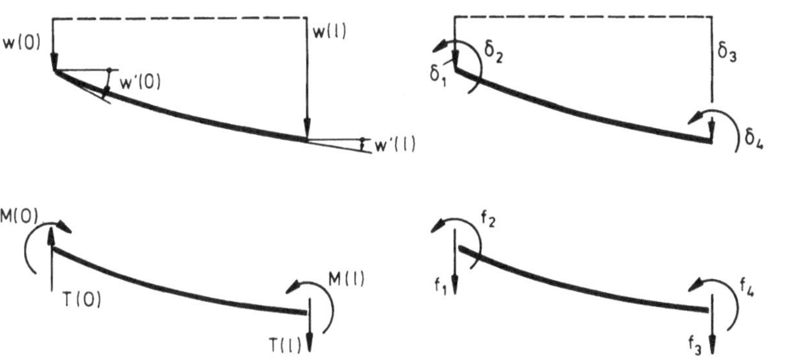

Figure 9.5

Furthermore, the homogeneous solution in Eq. (9.6) can be expanded in terms of the end displacements δ_i giving

$$w(x) = \delta_1 \chi_1(x) + \delta_2 \chi_2(x) + \delta_3 \chi_3(x) + \delta_4 \chi_4(x) \tag{9.7}$$

where

$$\chi_1(x) = \frac{1}{m}\left[\sin\varepsilon\left(b_3(x) - b_1(x) - \varepsilon\right) + (1 - \cos\varepsilon)\left(b_2(x) + 1\right)\right]$$

$$\chi_2(x) = \frac{1}{m}\frac{l}{\varepsilon}\left[\sin\varepsilon\left(\varepsilon b_1(x) - b_2(x) + 1\right) + (\cos\varepsilon - 1)\left(b_1(x) + b_3(x)\right)\right.$$
$$\left. + \varepsilon\cos\varepsilon\left(b_2(x) - 1\right)\right]$$

$$\chi_3(x) = \frac{1}{m}[\sin\varepsilon\,(b_1(x) - b_3(x)) + (1 - \cos\varepsilon)\,(1 - b_2(x))]$$

$$\chi_4(x) = \frac{1}{m}\frac{l}{\varepsilon}\,[(\sin\varepsilon - \varepsilon)\,(b_2(x) - 1) + (1 - \cos\varepsilon)\,(b_1(x) - b_3(x))]$$

and

$$m = [2(1 - \cos\varepsilon) - \varepsilon\sin\varepsilon],$$

$$b_1(x) = \sin\left(\frac{\varepsilon x}{l}\right), \quad b_2(x) = \cos\left(\frac{\varepsilon x}{l}\right), \quad b_3(x) = \frac{\varepsilon x}{l},$$

The thus modified function in (9.7) has the properties

$$Dw = EI\,w^{IV} - \lambda P\,w'' = 0$$

$$w(0) = \delta_1, \quad w'(0) = -\delta_2, \quad w(l) = \delta_3, \quad w'(l) = -\delta_4$$

If we substitute two such homogeneous solutions

$$w = \chi_i\delta_i = \boldsymbol{\chi}^T\boldsymbol{\delta}, \quad \hat{w} = \chi_i\hat{\delta}_i = \boldsymbol{\chi}^T\hat{\boldsymbol{\delta}}$$

into the first identity $G(w, \hat{w})$, see Eq. (9.2), then we obtain

$$G(w, \hat{w}) = \boldsymbol{f}^T\hat{\boldsymbol{\delta}} - \boldsymbol{\delta}^T\left(EI\int_0^l \boldsymbol{\chi}''\,\boldsymbol{\chi}''^T\,dx + \lambda P\int_0^l \boldsymbol{\chi}'\,\boldsymbol{\chi}'^T\,dx\right)\hat{\boldsymbol{\delta}}$$

$$= \boldsymbol{f}^T\hat{\boldsymbol{\delta}} - \boldsymbol{\delta}^T\boldsymbol{K}\hat{\boldsymbol{\delta}} = 0 \quad \forall\,\boldsymbol{\delta}, \hat{\boldsymbol{\delta}} \in \mathbb{R}^4$$

From whence follows that $\boldsymbol{f}^T - \boldsymbol{\delta}^T\boldsymbol{K} = \boldsymbol{0}^T$ or, because $\boldsymbol{K} = \boldsymbol{K}^T$, that

$$\boldsymbol{K}\boldsymbol{\delta} = \boldsymbol{f}$$

The stiffness matrix $\boldsymbol{K}$ has the elements

$$\boldsymbol{K} = \frac{EI}{l^3}\begin{bmatrix} 2(A' + B') - \varepsilon^2 & -(A' + B')l & -2(A' + B') + \varepsilon^2 & -(A' + B')l \\ & A'l^2 & (A' + B')l & B'l^2 \\ & & 2(A' + B') - \varepsilon^2 & (A' + B')l \\ \text{sym.} & & & A'l^2 \end{bmatrix} \tag{9.8}$$

where

$$A' = \frac{\varepsilon(\sin\varepsilon - \varepsilon\cos\varepsilon)}{2(1 - \cos\varepsilon) - \varepsilon\sin\varepsilon}, \qquad B' = \frac{\varepsilon(\varepsilon - \sin\varepsilon)}{2(1 - \cos\varepsilon) - \varepsilon\sin\varepsilon} \qquad \varepsilon^2 > 0$$

and

$$A' = \frac{\varepsilon(\sinh|\varepsilon| - \varepsilon\cosh|\varepsilon|)}{2(\cosh|\varepsilon| - 1) - \varepsilon\sinh|\varepsilon|}, \qquad B' = \frac{\varepsilon(\varepsilon - \sinh|\varepsilon|)}{2(\cosh|\varepsilon| - 1) - \varepsilon\sinh|\varepsilon|} \qquad \varepsilon^2 < 0$$

This matrix K, the exact second-order stiffness matrix of a beam, is not very popular because its elements are transcendental functions of ε. In practical applications, especially in frame analysis, it is far easier to work with a "polynomial" approximation of K which is obtained as follows:

Instead of formulating the first identity with two homogeneous solutions of the original Eq. (9.4) the first identity is formulated with two homogeneous solutions

$$w = \boldsymbol{\Psi}^T \boldsymbol{\delta}, \quad \hat{w} = \boldsymbol{\Psi}^T \hat{\boldsymbol{\delta}} \tag{9.9}$$

of the simple beam-equation $EI w^{IV} = 0$.

These functions, see Eq. (3.17), are not homogeneous solutions of Eq. (9.4)

$$Dw = EI\, \boldsymbol{\Psi}^{IV^T} \boldsymbol{\delta} - \lambda P\, \boldsymbol{\Psi}^{''T} \boldsymbol{\delta} = -\lambda P\, \boldsymbol{\Psi}^{''T} \boldsymbol{\delta} \neq 0$$

Consequently in the formulation of the identity $G(w, \hat{w})$ with two such functions $\{w, \hat{w}\}$

$$G(w, \hat{w}) = -\lambda P\, \boldsymbol{\delta}^T \left(\int_0^l \boldsymbol{\Psi}'' \boldsymbol{\Psi}^T \, dx \right) \hat{\boldsymbol{\delta}} + f^T \hat{\boldsymbol{\delta}} - \boldsymbol{\delta}^T \left(EI \int_0^l \boldsymbol{\Psi}'' \boldsymbol{\Psi}^{''T} \, dx \right.$$

$$\left. + \lambda P \int_0^l \boldsymbol{\Psi}' \boldsymbol{\Psi}^{'T} \, dx \right) \hat{\boldsymbol{\delta}} = p^T \hat{\boldsymbol{\delta}} + f^T \hat{\boldsymbol{\delta}} - \boldsymbol{\delta}^T \tilde{K} \hat{\boldsymbol{\delta}} = 0 \quad \forall\, \boldsymbol{\delta}, \hat{\boldsymbol{\delta}} \in R^4 \tag{9.10}$$

appears a vector p

$$p^T = -\lambda P\, \boldsymbol{\delta}^T \int_0^l \boldsymbol{\Psi}'' \boldsymbol{\Psi}^T \, dx$$

which is caused by the "residuum" of w with respect to the operator D.

The scalar $p^T \hat{\boldsymbol{\delta}}$ is the virtual work done in the domain while the two other terms, $f^T \hat{\boldsymbol{\delta}}$ and $\boldsymbol{\delta}^T \tilde{K} \hat{\boldsymbol{\delta}}$, represent the virtual work on the boundary and the internal virtual strain energy, resp.

The kernel of the quadratic form $\boldsymbol{\delta}^T \tilde{K} \boldsymbol{\delta}$, the matrix $\tilde{K}$, is the sum of two matrices

$$\tilde{K} = \frac{EI}{l^3} \begin{bmatrix} 12 & -6l & -12 & -6l \\ & 4l^2 & 6l & 2l^2 \\ & & 12 & 6l \\ \text{sym.} & & & 4l^2 \end{bmatrix} + \lambda P \begin{bmatrix} \frac{6}{5l} & -\frac{1}{10} & -\frac{6}{5l} & -\frac{1}{10}l \\ & \frac{2}{15}l & \frac{1}{10} & -\frac{11}{30} \\ & & \frac{6}{5l} & \frac{1}{10} \\ \text{sym.} & & & \frac{2}{15}l \end{bmatrix} \tag{9.11}$$

Of which the first one is the first-order stiffness matrix.

Choosing in the first identity for the vector $\hat{\boldsymbol{\delta}}$ consecutively the unit vectors $e_1 = \{1,0,0,0\}$ etc. we learn that

$$\tilde{K}\boldsymbol{\delta} = p + f$$

What, in the eyes of an engineer, spoils the picture is the vector p. To camouflage its existence the engineer introduces the vector

$$\tilde{f} = p + f$$

the vector of the so-called *"modified end forces"*, see e.g. [C4] that is the engineer works with the formula

$$\tilde{K}\,\delta = \tilde{f}$$

We close with some remarks.

1) The exact stiffness matrix $K = K(\lambda)$ in Eq. (9.8) is a function of the parameter λ. The first two terms in the expansion of this matrix at $\lambda_0 = 0$

$$K(\lambda) = K(\lambda_0) - (\lambda - \lambda_0)\frac{d}{d\lambda}\,K(\lambda)|_{\lambda = \lambda_0} + \dots$$

are, see [L&T] p. 292, just the two matrices in Eq. (9.11).

2) The approximate nature of the stiffness matrix $\tilde{K}$ in Eq. (9.11) is also signaled by the fact that its determinant is a fourth-order polynominal in λ.

According to the fundamental theorem of algebra such a polynomial has exactly four zeros and this contradicts the fact that a beam has an infinite number of eigenmodes and, hence, eigenvalues λ_i.

3) The system A associated with the differential equation (9.4) is

$$w'' - \varepsilon = 0$$

$$EI\varepsilon + M = 0$$

$$-M'' + \frac{\lambda P}{EI}\,M = p$$

4) The fundamental solutions of other variants of Eq. (9.1) can be found in [F3] p. 31–3 (beam on elastic support) and p. 31–7 (beam on elastic support with axial force). The associated stiffness matrices are listed in [A2].

9.2 Stability

Two criteria are used in the literature to define the stability of elastic systems under conservative loads λP.

  a) *static criterion*

  b) *energy criterion*

If the criterion is static then it is understood that the behaviour of the system is governed by a differential equation which formulates, pointwise, the equilibrium condition and instability, $\lambda = \lambda_c$, is reached when the solution of the *bvp* is no longer unique.

The energy criterion defines the equilibrium position u as that position u which renders the potential energy, Π_1, stationary in the class R_1, the class of all displacements satisfying the geometric boundary conditions. Instability, $\lambda = \lambda_c$, is reached if there exists a second function u which renders Π_1 stationary in R_1.

In classical mechanics the two criteria coincide because what is introduced as potential energy functional is just the basic functional (restricted to R_1) of the homogeneous *bvp*.

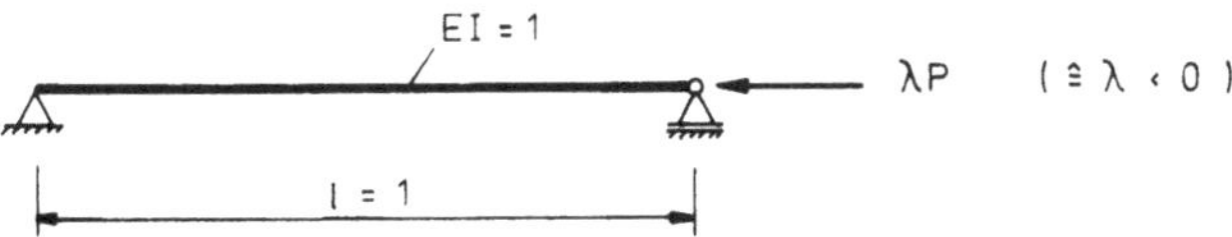

Figure 9.6

Consider, e. g., the beam in Fig. 9.6. The static criterion requires that every equilibrium configuration, as e. g. the unbuckled configuration $w_0 = 0$, is a solution of the homogeneous *bvp*

$$EI\,w^{IV} - \lambda P\,w'' = 0 \qquad 0 < x < 1$$
$$w(0) = w(1) = w''(0) = w''(1) = 0 \tag{9.12}$$

As every solution of this regular homogeneous *bvp* is also a stationary point of the functional

$$\Pi_1(w) = \frac{1}{2}\int_0^1 \left[\frac{M^2}{EI} + \lambda P\,w'^2\right] dx = \frac{1}{2}\,E(w,\,w)$$

the restriction of the basic functional $\Pi(w)$ to the class

$$R_1 = \{w \in C^2 \,|\, w(0) = w(1) = 0\}$$

we must require that every non-trivial solution, $\bar{w}$, satisfies the condition

$$\delta\Pi_1(\bar{w},\,\hat{w}) = E(\bar{w},\,\hat{w}) = 0 \qquad \forall\,\hat{w} \in R_{1,0} \tag{9.13}$$

The very same functional, $\Pi_1(w)$, is now also labeled the potential energy functional of the beam and, therewith, is also the basis of the energy criterion.

According to the energy criterion the unbuckled configuration, $w_0 = 0$, is an equilibrium configuration because the first variation of the potential energy at $w_0 = 0$ is zero

$$\delta\Pi_1(w_0,\,\hat{w}) = \int_0^1 \left[\frac{M_0\,\hat{M}}{EI} + \lambda P\,w_0'\,\hat{w}'\right] dx = E(w_0,\,\hat{w}) = 0 \qquad \forall\,\hat{w} \in R_{1,0}$$

with respect to all virtual displacements $\hat{w}$.

A second equilibrium configuration exists if there is a function $\bar{w}$ in $R_{1,0} = R_1$ such that

$$\delta \Pi_1 (w_0 + \bar{w}, \hat{w}) = 0 \qquad \forall\, \hat{w} \in R_{1,0}$$

Because the potential energy at $w = w_0 + \bar{w}$ expresses as

$$\Pi_1 (w_0 + \bar{w}) = \Pi_1 (w_0) + \delta \Pi_1 (w_0, \bar{w}) + \frac{1}{2}\, \delta^2\, \Pi_1 (w_0, \bar{w}) = \frac{1}{2}\, \delta^2\, \Pi_1 (w_0, \bar{w})$$

this equilibrium condition is equivalent with

$$\delta \{\Pi_1 (w_0 + \bar{w})\} = \delta \left\{ \frac{1}{2}\, \delta^2\, \Pi_1 (w_0, \bar{w}) \right\} = 0 \qquad \forall\, \hat{w} \in R_{1,0} \tag{9.14}$$

Or as textbooks formulate it: "The first variation of the second variation of the potential energy at w_0 with respect to $\bar{w}$ must be zero with respect to all virtual displacements $\hat{w}$."

In this form, Eq. (9.14), the energy criterion usually appears in the literature.

Luckily (a little bit of simple algebra does this, see section 4.3) it can be demonstrated that the second variation of the potential energy at w_0 in the direction of $\bar{w}$ is just the energy of $\bar{w}$ itself

$$\frac{1}{2}\, \delta^2\, \Pi_1 (w_0, \bar{w}) = \frac{1}{2}\, E(\bar{w}, \bar{w})$$

and we, therefore, can replace the strange expression on the right-hand side of Eq. (9.14) by

$$\delta \left\{ \frac{1}{2}\, \delta^2\, \Pi_1 (w_0, \bar{w}) \right\} = \frac{d}{d\varepsilon} \left(\frac{1}{2}\, E(\bar{w} + \varepsilon\, \hat{w}, \bar{w} + \varepsilon\, \hat{w}) \right) \Bigg|_{\varepsilon = 0} = E(\bar{w}, \hat{w})$$

which is just the same expression as in Eq. (9.13) that is the two criteria coincide.

Let us perform the calculation of the critical value $\lambda = \lambda_c$ of the simple beam in Fig. 9.6 when $\lambda < 0$ (compression).

In this case the solutions of Eq. (9.12) have the form

$$w = c_1 \sin (\varepsilon\, x) + c_2 \cos (\varepsilon\, x) + c_3\, x + c_4$$

and on account of the homogeneous boundary conditions the coefficients c_i are subject to the condition that they satisfy the system

$$\begin{bmatrix} 0 & 1 & 0 & 1 \\ 0 & \varepsilon^2 & 0 & 0 \\ \sin \varepsilon & \cos \varepsilon & 1 & 1 \\ \varepsilon^2 \sin \varepsilon & \varepsilon^2 \cos \varepsilon & 0 & 0 \end{bmatrix} \begin{bmatrix} c_1 \\ c_2 \\ c_3 \\ c_4 \end{bmatrix} = \begin{bmatrix} 0 \\ 0 \\ 0 \\ 0 \end{bmatrix}$$

which, due to $c_2 = c_4 = 0$, reduces to

$$\begin{bmatrix} \sin \varepsilon & 1 \\ \varepsilon^2 \sin \varepsilon & 0 \end{bmatrix} \begin{bmatrix} c_1 \\ c_3 \end{bmatrix} = \begin{bmatrix} 0 \\ 0 \end{bmatrix}$$

For a non-trivial solution, $\{c_1, c_3\}^T \neq \{0, 0\}^T$, to exist the determinant must be zero, that is we must have

$$\sin \varepsilon = 0$$

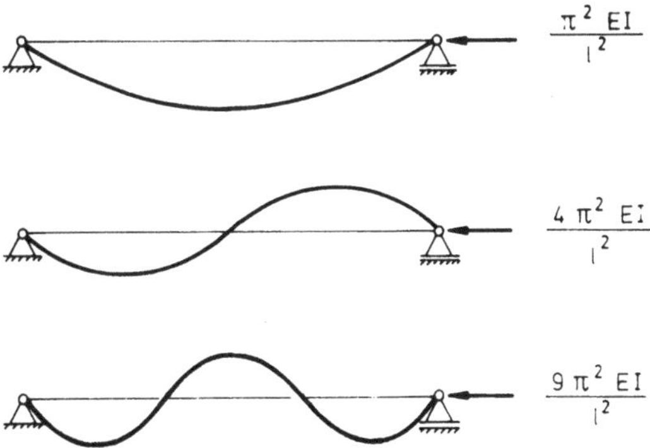

Figure 9.7

Substituting the positive zeros $\varepsilon_n = n\,\pi$, $n = 1, 2, 3 \ldots$, into Eq. (9.15) we obtain the system

$$\begin{bmatrix} 0 & 1 \\ 0 & 0 \end{bmatrix} \begin{bmatrix} c_1 \\ c_3 \end{bmatrix} = \begin{bmatrix} 0 \\ 0 \end{bmatrix}$$

which has for all n the solutions $c_1 = $ arbitrary, $c_3 = 0$.

Hence, the eigenfunctions corresponding to the eigenvalue ε_n are

$$w_n = c_1 \sin n\pi x$$

The smallest eigenvalue is $\varepsilon_1 = \pi$. To it corresponds the axial force

$$\lambda P = -\pi^2$$

To calculate the end couples of the eigenfuction we can substitute the data of the beam into the second-order stiffness matrix calculated in section 9.1. Because of $\delta_1 = \delta_3 = 0$ this matrix simplifies to

$$\begin{bmatrix} A' & B' \\ B' & A' \end{bmatrix} \begin{bmatrix} \delta_2 \\ \delta_4 \end{bmatrix} = \begin{bmatrix} f_2 \\ f_4 \end{bmatrix}$$

and with $\varepsilon = \varepsilon_1 = \pi$ follows

$$A' = B' = \frac{\pi^2}{4}$$

In the case of the eigenfunction w_1 we have, see Fig. 9.8,

$$\delta_2 = -\delta_4$$

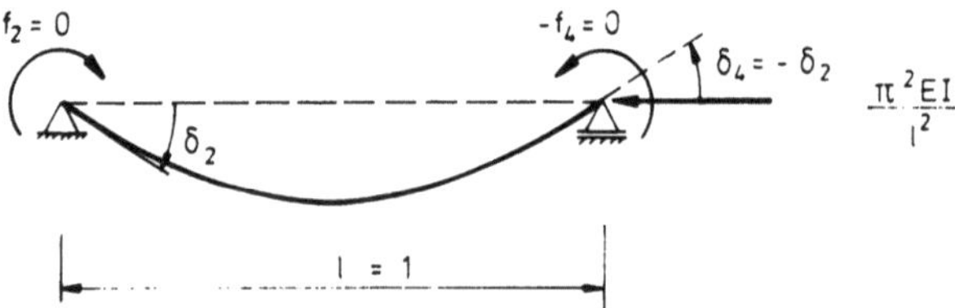

Figure 9.8

and, hence, the couples f_2 and f_4 at the ends are zero if the axial load is the critical load.

$$\frac{\pi^2}{4} \begin{bmatrix} 1 & 1 \\ 1 & 1 \end{bmatrix} \begin{bmatrix} \delta_2 \\ -\delta_2 \end{bmatrix} = \begin{bmatrix} 0 \\ 0 \end{bmatrix}$$

That is, zero couples are sufficient to rotate the ends in the direction of the first eigenform, w_1. No work is done in this case

$$f_2 \delta_2 + f_4 \delta_4 = 0$$

In agreement with this the internal energy of the eigenform w_1 (as of any eigenform w_n) is zero because, see Eq. (9.2),

$$G(w_1, w_1) = \int_0^l D\overset{\circ}{w_1}\, w_1\, dx + [(V + \lambda_1 Pw_1')\overset{\circ}{w_1} - M\overset{\circ}{w_1'}]_0^l - E(w_1, w_1)$$
$$= -E(w_1, w_1) = 0$$

For a comparison let us calculate in addition an approximation of the critical load $\lambda_c P = -\pi^2$ the way the engineer does it in case he has to analyze complex framed structures, i. e. by approximating the eigenfunction in terms of the four homogeneous simple beam-solutions

$$w = \sum_{i=1}^{4} \delta_i \psi_i$$

The first identity of such an expansion reads, as demonstrated above,

$$G(w, w) = (p^T + f^T)\,\delta - \delta^T \tilde{K}\,\delta = \tilde{f}^T \delta - \delta^T \tilde{K}\,\delta = 0$$

The geometric boundary conditions, $\delta_1 = \delta_3 = 0$ (vertical deflection), can be satisfied, too. But if we would let the end couples be zero, $f_2 = f_4 = 0$, then we would be left with nonzero forces p_i. We, therefore, compromise, that is we set the "modified end couples" equal to zero, $\tilde{f}_2 = f_2 + p_2 = \tilde{f}_4 = f_4 + p_4 = 0$.

It thus results, as in the case of the exact expansion, that the external work

$$\tilde{f}^T \delta = \tilde{f}_1\, 0 + 0\, \delta_2 + \tilde{f}_3\, 0 + 0\, \delta_4 = 0$$

is zero and, hence, the internal energy zero as well.

$$\delta^T \tilde{K} \delta = 0$$

The smallest parameter λ at which this happens (not counting the neutral position $\delta = 0$) is

$$\lambda_c P = -12$$

In which case the beam is in the buckled position $\delta = \{0, \delta_2, 0, -\delta_2\}$ (eigenvector). The approximation, see Fig. 9.9,

$$w = \delta_2(\psi_2 - \psi_4) = \delta_2(x^2 - x)$$

corresponds to a load case with end couples and lateral forces, the residuum of w with respect to the operator D.

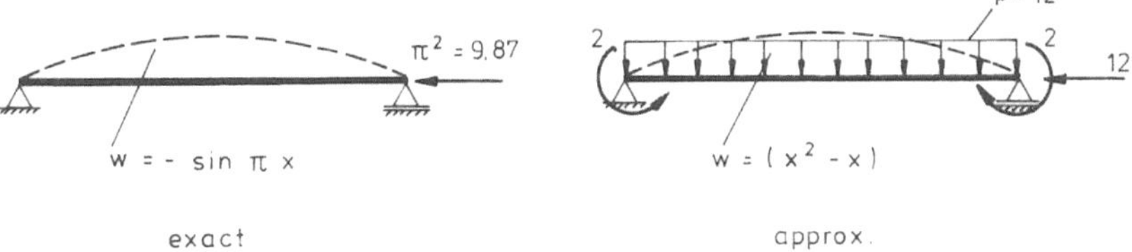

Figure 9.9

In closing this section, we think, we owe the attentive reader yet an explanation:

How is it, so he might ask, that the eigenfunction $w = \sin \pi x$ and also its approximation $w = x^2 - x$ have zero internal energy though they are not rigid-body movements?

Is there something wrong with mechanics? No—clearly not! The truth is simply that the second-order beam equation alone does not completely describe the behaviour of the beam. The horizontal displacement u must be taken into account, too.

This leads to the nonlinear system of differential equations, see section 10.5,

$$-EA\left(u' + \frac{1}{2}(w')^2\right)' = 0$$

$$EI\,w'^{IV} - \left(EA\left(u' + \frac{1}{2}w'\right)^2 w'\right)' = 0$$

which at the critical load, at $N = EA\left(u' + 1/2(w')^2\right) = -\pi^2$ has the solution

$$w(x) = \sin \pi x$$

$$u(x) = -\pi(1.25\,\pi x + 0.125 \sin 2\pi x)$$

The energy balance of the buckled beam, therefore, reads, see Eq. (10.23)

$$\frac{1}{2}\int_0^l \left(\frac{N^2}{EA} + \frac{M^2}{EI}\right) dx = \frac{1}{2}\,N(1)\,u(1) - \frac{1}{2}\int_0^l \frac{1}{2}(w')^2\,dx$$

$$\frac{3}{4}\,\pi^4 = \frac{5}{8}\,\pi^4 + \frac{1}{8}\,\pi^4$$

and, clearly, the internal energy and, therefore, the external work as well are not zero. So nothing is wrong.

Please note that the external work is not simply $1/2\,N(1)\,u(1)$ as in linear mechanics but that it is supplemented by a secondary quantity. We shall discuss this at length in section 10.7.

9.3 Lateral Buckling of Beams

We consider the system

$$EI_z\,v^{IV} + P\,v'' + (M_y\varphi)'' = p_y$$
$$M_y v'' + EC_T\varphi^{IV} - (GJ_T - P\,i_p^2)\,\varphi'' = m_x \tag{9.16}$$

which governs ($p_y = m_x = 0$) the lateral buckling of straight slender beams whose cross sections have two planes of symmetry.

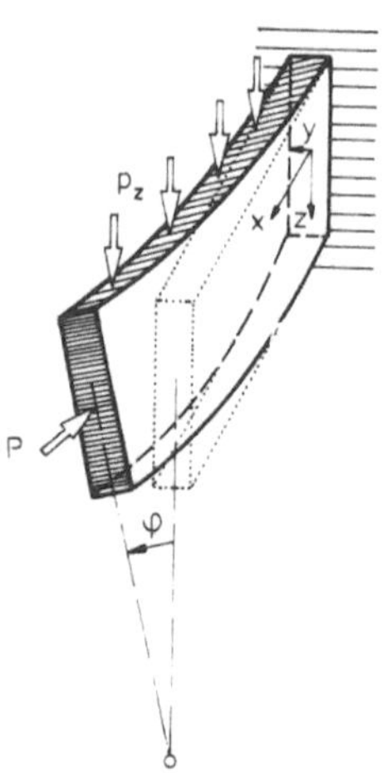

Figure 9.10

The configuration variables are the lateral displacement v and the rotation φ, see Fig. 9.10.

The single quantities which constitute the coefficients of the system (9.16) are the following:

P = axial force, A = cross-sectional area,

E = Young's modulus, M_y = bending moment,

I_z = moment of inertia G = shear modulus,

     about the z-axis, C_T = warping rigidity

I_T = torsional rigidity, I_p = $\int r^2 \, dA,\ i_p^2 = A^{-1} I_p$

The first identity of the system (9.16) is the expression

$$p: v, \varphi \in C^4, \hat{v}, \hat{\varphi} \in C^2$$

$$q: G(v, \varphi; \hat{v}, \hat{\varphi}) = \int_0^l \left[\left(E I_z \, v^{IV} + P \, v'' + (M_y \, \varphi)'' \right) \hat{v} + \left(M_y \, v'' + E C_T \, \varphi^{IV} \right. \right.$$

$$\left. - (G J_T - P \, i_p^2) \, \varphi'' \right) \hat{\varphi} \right] dx + \left[- E I_z \, v''' \, \hat{v} + E I_z \, v'' \, \hat{v} - P v' \, \hat{v} - (M_y \, \varphi') \, \hat{v} \right.$$

$$+ M_y \, \varphi \, \hat{v}' - E C_T \, \varphi''' \, \hat{\varphi}' + E C_T \, \varphi'' \, \hat{\varphi}'' - (G J_T - P \, i_p^2) \, \varphi' \, \hat{\varphi} \Big]_0^l$$

$$- E(v, \varphi; \hat{v}, \hat{\varphi}) = 0$$

where

$$E(v, \varphi; \hat{v}, \hat{\varphi}) = \int_0^l \left(E I_z \, v'' \, \hat{v}'' - P \, v' \, \hat{v}' + M_y (\varphi \, \hat{v}'' + \hat{\varphi} \, v'') + E C_T \, \varphi'' \, \hat{\varphi}'' \right.$$

$$\left. - (G J_T - P \, i_p^2) \, \varphi' \, \hat{\varphi}' \right) dx$$

is the symmetric energy.

Other instabilities which occur in beams are torsional buckling or combined torsional and flexural buckling. The equations which govern these modes are somewhat more lengthy but they pose no additional difficulties when it comes to the formulation of the first identity.

9.4 The Kirchhoff Plate

Consider a plate which is loaded at its edge by in-plane forces and let $S = [\sigma_{ij}] \in C^1(\Omega) \cap C(\bar{\Omega})$ the associated stress tensor which satisfies in Ω the equation

$$\operatorname{div} S = 0 \tag{9.17}$$

In addition to being acted upon by the in-plane forces let the plate be loaded with forces orthogonal to its plane. The deflection w in the direction of these forces satisfies the equation

$$K \Delta \Delta w - N_1 \, w_{,11} - 2 N_{12} \, w_{,12} - N_2 \, w_{,22} = p$$

where the coefficients N_{ij} are the stress-resultants (h = plate thickness) of the in-plane stresses

$$N_1 = h\sigma_{11}, \quad N_{12} = h\sigma_{12} = h\sigma_{21} = N_{21}, \quad N_2 = h\sigma_{22}$$

Because of Eq. (9.17) we have

$$N_{1,1} + N_{12,2} = 0 \quad N_{2,2} + N_{21,2} = 0$$

and, therefore, the equation can also be written as

$$K\Delta\Delta w - (N_1\,w_{,1})_{,1} - (N_{12}\,w_{,1})_{,2} - (N_2\,w_{,2})_{,2} - (N_{21}\,w_{,2})_{,1} = p$$

or if we introduce the abbreviations

$$D_1\,w = K\Delta\Delta w \tag{9.18}$$

$$D_2\,w = (N_1\,w_{,1})_{,1} + (N_{12}\,w_{,1})_{,2} + (N_2\,w_{,2})_{,2} + (N_{21}\,w_{,2})_{,1} \tag{9.19}$$

as

$$D w = D_1\,w - D_2\,w = p \tag{9.20}$$

The first identity of the operator D_2 is

$$p:\ w, \hat{w} \in C^2 \times C^1,\ S \in C^1(\Omega) \cap C(\bar{\Omega})$$

$$q:\ G(w,\hat{w}) = \int_\Omega D_2\,w\,\hat{w}\,d\Omega - \int_\Gamma (N_1\,w_{,1}\,n_1 + N_{12}\,w_{,1}\,n_2 + N_2\,w_{,2}\,n_2$$
$$+ N_{21}\,w_{,2}\,n_1)\,\hat{w}\,ds + E_2(w,\hat{w}) = 0$$

where $E_2(w,\hat{w})$ is the symmetric expression

$$E_2(w,\hat{w}) = \int_\Omega (N_1\,w_{,1}\,\hat{w}_{,1} + N_{12}\,w_{,1}\,\hat{w}_{,2} + N_{21}\,w_{,2}\,\hat{w}_{,1} + N_2\,w_{,2}\,\hat{w}_{,2})\,d\Omega$$
$$\tag{9.21}$$

Combining this identity with the first identity of the operator $K\Delta\Delta$ formulated in chapter 2, Eq. (2.18), we obtain the first identity of the operator $D = D_1 - D_2$

$$p:\ w, \hat{w} \in C^4 \times C^2,\ S \in C^1(\Omega) \cap C(\bar{\Omega})$$

$$q:\ G(w,\hat{w}) = \int_\Omega D w\,\hat{w}\,d\Omega + \int_\Gamma \left[V_n\,\hat{w} - M_n\,\frac{\partial\hat{w}}{\partial n} \right] ds + [[M_{nt}\,\hat{w}]]$$

$$+ \int_\Gamma (N_1\,w_{,1}\,n_1 + N_{12}\,w_{,1}\,n_2 + N_{21}\,w_{,2}\,n_1$$

$$+ N_2\,w_{,2}\,n_2)\,\hat{w}\,ds - E(w,\hat{w}) = 0$$

with

$$E(w, \hat{w}) = E_1(w, \hat{w}) + E_2(w, \hat{w})$$

where $E_1(w, \hat{w})$ is the energy of the operator $K \Delta \Delta$, see Eq. (2.12), and $E_2(w, \hat{w})$ the energy of the operator D_2, see Eq. (9.21).

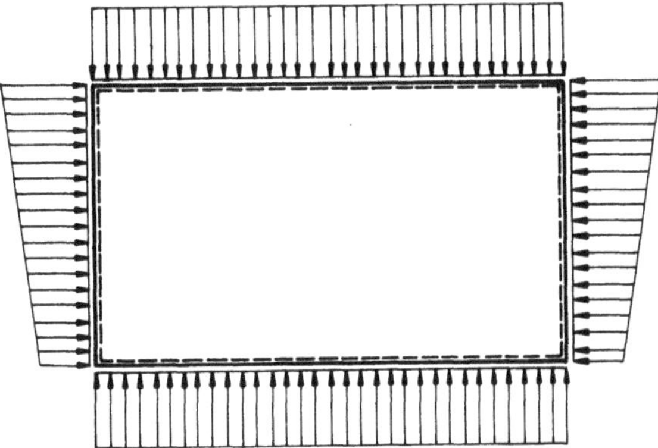

Figure 9.11

As an application of this identity consider the problem of a plate loaded at its edge by in-plane forces, see Fig. 9.11, and let λN_1, λN_2, λN_{12} the corresponding in-plane stress resultants.

We want to know: for which values of λ does the homogeneous bvp

$$D_1 w - \lambda D_2 w = 0 \quad \text{in } \Omega, \qquad w = M_n = 0 \quad \text{on } \Gamma$$

have a non-trivial solution $w \neq 0$, that is when does the plate buckle?

The restriction of the basic functional Π of this homogeneous bvp to the class R_1 is

$$\Pi_1(w) = \frac{1}{2} E(w, w)$$

and because of

$$G(w, w) = \int_\Omega Dw \, w \, d\Omega - E(w, w) = 0$$

every solution $w \in C^4(\bar{\Omega})$ of the homogeneous bvp is also a zero of $\Pi_1(w)$

$$\Pi_1(w) = \frac{1}{2} E(w, w) = \frac{1}{2} \int_\Omega Dw \, w \, d\Omega = \frac{1}{2} \int_\Omega 0 \, w \, d\Omega = 0$$

Hence, the potential energy is zero in the buckled position (as in the neutral position). A second solution $w \neq 0$, therefore, can only exist if $E(w, w)$ is not positive (or negative) definite on $R_1 - \{0\}$.

Now it is

$$E(w, w) = E_1(w, w) + \lambda E_2(w, w)$$

and, furthermore,

$$E_1(w, w) > 0, \quad E_2(w, w) > 0 \quad \forall\, w \in R_1 - \{0\}$$

Hence, the energy is positive definite whenever $\lambda > 0$ (tension)

$$E(w, w) > 0 \quad \forall\, w \in R_1 - \{0\}$$

A second zero can, therefore, only exist if $\lambda < 0$ (compression).

9.5 Nonconservative Problems

We consider the same beam as in Fig. 9.1 only that this time the longitudinal forces $p_x(x)$ follow the movement of the axis, that is they remain tangential to the axis, see Fig. 9.12.

The deflection of the beam in this case satisfies the differential equation, see [P2] p. 39 (1.219),

$$EI\, w^{IV} - \lambda L w'' + c w = p_z$$

where

$$L(x) = P + \int_0^x p_x(y)\, dy$$

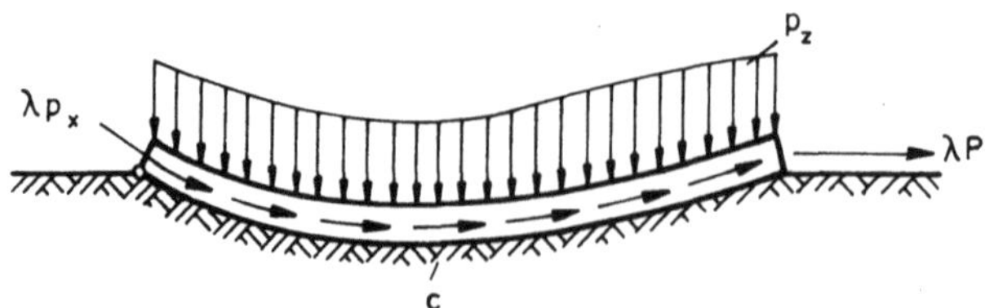

Figure 9.12

The first identity of the operator

$$D w = EI\, w^{IV} - \lambda L(x)\, w'' + c w$$

is

$$G(w, \hat{w}) = \int_0^l D w \hat{w}\, dx + [(V + \lambda L w')\, \hat{w} - M \hat{w}']_0^l - E(w, \hat{w}) = 0$$

where

$$E(w, \hat{w}) = \int_0^l \left[\frac{M \hat{M}}{EI} - \lambda w'\, L \hat{w}' - w\, c\, \hat{w} - \lambda w'\, p_x \hat{w} \right] dx$$

This integral $E(w, \hat{w})$, due to the last term $w' p_x \hat{w}$ is no longer symmetric.

In the symmetric case the functional associated with $E(w, \hat{w})$ is simply $F(w)$ $= 1/2\, E(w, w)$. Now this is no longer true. $F(w) = 1/2\, E(w, w)$ and $E(w, \hat{w})$ do not match and even when we try other variants we soon recognize that there is no functional which has $E(w, \hat{w})$ as its first variation. The essential condition, see section 4.12, is violated.

If we try to shift the operator D fully onto $\hat{w}$ then we obtain the following result

$$B(w, \hat{w}) = \int_0^l D w\, \hat{w}\, dx + [(V + \lambda L w')\, \hat{w} - M \hat{w}']_0^l - [w(\hat{V} + \lambda L \hat{w}')$$

$$- w'\, \hat{M}]_0^l - [w\, p_x \hat{w}]_0^l - \int_0^l w\, D^* \hat{w}\, dx$$

where

$$D^* w = EI\, w^{IV} - \lambda (L w)'' + c w$$

is not the operator D we started with.

The operator D^* is called the formal adjoint operator. If $D = D^*$, as in the previous chapters, then we say that D is formally self-adjoint. If D is a linear operator of the form

$$D w = \sum_{i=0}^m (f_i w^{(i)})^{(i)}, \qquad w^{(i)} = \frac{d^i w}{dx^i}$$

with coefficients $f_i(x) \in C^i[a, b]$ then the formal adjoint operator D is

$$D^* w = \sum_{i=0}^{2m} (-1)^i (f_i w)^{(i)}$$

The self-adjoint operators are operators with even degree and they all can be expressed in the "self-adjoint" form

$$D w = \sum_{i=0}^m (g_i w^{(i)})^{(i)} \tag{9.22}$$

As an example consider Eq. (9.1) which can be written as

$$Dw = (EIw'')'' + (-\lambda Lw')' + cw = p$$

where

$$g_2 = EI, \quad g_1 = -\lambda L, \quad g_0 = c$$

The energy associated with a formally self-adjoint operator as in Eq. (9.22) has the form

$$E(w, \hat{w}) = \int_0^l \sum_{i=0}^m g_i \, w^{(i)} \, \hat{w}^{(i)} \, dx$$

and, therefore, is always symmetric while the energy of a not formally self-adjoint operator is not.

Formally self-adjoint operators play an important role in structural mechanics. Betti's principle

$$W_{1,2} = W_{2,1}$$

and the symmetry of the fundamental solutions (*Maxwell's principle*)

$$g_0(y, x) = g_0(x, y)$$

are based on the fact that $D = D^*$. Maxwell's principle, e. g., no longer applies in the case of the beam in Fig. 9.13.

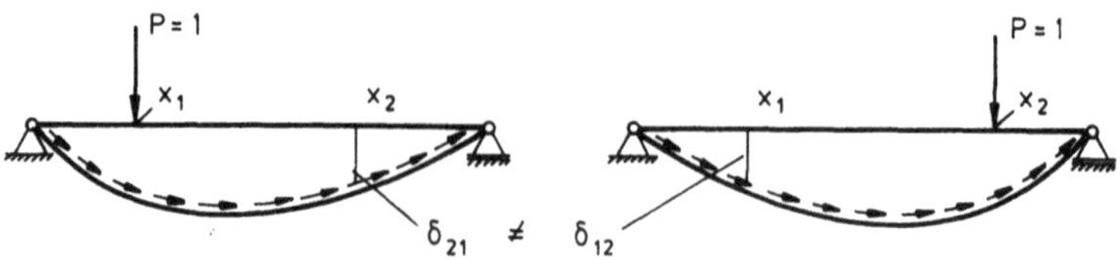

Figure 9.13

9.6 Initial Value Problems

With due allowance by the reader we will digress in this section shortly into the time domain to talk about another class of problems, namely initial value problems, which also do not allow the formulation of variational principles (at least as long as we concentrate on the classical concepts).

Consider the ball with mass m in Fig. 9.14 which we hold fixed at $x = x_0$ and which we let go at time $t = 0$.

The movement $x(t)$ of the ball that follows is the solution of the initial value problem

$$-\left[\ddot{x}(t)+\frac{c}{m}\,x\right]=0 \quad 0<t, \quad x(0)=x_0, \quad \dot{x}(0)=0$$

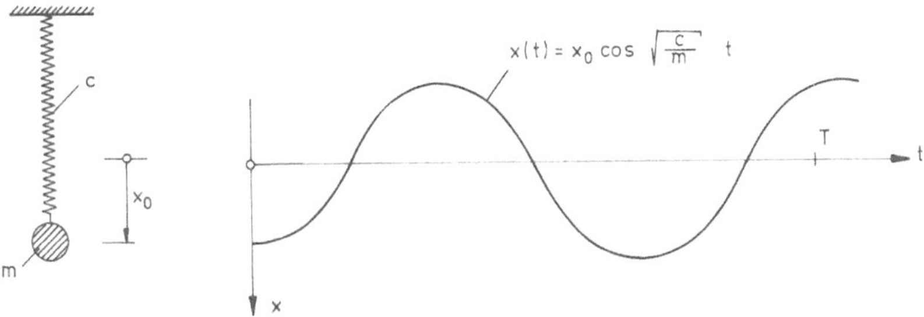

Figure 9.14

The first identity of the governing operator in the domain $[0, T]$ reads

$$G(x, \hat{x})=\int_0^T -\left(\ddot{x}+\frac{c}{m}\,x\right)\hat{x}\,dt+[\dot{x}\hat{x}]_0^T-\int_0^T\left(\dot{x}\dot{\hat{x}}-\frac{c}{m}\,x\hat{x}\right)dt$$

and, clearly, the initial value problem is not regular with respect to this identity because two conjugated boundary terms, x and $\dot{x}$, are prescribed on the same part of the boundary, namely at $t=0$.

This is why "a variational formulation for the entire class of initial value problems cannot be achieved as long as we persist in using the procedures of the classical calculus of variations... Even the fundamental principle of Hamilton must artificially be converted into the treatment of motions between two terminal configurations, instead of considering the initial conditions alone", [N3] p. 244. In this respect see also Gurtin's remark on Hamilton's principle, [G2] p. 226.

Gurtin bypasses these difficulties by making use of convolutions.

Consider the simplest initial value problem

$$u'(t)=f(t) \quad 0<t \quad u(0)=0$$

To the operator $D=d/dt$ belongs the identity, see [N3] p. 245,

$$p: u, \hat{u} \in C^1[0, T]$$

$$q: C(u, \hat{u})=\int_0^T u'\,\hat{u}(T-t)\,dt-u(T)\,\hat{u}(0)+u(0)\,\hat{u}(T)$$

$$-\int_0^T u(T-t)\,\hat{u}'(t)\,dt=0$$

Let $u = u_s$ and $\hat{u}$ an admissible virtual displacement, $\hat{u}(0) = 0$, from C^1. At the time mark T we then have

$$(-1)\, C(u_s, \hat{u}) = -\int_0^T f(t)\, \hat{u}(T-t)\, dt + \int_0^T u_s(T-t)\, \hat{u}'\, dt = 0$$

As this expression is (this is easy to demonstrate) the first variation of the functional

$$\Pi(u) = \frac{1}{2}\int_0^T u(T-t)\, u'(t)\, dt - \int_0^T f(t)\, u(T-t)\, dt$$

we conclude that u_s is a stationary point of $\Pi(u)$ with respect to all admissible virtual displacements.

As an application consider the following problem of a dynamically loaded bar ($\mu = $ mass per unit length)

$$-EA\, u''(x, t) + \mu\, \ddot{u}(x, t) = p(x, t) \quad 0 < x < l, \quad 0 < t$$

$$\text{init. cond.}: u(x, 0) = \bar{u}(x)\,; \ \dot{u}(x, 0) = \bar{\bar{u}}(x)$$

$$\text{bound. cond.}: u(0, t) = N(l, t) = 0 \quad 0 \leqslant t$$

To this problem belongs the identity

$$p: u(x, t) \in C^2[0, l] \times C^2[0, T],\ \hat{u} \in C^1 \times C^1$$

$$q: C(u, \hat{u}) = \int_0^l (-EA\, u'' + \mu\, \ddot{u}) * \hat{u}\, dx + [N * \hat{u}]_0^l - \int_0^l \mu\, [\dot{u}(x, T)\, \hat{u}(x, 0)$$

$$-\dot{u}(x, 0)\, \hat{u}(x, T)]\, dx - \int_0^l \left[\frac{N * \hat{N}}{EA} - \mu\, \dot{u} * \hat{\dot{u}}\right] dx = 0$$

where

$$u * \hat{u} = \int_0^T u(t)\, \hat{u}(T-t)\, dt \quad T > 0$$

denotes the convolution of two functions.

If we let the function u in this identity the solution of the initial value problem, $u = u_s$, and $\hat{u}$ an admissible virtual displacement

$$\hat{u} \in R_{1,0}(T) = \{\hat{u} \in C^1 \times C^1 | \hat{u}(x, 0) = 0\,; \ \hat{u}(0, t) = 0,\ 0 \leqslant t \leqslant T\}$$

then this identity becomes

$$C(u_s, \hat{u}) = \int_0^l p * \hat{u}\, dx + \int_0^l \mu\, \bar{u}(x)\, \hat{u}(x, T)\, dx - \int_0^l \left[\frac{N * \hat{N}}{EA} - \mu\, \dot{u} * \hat{\dot{u}}\right] dx = 0$$

$$\forall\, \hat{u} \in R_{1,0}$$

that is u_s is a stationary point of the functional

$$\Pi_T(u) = \frac{1}{2} \int_0^l \left[\frac{N * N}{EA} - \mu \dot{u} * \dot{u} \right] dx - \int_0^l \mu \bar{\bar{u}}(x)\, u(x, T)\, dx - \int_0^l p * u\, dx$$

A further variant of this approach is obtained if we apply one more convolution. To this end let i denote the function $i(t) = t$. Given the initial data $\bar{u}$ and $\bar{\bar{u}}$ it follows easily that a first convolution

$$t \xrightarrow{*} T$$

of both sides of the partial differential equation with respect to i

$$i * (-EA\, u'' + \mu \ddot{u}) = i * p$$

renders

$$-i * EA\, u'' + \mu u(x, T) = \mu[T\bar{\bar{u}}(x) + \bar{u}(x)] + i * p$$

which means "that the equation of motion and the appropriate initial conditions are together equivalent to a single integral-differential equation", [G 2] p. 65.

To the left-hand side belongs, under the same assumptions on u and $\hat{u}$ as above, the identity

$$C(u, \hat{u}) = \int_0^l [-i * EAu'' + \mu u(x, T)] * \hat{u}\, dx + [i * N * \hat{u}]_0^l$$

$$- \int_0^l \left[\frac{i * N * \hat{N}}{EA} + \mu u(x, T) \right] * \hat{u}\, dx = 0$$

where it is understood that the second convolution now takes us from T to τ

$$t \xrightarrow{*} T \xrightarrow{*} \tau$$

And as above results that u_s, the solution of the initial value problem, is a stationary point of the functional

$$\Pi_\tau(u) = \frac{1}{2} \int_0^l \left[\frac{i * N * N}{EA} + \mu u * u \right] dx - \int_0^l [\mu\{T\bar{\bar{u}}(x) + \bar{u}(x)\}$$

$$+ i * p] * u\, dx$$

A functional which now contains all the initial data of the problem.

9.7 Vibrations

An engineer's approach to the determination of the natural frequencies of framed structures is virtually a copy of his approach to the determination of the critical load of such structures.

The lateral vibration $w(x, t)$ of a single beam with mass μ per unit length is the solution of the equation

$$EI\, w^{IV}(x, t) + \mu\, \ddot{w}(x, t) = 0 \quad 0 < x < l, \quad 0 < t \tag{9.23}$$

Substituting

$$w(x, t) = w(x) \cos(\omega t + \varphi)$$

into this equation renders the ordinary differential equation

$$EI\, w^{IV} - \omega^2\, \mu\, w = 0 \quad 0 < x < l$$

whose general solution is

$$w(x) = a_1 \cos(\lambda x) + a_2 \sin(\lambda x) + a_3 \cosh(\lambda x) + a_4 \sinh(\lambda x)$$

$$\lambda = \left(\frac{\mu\, \omega^2}{EI}\right)^{1/4}$$

Let us assume

$$w(x) = \sum_{i=1}^{4} \theta_i(x)\delta_i = \boldsymbol{\Theta}^T \boldsymbol{\delta}$$

is the "normalized" expansion of w, that is the parameters δ_i are unit end displacements/rotations.

Substituting two such homogeneous solutions

$$w = \boldsymbol{\Theta}^T \boldsymbol{\delta} \quad \hat{w} = \boldsymbol{\Theta}^T \hat{\boldsymbol{\delta}}$$

into the first identity

$$G(w, \hat{w}) = \int_0^l (EI\, w^{IV} \overset{\circ}{\diagup} \mu\, \omega^2\, w)\, \hat{w}\, dx + [V\hat{w} - M\hat{w}']_0^l$$

$$- \int_0^l (EI\, w''\, \hat{w}'' - \mu\, \omega^2\, w\, \hat{w})\, dx = \boldsymbol{f}^T \hat{\boldsymbol{\delta}} - \boldsymbol{\delta}^T \boldsymbol{K} \hat{\boldsymbol{\delta}} = 0 \tag{9.24}$$

we obtain the exact stiffness matrix

$$K_{ij} = \int_0^l (EI\theta_i''\, \theta_j'' - \mu\, \omega^2\, \theta_i\, \theta_j)\, dx$$

which is a transcendental function of ω.

This property renders the application of K in the determination of the natural frequencies of framed structures, structures which consist of many single beams, practically useless.

The engineer, therefore, replaces K by an approximation $\tilde{K}$ he obtains with the homogeneous solutions, ψ_i, of the simple beam-equation.

Formulating the first identity (9.24) with two such functions

$$w = \boldsymbol{\Psi}^T \boldsymbol{\delta} \qquad \hat{w} = \boldsymbol{\Psi}^T \hat{\boldsymbol{\delta}}$$

he obtains

$$G(w, \hat{w}) = \boldsymbol{p}^T \hat{\boldsymbol{\delta}} + \boldsymbol{f}^T \hat{\boldsymbol{\delta}} - \boldsymbol{\delta}^T \tilde{K} \hat{\boldsymbol{\delta}} = 0 \qquad \forall \, \boldsymbol{\delta}, \hat{\boldsymbol{\delta}} \in \mathbb{R}^4$$

where

$$p_i = \int_0^l (EI\, w^{IV} - \mu \omega^2\, w)\, \psi_i\, dx = \int_0^l (-\mu \omega^2\, w)\, \psi_i\, dx$$

$$f_i = [V\, \psi_i - M\, \psi_i']_0^l$$

and where ($\mu = \text{constant}$)

$$\tilde{K}_{ij} = \int_0^l EI\, \psi_i''\, \psi_j''\, dx - \int_0^l \mu \omega^2\, \psi_i\, \psi_j\, dx$$

$$\hat{=} \frac{EI}{l^3} \begin{bmatrix} 12 & -6l & -12 & -6l \\ & 4l^2 & 6l & 2l^2 \\ & & 12 & 6l \\ \text{sym.} & & & 4l^2 \end{bmatrix} - \frac{\mu l \omega^2}{420} \begin{bmatrix} 156 & -22l & 54 & 13l \\ & 4l^2 & -13l & -3l^2 \\ & & 156 & 22l \\ \text{sym.} & & & 4l^2 \end{bmatrix}$$

is the approximate stiffness matrix. It is the sum of the first-order stiffness matrix and the so-called "*consistent mass matrix*".

Finally the engineer adds p and f

$$\tilde{f} = p + f$$

and calls $\tilde{f}$ the vector of the modefied end forces.

Fig. 9.15 compares the exact eigenfunction of a hinged beam, $w = \boldsymbol{\Theta}^T \boldsymbol{\delta}$, where $\omega = \pi^2$ and $\boldsymbol{\delta} = \{0, \delta_2, 0, -\delta_2\}^T$ are the non-trivial solutions of the problem

$$G(w, w) = \boldsymbol{f}^T \boldsymbol{\delta} - \boldsymbol{\delta}^T K(\omega)\, \boldsymbol{\delta} = 0 \qquad \boldsymbol{f} = \{f_1, 0, f_3, 0\}^T$$

with its approximation, $w = \boldsymbol{\Psi}^T \boldsymbol{\delta}$, where $\tilde{\omega} = 10.95$ and $\boldsymbol{\delta} = \{0, \delta_2, 0, -\delta_2\}^T$ are the non-trivial solutions of the problem

$$G(w, w) = \tilde{\boldsymbol{f}}^T \boldsymbol{\delta} - \boldsymbol{\delta}^T \tilde{K}(\tilde{\omega})\, \boldsymbol{\delta} = 0 \qquad \tilde{\boldsymbol{f}} = \{\tilde{f}_1, 0, \tilde{f}_3, 0\}^T$$

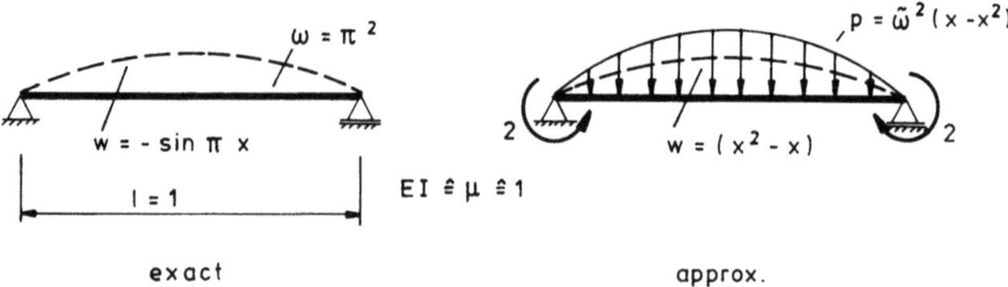

Figure 9.15

9.8 Hamilton's Principle

As in the case of the second-order beam equation the fact that the amplitude, $w(x)$, of the exact eigenfunction satisfies

$$f^T \delta = \delta^T K \delta = 0$$

does not mean that the vibrating beam is free of work and energy.

The full expression of the internal energy is obtained if we integrate $D w \, \hat{w}$ (D is the original partial differential equation, (9.23)) over the full domain, the rectangle

$$\Omega = [0, l] \times [0, T]$$

where $T > 0$ is an arbitrary time mark.

Doing integration by parts once with respect to x—in the integral $(EI w^{IV}, \hat{w})_{(0, l)}$—and once with respect to t—in the integral $(\mu \ddot{w}, \hat{w})_{(0, T)}$—we obtain

$$p : w(x, t) \in C^4[0, l] \times C^2[0, T], \; \hat{w}(x, t) \in C^2[0, l] \times C^1[0, T]$$

$$q : G(w, \hat{w}) = \int_0^T \int_0^l (EI w^{IV} + \mu \ddot{w}) \, \hat{w} \, dx \, dt + \int_0^T [V \hat{w} - M \hat{w}']_0^l \, dt$$

$$- \int_0^l [\mu \dot{w} \hat{w}]_0^T \, dx - \int_0^T \int_0^l \left(\frac{M \hat{M}}{EI} - \mu \dot{w} \dot{\hat{w}} \right) dx \, dt = 0 \qquad (9.25)$$

The last integral is the first variation of the internal energy

$$F(w) = \frac{1}{2} \int_0^T \int_0^l \left(\frac{M^2}{EI} - \mu \dot{w}^2 \right) dx \, dt$$

with respect to $\hat{w}(x, t)$.

Substituting the eigensolution

$$w(x, t) = \sin(\pi x) \cos(\pi^2 t)$$

into this identity we obtain at an arbitrary time $T > 0$

$$\frac{1}{2} G(w, w) = \frac{1}{2} \int_0^l [\mu \dot{w} w]_0^T \, dx - \frac{1}{2} \int_0^T \int_0^l \left(\frac{M^2}{EI} - \mu \dot{w}^2 \right) dx \, dt$$

$$= \frac{\pi^2}{4} \sin(2 \pi^2 T) - \frac{\pi^2}{4} \sin(2 \pi^2 T) = 0$$

and this means that the external eigenwork and the internal energy are not identical zero, both quantities oscillate with time.

The identity (9.25) will lead us now also to Hamilton's principle.

Assume that external loads, $p(x, t)$, force the hinged beam to vibrate with time. Let $w_0(x)$ be the shape of the elastic line and $\dot{w}_0(x)$ its speed at time $t = 0$. The full set of equations then reads

$$EI \, w^{IV} + \mu \ddot{w} = p \qquad 0 < x < l, \qquad 0 < t$$

$$w(x, 0) = w_0, \qquad \dot{w}(x, 0) = \dot{w}_0, \qquad 0 \leqslant x \leqslant l$$

$$w(0, t) = w(l, t) = M(0, t) = M(l, t) = 0 \qquad 0 \leqslant t$$

Substituting the solution $w = w_s$ and a geometrically admissible function $\hat{w}$, that is a function from

$$R_{1,0}(T) = \{\hat{w}(x,t) \in C^2[0,l] \times C^1[0,T] \,|\, \hat{w}(x,0) = \hat{w}(x,T) = 0, 0 \leqslant x \leqslant l,$$

$$\hat{w}(0, t) = \hat{w}(l, t) = 0, 0 \leqslant t \leqslant T\}$$

into the first identity renders the expression

$$(-1) G(w_s, \hat{w}) = \int_0^T \int_0^l p \hat{w} \, dx \, dt - \int_0^T \int_0^l \left(\frac{M \hat{M}}{EI} - \mu \dot{w} \dot{\hat{w}} \right) dx \, dt = 0$$

$$\forall \, \hat{w} \in R_{1,0}(T)$$

which is the first variation of the functional

$$\Pi_1(w) = \frac{1}{2} \int_0^T \int_0^l \left(\frac{M^2}{EI} - \mu \dot{w}^2 \right) dx \, dt - \int_0^T \int_0^l p w \, dx \, dt$$

$$= U - T + A = L$$

the so-called Lagrangian, see [F4] p. 318.

We have, thus, found Hamilton's principle:

"The time integral of the Lagrangian function over a time interval 0 to T is stationary for the 'actual' motion with respect to all admissible virtual displacements which vanish, first, at instants of time 0 and T and at all points of the body, and second, over Γ_1, where the displacements are prescribed, throughout the entire time interval."

Note that we do not claim that the solution $w(x, t)$ renders the Lagrangian an extremum. For this to be true the internal energy must be positive definite on $R_{1,0}(T)$. But this is in no way obvious.

10 Nonlinear Theory of Elasticity

We extend in this chapter our formulations to the nonlinear theory of elasticity (geometric and physical nonlinearities) and the large displacement analysis of beams and plates (geometric nonlinearities)

10.1 The Differential Equations

All indices have the range $\{1, 2, 3\}$ and summation is implied if an index appears twice.

The capital letters $E = [\varepsilon_{ij}]$ and $S = [\sigma_{ij}]$ denote matrices and the lower case letter $u = \{u_i\}$ a vector. All three, u, E, S, are functions of x.

The matrix $\nabla u = [u_{i,j}]$ is the gradient of the vector-valued function $u = \{u_i\}$. The symbol $E(\ )$ is an operator whose arguments are the displacement fields u. The result is a symmetric matrix

$$E(u) = \frac{1}{2}\,(\nabla u + \nabla u^T + \nabla u^T \nabla u) \tag{10.1}$$

with the elements

$$\varepsilon_{ij}(u) = \frac{1}{2}\,(u_{i,j} + u_{j,i} + u_{k,i}u_{k,j})$$

We assume the material to be hyperelastic, i.e. we assume that there exists a (sufficiently often differentiable) function

$$W(E) = W(\varepsilon_{11}, \varepsilon_{12}, \ \ldots \ \varepsilon_{33})$$

termed the specific strain energy whose derivatives are the stresses

$$\sigma_{ij} = \sigma_{ij}(\varepsilon_{11}, \varepsilon_{12}, \ \ldots \varepsilon_{33}) = \frac{\partial W(\varepsilon_{11}, \varepsilon_{12}, \ \ldots \ \varepsilon_{33})}{\partial \varepsilon_{ij}} \tag{10.2}$$

and that these equations are invertible, that is the strains ε_{ij} can be expressed in terms of the stresses.

By the symbol

$$W'(E) = [\partial W/\partial \varepsilon_{ij}]$$

we denote the symmetric 3×3 matrix formed with the derivatives of W.

The scalar-valued function

$$W^*(S) = S \cdot E - W(E)$$

(replace the tensor E on the right-hand side by the tensor S according to the postulated inverse of Eq. (10.2)) is termed the specific complementary strain energy.

In the linear theory (isotropic material is a hyperelastic material) and in the geometrically nonlinear theory it is, e. g.,

$$W = \frac{1}{2} C[E] \cdot E, \quad S = C[E], \quad E = C^{-1}[S]$$

and, hence,

$$W^*(S) = S \cdot C^{-1}[S] - \frac{1}{2} S \cdot C^{-1}[S] = \frac{1}{2} S \cdot C^{-1}[S]$$

Next, we formulate the system A of the nonlinear theory of elasticity.

In the triple $\{u, E, S\}$ we let E the Lagrangian-strain tensor and S the second Piola-Kirchhoff-stress tensor.

If p are the volume forces then the elastic state $\Sigma = \{u, E, S\}$ satisfies at every interior point x of Ω (the domain of the undeformed body) the equations

$$E(u) - E = 0_{(3 \times 3)} \qquad \frac{1}{2}(u_{i,j} + u_{j,i} + u_{k,i}u_{k,j}) - \varepsilon_{ij} = 0$$

$$W'(E) - S = 0_{(3 \times 3)} \quad \hat{=} \qquad \frac{\partial W}{\partial \varepsilon_{ij}} - \sigma_{ij} = 0 \qquad (10.3)$$

$$-\operatorname{div}(S + \nabla u \, S) = p_{(3)} \qquad -(\sigma_{ij} + u_{i,k}\sigma_{kj})_{,j} = p_i$$

These equations are supplemented by boundary conditions of displacement type, $u = \bar{u}$ on Γ_1, and force type, $\tau(S, u) = \bar{t}$ on Γ_2, where

$$\tau(S, u) = (S + \nabla u \, S) \, n \qquad \tau_i = (\sigma_{ij} + u_{i,k}\sigma_{kj}) \, n_j$$

is the traction vector on the boundary.

10.2 The first Identity

To start we choose among all symmetric elastic states

$$\mathscr{S} = \{\Sigma = \{u, E, S\} | E \text{ and } S \text{ symmetric}\}$$

the two subsets

$$\mathscr{S}^a = \{\Sigma \in \mathscr{S} \,|\, \pmb{u} \in C^2(\bar{\Omega}),\ \pmb{E} \in C(\bar{\Omega}),\ \pmb{S} \in C^1(\bar{\Omega})\}$$

$$\mathscr{S}^b = \{\Sigma \in \mathscr{S} \,|\, \pmb{u} \in C^1(\bar{\Omega}),\ \pmb{E} \in C(\bar{\Omega}),\ \pmb{S} \in C(\bar{\Omega})\}$$

and form with two such states $\Sigma, \hat{\Sigma} \in \mathscr{S}^a \times \mathscr{S}^b$ the expression

$$\langle A(\Sigma), \hat{\Sigma} \rangle = \int\limits_{\Omega} \left(E(\pmb{u}) - \pmb{E} \right) \cdot \hat{\pmb{S}}\, d\Omega + \int\limits_{\Omega} \left(W'(\pmb{E}) - \pmb{S} \right) \cdot \hat{\pmb{E}}\, d\Omega$$
$$+ \int\limits_{\Omega} - \operatorname{div}(\pmb{S} + \nabla \pmb{u}\, \pmb{S}) \cdot \hat{\pmb{u}}\, d\Omega \qquad (10.4)$$

Consider now the auxiliary lemma

$$p:\ \pmb{u} \in C^2, \hat{\pmb{u}}, \pmb{S} \in C^1,\ \pmb{S}\ symmetric$$

$$q:\ \int\limits_{\Omega} - \operatorname{div}(\pmb{S} + \nabla \pmb{u}\, \pmb{S}) \cdot \hat{\pmb{u}}\, d\Omega = - \int\limits_{\Gamma} \tau(\pmb{S}, \pmb{u}) \cdot \hat{\pmb{u}}\, ds + \int\limits_{\Omega} E_{\pmb{u}}(\hat{\pmb{u}}) \cdot \pmb{S}\, d\Omega \qquad (10.5)$$

The matrix in this lemma

$$E_{\pmb{u}}(\hat{\pmb{u}}) = \frac{1}{2} \left(\nabla \hat{\pmb{u}} + \nabla \hat{\pmb{u}}^T + \nabla \pmb{u}^T \nabla \hat{\pmb{u}} + \nabla \hat{\pmb{u}}^T \nabla \pmb{u} \right)$$

with the elements

$$\varepsilon_{ij_{\pmb{u}}}(\hat{\pmb{u}}) = \frac{1}{2} \left(\hat{u}_{i,j} + \hat{u}_{j,i} + u_{k,i} \hat{u}_{k,j} + \hat{u}_{k,i} u_{k,j} \right)$$

is the Gateaux differential of the matrix $E(\pmb{u})$

$$\frac{d}{d\varepsilon} \left\{ E(\pmb{u} + \varepsilon \hat{\pmb{u}}) \right\} \Big|_{\varepsilon = 0} = E_{\pmb{u}}(\hat{\pmb{u}})$$

According to this lemma we commit no error if we replace the last integral in $\langle A(\Sigma), \hat{\Sigma} \rangle$ by the right-hand side of Eq. (10.5). Hence, Eq. (10.4) is equivalent with

$$\langle A(\Sigma), \hat{\Sigma} \rangle = \int\limits_{\Omega} \left(E(\pmb{u}) - \pmb{E} \right) \cdot \hat{\pmb{S}}\, d\Omega + \int\limits_{\Omega} \left(W'(\pmb{E}) - \pmb{S} \right) \cdot \hat{\pmb{E}}\, d\Omega$$
$$+ \int\limits_{\Omega} E_{\pmb{u}}(\hat{\pmb{u}}) \cdot \pmb{S}\, d\Omega - \int\limits_{\Gamma} \tau(\pmb{S}, \pmb{u}) \cdot \hat{\pmb{u}}\, ds \qquad (10.6)$$

The sum of the three domain integrals in this equation

$$E(\Sigma, \hat{\Sigma}) = \int\limits_{\Omega} \left(E(\pmb{u}) - \pmb{E} \right) \cdot \hat{\pmb{S}}\, d\Omega + \int\limits_{\Omega} \left(W'(\pmb{E}) - \pmb{S} \right) \cdot \hat{\pmb{E}}\, d\Omega + \int\limits_{\Omega} E_{\pmb{u}}(\hat{\pmb{u}}) \cdot \pmb{S}\, d\Omega$$

is the first variation of the functional

$$F(\Sigma) = \int_{\Omega} \big(E(u) - E\big) \cdot S \, d\Omega + \int_{\Omega} W(E) \, d\Omega$$

at Σ in the direction of $\hat{\Sigma}$.

$$E(\Sigma, \hat{\Sigma}) = \delta F(\Sigma, \hat{\Sigma}) = \frac{d}{d\varepsilon} F(\Sigma + \varepsilon \hat{\Sigma})\Big|_{\varepsilon = 0} \tag{10.7}$$

Regrouping Eq. (10.6) we have, thus, found the first identity

$$p: \ \Sigma, \hat{\Sigma} \in \mathscr{S}^a \times \mathscr{S}^b$$

$$q: \ G(\Sigma, \hat{\Sigma}) = \langle A(\Sigma), \hat{\Sigma} \rangle + \int_{\Gamma} \tau(S, u) \cdot \hat{u} \, ds - E(\Sigma, \hat{\Sigma}) = 0 \tag{10.8}$$

and, hence, also the principle of virtual displacements

$$p: \ \Sigma \in \mathscr{S}^a$$

$$q: \ G(\Sigma, \hat{\Sigma}) = 0 \quad \forall \, \hat{\Sigma} \in \mathscr{S}^b$$

If we interchange the places of Σ and $\hat{\Sigma}$ this results in

$$p: \ \Sigma \in \mathscr{S}^b$$

$$q: \ G(\hat{\Sigma}, \Sigma) = \langle A(\hat{\Sigma}), \Sigma \rangle + \int_{\Gamma} \tau(\hat{S}, \hat{u}) \cdot u \, ds - E(\hat{\Sigma}, \Sigma) = 0 \quad \forall \, \hat{\Sigma} \in \mathscr{S}^a$$

But this statement cannot be interpreted as the principle of virtual forces because the symmetry condition

$$E(\Sigma, \hat{\Sigma}) \neq E(\hat{\Sigma}, \Sigma) \tag{10.9}$$

is violated and, hence, $E(\hat{\Sigma}, \Sigma)$ is not, contrary to the left-hand side, the first variation of the functional $F(\Sigma)$

$$\frac{d}{d\varepsilon} F(\Sigma + \varepsilon \hat{\Sigma})\Big|_{\varepsilon = 0} \neq E(\hat{\Sigma}, \Sigma) \tag{10.10}$$

Consequently the principle of virtual forces does not apply in nonlinear elasticity. Next let us turn to Betti's principle, corollary 3. In linear mechanics we formulated

$$B(\Sigma, \hat{\Sigma}) = G(\Sigma, \hat{\Sigma}) - G(\hat{\Sigma}, \Sigma) = 0$$

If we do the same in nonlinear elasticity then, due to the violated symmetry condition, see Eq. (10.9), the energy terms $E(\Sigma, \hat{\Sigma})$ and $E(\hat{\Sigma}, \Sigma)$ do not drop out.

Hence we cannot interpret the equation $B(\Sigma, \hat{\Sigma}) = 0$ as a statement of Betti's principle. This principle does not apply in nonlinear elasticity.

Then what about corollary 4, the principle "eigenwork = int. energy"?

If we multiply the first identity on the diagonal, $G(\Sigma, \Sigma)$, with $1/2$ we obtain

$$G(\Sigma, \Sigma) = \frac{1}{2} \langle A(\Sigma), \Sigma \rangle + \frac{1}{2} \int_{\Gamma} \tau(S, u) \cdot u \, ds - \frac{1}{2} E(\Sigma, \Sigma) = 0$$

But since $1/2 \, E(\Sigma, \Sigma)$ is not the internal energy $F(\Sigma)$ this equation cannot be interpreted as the statement that "the external eigenwork is stored as internal energy".

Instead this equation expresses the fact that the first variation of the internal energy at Σ in the direction of Σ itself is equal to the virtual work done by the external forces of Σ acting through Σ. It is simply the same statement as $G(\Sigma, \hat{\Sigma})$ only that $\hat{\Sigma} = \Sigma$.

Why this is so and how the mathematical expression of the principle "eigenwork = int. energy" formulates in nonlinear elasticity is discussed in section 10.7.

What about the equilibrium conditions, corollary 5, which in linear mechanics read

$$G(u, r) = 0 \quad \forall \, r = \text{rigid-body movements?}$$

These conditions, naturally, also apply in nonlinear elasticity. Only that the rigid-body movements are now the displacement fields of the form

$$r = a + Q x, \quad \det Q = 1$$

That is, translations and genuine rotations. The rotations must no longer be infinitesimal small as in the linear theory.

Remember that the rigid-body movements are those displacement fields which have zero strain energy,

$$u = \text{rigid-body movement} \ \rightarrow \ \frac{1}{2} E(u, u) = 0$$

This is, if the stress-strain relations are "reasonable", equivalent with the fact that the strain tensor $E(u)$ is zero. The zeros of the strain tensor $E(u)$ defined in Eq. (10.1) are, exactly, the displacement fields of the form $u = a + Q x$, see [G 2] p. 30.

We collect our results in the following table 10.1

principle		*linear*	*nonlinear*
of virtual displacements	$G(u, \hat{u}) = 0$	*yes*	*yes*
of virtual forces	$G(\hat{u}, u) = 0$	*yes*	*no*
"eigenwork = int. energy"	$\frac{1}{2} G(u, u) = 0$	*yes*	*(no)*
Betti	$B(u, \hat{u}) = 0$	*yes*	*no*
Equilibrium	$G(u, r) = 0$	*yes*	*yes*

The (no) is to indicate that this principle holds in nonlinear mechanics but that $1/2$ $G(u, u) = 0$ is not its expression.

10.3 Energy Principles

In this section we shall derive, by way of the form $V(\Sigma, \hat{\Sigma})$, the energy principles associated with regular $bvps$ of the operator $A(\Sigma)$.

We trust that the reader is familiar with section 4.12 where we explained the formalism which leads to the form $V(\Sigma, \hat{\Sigma})$ and therewith to the basic functional $\Pi(\Sigma)$ of a regular bvp.

To start consider the first identity of the operator $A(\Sigma)$

$$G(\Sigma, \hat{\Sigma}) = \langle A(\Sigma), \hat{\Sigma} \rangle + \int_\Gamma \tau(S, u) \cdot \hat{u} \, ds - E(\Sigma, \hat{\Sigma}) = 0$$

of which the form $V(\Sigma, \hat{\Sigma})$ is essentially an "updated" version. That is to obtain $V(\Sigma, \hat{\Sigma})$ we must add to $(-1) G(\Sigma, \hat{\Sigma})$ and subtract from it (simultaneously) the missing terms of the first variation of the functional

$$R(S, u) = \int_\Gamma \tau(S, u) \cdot u \, ds = \int_\Gamma (S + \nabla u S) \, n \cdot u \, ds$$

the boundary integral in $G(\Sigma, \Sigma)$.

The first variation of this functional is the expression

$$\delta R(S, u; \hat{S}, \hat{u}) = \int_\Gamma \tau_S(\hat{S}, u) \cdot u \, ds + \int_\Gamma \tau_u(S, \hat{u}) \cdot u \, ds + \int_\Gamma \tau(S, u) \cdot \hat{u} \, ds$$

where the terms

$$\tau_S(\hat{S}, u) := \frac{d}{d\eta} \left(S + \eta \hat{S} + \nabla u (S + \eta \hat{S}) \right) n \Big|_{\eta = 0} = (\hat{S} + \nabla u \hat{S}) \, n \qquad (10.11)$$

$$\tau_u(S, \hat{u}) := \frac{d}{d\varepsilon} \left(S + (\nabla u + \varepsilon \nabla \hat{u}) \, S) \, n \right) \Big|_{\varepsilon = 0} = \nabla \hat{u} \, S n \qquad (10.12)$$

are the Gateaux differentials of $\tau(S, u) = (S + \nabla u S) \, n$ in the direction of $\hat{S}$ and $\hat{u}$, resp.

Missing, therefore, are in $G(\Sigma, \hat{\Sigma})$ the first two integrals of the expression $\delta R(S, u; \hat{S}, \hat{u})$. Hence the form $V(\Sigma, \hat{\Sigma})$ is obtained if we add to $(-1) G(\Sigma, \hat{\Sigma})$ the zero sum

$$N(\Sigma, \hat{\Sigma}) = \int_\Gamma [\tau_S(\hat{S}, u) \cdot u + \tau_u(S, \hat{u}) \cdot u] \, ds - \int_\Gamma [\tau_S(\hat{S}, u) \cdot u$$

$$+ \tau_u(S, \hat{u}) \cdot u] \, ds = 0$$

and place the curly brackets $\{\ldots \ldots\}$,

$$V(\Sigma, \hat{\Sigma}) = (-1)\, G(\Sigma, \hat{\Sigma}) + N(\Sigma, \hat{\Sigma}) = \left\{ - <A(\Sigma), \hat{\Sigma}> - \int_{\Gamma} \tau(S, u) \cdot \hat{u}\, ds \right.$$

$$\left. + \int_{\Gamma} [\tau_S(\hat{S}, u) \cdot u + \tau_u(S, \hat{u}) \cdot u]\, ds \right\} - \int_{\Gamma} [\tau_S(\hat{S}, u) \cdot u$$

$$+ \tau_u(S, \hat{u}) \cdot u]\, ds + E(\Sigma, \hat{\Sigma}) = 0$$

(For clarity we have underlined the integrals of the first identity).

Consider now the mixed *bvp*

$$A(\Sigma) = \{0_{(3 \times 3)}, \, 0_{(3 \times 3)}, \, p_{(3)}\}^T \quad \text{in } \Omega \, .$$

$$u = \bar{u} \text{ on } \Gamma_1, \ \tau(S, u) = \bar{t} \text{ on } \Gamma_2, \ \Gamma = \Gamma_1 \cup \Gamma_2 \tag{10.13}$$

and assume the solution Σ is in $\mathscr{S}^a$.

Substituting the data of the *bvp* (10.13) into $V(\Sigma, \hat{\Sigma})$ and collecting terms we obtain

$$\delta \Pi(\Sigma, \hat{\Sigma}) = - \int_{\Omega} p \cdot \hat{u}\, d\Omega + \int_{\Gamma_1} [\tau_S(\hat{S}, u) + \tau_u(S, \hat{u})] \cdot (\bar{u} - u)\, ds$$

$$- \int_{\Gamma_1} \tau(S, u) \cdot \hat{u}\, ds - \int_{\Gamma_2} \bar{t} \cdot \hat{u}\, ds + E(\Sigma, \hat{\Sigma}) = 0 \tag{10.14}$$

which is the first variation of **the basic functional**

$$\Pi(\Sigma) = \int_{\Omega} (E(u) - E) \cdot S\, d\Omega + \int_{\Omega} W(E)\, d\Omega - \int_{\Omega} p \cdot u\, d\Omega$$

$$+ \int_{\Gamma_1} \tau(S, u) \cdot (\bar{u} - u)\, ds - \int_{\Gamma_2} \bar{t} \cdot u\, ds \tag{10.15}$$

Hence, our first result is: if $\Sigma \in \mathscr{S}^a$ is a solution of the *bvp* then Σ is also a stationary point of the basic functional, that is

$$\delta \Pi(\Sigma, \hat{\Sigma}) = 0 \quad \forall\, \hat{\Sigma} \in \mathscr{S}^b$$

To obtain the converse result we replace in Eq. (10.14) the integral $E(\Sigma, \hat{\Sigma})$ according to Eq (10.6). This renders the result

$$- \int_{\Omega} (p + \operatorname{div}(S + \nabla u\, S)) \cdot \hat{u}\, d\Omega + \int_{\Gamma_1} (\tau_S(\hat{S}, u) + \tau_u(S, \hat{u})) \cdot (\bar{u} - u)\, ds$$

$$- \int_{\Gamma_2} (\bar{t} - \tau(S, u)) \cdot \hat{u}\, ds - \int_{\Omega} (S - W'(E)) \cdot \hat{E}\, d\Omega$$

$$- \int_{\Omega} (E - E(u)) \cdot \hat{S}\, d\Omega = 0 \quad \forall\, \hat{\Sigma} \in \mathscr{S}^b \tag{10.16}$$

Now, let $\Sigma \in \mathcal{S}^a$ be an elastic state which satisfies Eq. (10.14) and, therefore, also Eq. (10.16) for all $\hat{\Sigma} \in \mathcal{S}^b$. Then, by virtue of lemma 1, 2 and 3 (see chapter 4 and 7) the elastic state Σ satisfies the equations

$$\operatorname{div}(S + \nabla u\, S) + p = 0, \quad E - E(u) = 0, \quad W'(E) - S = 0 \quad \text{in } \Omega$$

$$\bar{t} - \tau(S, u) = 0 \quad \text{in } \Gamma_2$$

That is, Σ is very nearly a solution of the bvp. What is only missing is the statement that $u = \bar{u}$ on Γ_1. To draw this conclusion we need an additional lemma

Lemma 4

If the equation

$$\int\limits_{\Gamma_1} \left(\tau_S(\hat{S}, u) + \tau_u(S, \hat{u}) \right) \cdot (\bar{u} - u)\, ds = 0$$

is satisfied for all $\hat{S}$ and for all $\hat{u} \in C^\infty$ which satisfy $\hat{u} = 0$ on Γ_2 then $u = \bar{u}$ on Γ_1.

Though the proof of this lemma is yet open we dare to give our (preliminary) result the status of a principle

Basic principle

The elastic state $\Sigma \in \mathcal{S}^a$ solves the bvp (10.13) if and only if $\Sigma \in \mathcal{S}^a$ is a stationary point of the basic functional (10.15) with respect to all variations $\hat{\Sigma} \in \mathcal{S}^b$.

To formulate further energy principles we consider, as in the linear theory, the following subsets of $\mathcal{S}$

$$\mathcal{S}_0 = \{\Sigma \in \mathcal{S}^a \,|\, W'(E) - S = 0\}$$

$$\mathcal{S}_1 = \{\Sigma \in \mathcal{S}^b \,|\, W'(E) - S = 0,\ E(u) - E = 0,\ \bar{u} = u \text{ on } \Gamma_1\}$$

$$\mathcal{S}_2 = \{\Sigma \in \mathcal{S}^a \,|\, W'(E) - S = 0,\ -\operatorname{div}(S + \nabla u\, S) = p,\ \tau(S, u) = \bar{t} \text{ on } \Gamma_2\}$$

The restrictions of the basic functional to these classes are

$$\Pi|_{\mathcal{S}_0} = \Pi_0(S, u) = \int\limits_\Omega E(u) \cdot S\, d\Omega - \int\limits_\Omega W^*(S)\, d\Omega - \int\limits_\Omega p \cdot u\, d\Omega$$

$$- \int\limits_{\Gamma_2} \bar{t} \cdot u\, ds + \int\limits_{\Gamma_1} \tau(S, u) \cdot (\bar{u} - u)\, ds$$

$$\Pi|_{\mathcal{S}_1} = \Pi_1(u) = \int\limits_\Omega W(E(u))\, d\Omega - \int\limits_\Omega p \cdot u\, d\Omega - \int\limits_{\Gamma_2} \bar{t} \cdot u\, ds$$

$$\Pi|_{\mathcal{S}_2} = \Pi_2(S, u) = - \int\limits_\Omega W^*(S)\, d\Omega$$

$$- \frac{1}{2} \int\limits_\Omega \nabla u^T \nabla u \cdot S\, d\Omega + \int\limits_{\Gamma_1} \tau(S, u) \cdot (\bar{u} - u)\, ds$$

The functional $\Pi_0(S, u)$ is the functional of a Hellinger-Reissner principle which, as in the linear theory, is a corollary of the basic principle.

The functional $\Pi_1(u)$ belongs to the

Principle of stationary value of potential energy

The elastic state $\Sigma \in \mathscr{S}^a \cap \mathscr{S}_1$ *is a solution of the bvp* (10.13) *if and only if* $\Sigma \in \mathscr{S}^a \cap \mathscr{S}_1$ *is a stationary point of the functional* $\Pi_1(u)$, *that is if* Σ *satisfies the variational equation*

$$\delta \Pi_1(u, \hat{u}) = 0 \quad \forall \, \hat{u} \in \mathscr{S}_{1,0}$$

where the variations $\hat{u}$ *belong to the class*

$$\mathscr{S}_{1,0} = \{ \Sigma \in \mathscr{S}^b \, | \, W'(E) - S = 0, \, E(u) - E = 0, \, u = 0 \text{ on } \Gamma_1 \}$$

The proof of this principle is standard (lemma 4 is not needed).

We do not speak of a minimum of potential energy because this would require further information about the function $W(E)$ as the reader will see for himself if he formulates the difference $\Pi_1(u + \hat{u}) - \Pi_1(u)$.

The functional $\Pi_2(S, u)$ is the restriction of the basic functional to the class $\mathscr{S}_2$, the class of all statically admissible elastic states. If the pair $\{\Sigma, \hat{\Sigma}\}$ is from $\mathscr{S}_2 \times \mathscr{S}_{2,0}$ then we have for all numbers ε the equality

$$\Pi_2(\Sigma + \varepsilon \hat{\Sigma}) = \Pi(\Sigma + \varepsilon \hat{\Sigma})$$

and, hence, as well

$$\left. \frac{d}{d\varepsilon} \Pi_2(\Sigma + \varepsilon \hat{\Sigma}) \right|_{\varepsilon = 0} = \left. \frac{d}{d\varepsilon} \Pi(\Sigma + \varepsilon \hat{\Sigma}) \right|_{\varepsilon = 0}$$

This means that the first variation, $\delta \Pi_2(\Sigma, \hat{\Sigma})$, of the functional $\Pi_2(\Sigma)$ can be calculated by substituting the pair $\{\Sigma, \hat{\Sigma}\}$ into the first variation, $\delta \Pi(\Sigma, \hat{\Sigma})$, of the basic functional, see Eq. (10.14).

This renders, if we consider in addition the equations

$$p = -\operatorname{div}(S + \nabla u S) \quad \text{in } \Omega, \quad \bar{t} = \tau(S, u) \quad \text{on } \Gamma_2$$

and Eq. (10.12), the result that the first variation of Π_2 vanishes with respect to all $\hat{\Sigma}$ in $\mathscr{S}_{2,0}$.

$$\delta \Pi_2(S, u; \hat{S}, \hat{u}) = -\int_\Omega (E - E(u)) \cdot \hat{S} \, d\Omega + \int_{\Gamma_1} (\tau_S(\hat{S}, u)$$

$$+ \tau_u(S, \hat{u})\,(\bar{u} - u))\, ds = 0 \quad \forall \, \hat{\Sigma} \in \mathscr{S}_{2,0}$$

With lemma 1 and lemma 4 it, thus, follows that every stationary point $\Sigma \in \mathscr{S}_2$ is geometrically compatible and that every such Σ solves the bvp (10.13). We formulate this as the

Principle of stationary value of complementary energy

The elastic state $\Sigma \in \mathscr{S}_2$ is a solution of the bvp if and only if $\Sigma \in \mathscr{S}_2$ is a stationary point of the functional Π_2, that is if Σ satisfies the equation

$$\delta \Pi_2(\Sigma, \hat{\Sigma}) = 0 \quad \forall \, \hat{\Sigma} \in \mathscr{S}_{2,0}$$

Here, $\mathscr{S}_{2,0}$ is the class

$$\mathscr{S}_{2,0} = \{\Sigma \in \mathscr{S}^a | W'(E) - S = 0, \quad -\operatorname{div}(S + \nabla u S) = 0, \ \tau(S, u) = 0 \text{ on } \Gamma_2\}$$

Note: in the literature it is usually $(-1)\,\Pi_2(S, u)$ and not $\Pi_2(S, u)$ which is termed complementary energy.

10.4 Incremental Procedures

Procedures which divide the loading path in a number of equilibrium states $i = 1, 2 \ldots N$ are only meaningful in nonlinear mechanics because the increments of a linear equation, e.g. $3x = y$, satisfy the same equation

$$3\Delta x = \Delta y$$

as the original quantities.

If the equation is nonlinear

$$y = g(x)$$

e.g. $g(x) = x^3$ then we expand $g(x)$ in a Taylor series

$$y + \Delta y = g(x) + g'(x)\,\Delta x + \ldots$$

and determine Δx approximately by

$$\Delta y = g'(x)\,\Delta x$$

This is the method of tangential stiffnesses.

If the equations involve operators which act on functions such as

$$E = E(u)$$

then we replace the partial derivatives by Gateaux derivatives

$$E + \Delta E = E(u) + E_u(\Delta u) + \dots$$

and determine Δu approximately so that the equation

$$\Delta E = E_u(\Delta u)$$

is satisfied.

We could, hence, perform a Gateaux expansion of the system (10.3) and formulate, thus, a system of differential equations for the increments Δu, ΔE, ΔS and then derive variational principles.

But there is a simpler approach:

We know that the first variation of the potential energy must vanish at the equilibrium point u

$$\delta \Pi_1(u, \hat{u}) = \delta F(u, \hat{u}) - \int_\Omega p \cdot \hat{u} \, d\Omega - \int_{\Gamma_2} \bar{t} \cdot \hat{u} \, ds = 0 \qquad \forall \, \hat{u} \in R_{1,0} \qquad (10.17)$$

Let u^0 the equilibrium point associated with p^0 and $\bar{t}^0$ and let Δu, Δp, $\Delta \bar{t}$ the increments which lead to

$$u = u^0 + \Delta u, \quad p = p^0 + \Delta p, \quad \bar{t} = \bar{t}^0 + \Delta \bar{t}$$

and assume we expand the single integrals in the first variation with respect to these increments

$$\delta F(u, \hat{u}) = \delta F(u^0, \hat{u}) + \mathrm{grad}_u \delta F(u^0, \hat{u}) \cdot \Delta u + \dots$$

$$\int_\Omega p \cdot \hat{u} \, d\Omega = \int_\Omega p^0 \cdot \hat{u} \, d\Omega + \int_\Omega \Delta p \cdot \hat{u} \, d\Omega$$

$$\int_{\Gamma_2} \bar{t} \cdot \hat{u} \, ds = \int_{\Gamma_2} \bar{t}^0 \cdot \hat{u} \, ds + \int_{\Gamma_2} \Delta \bar{t} \cdot \hat{u} \, ds$$

Then the first variation is, approximately,

$$\delta \Pi_1(u, \hat{u}) \simeq \delta \Pi_1(u^0, \hat{u}) + \mathrm{grad}_u \delta F(u^0, \hat{u}) \cdot \Delta u - \int_\Omega \Delta p \cdot \hat{u} \, d\Omega - \int_{\Gamma_2} \Delta \bar{t} \cdot \hat{u} \, ds$$

and because the first variation $\delta \Pi_1(u^0, \hat{u})$ is zero for all $\hat{u}$ the increment Δu (approximately) satisfies the equation

$$\mathrm{grad}_u \delta F(u^0, \hat{u}) \cdot \Delta u - \int_\Omega \Delta p \cdot \hat{u} \, d\Omega - \int_{\Gamma_2} \Delta \bar{t} \cdot \hat{u} \, ds = 0 \qquad \forall \, \hat{u} \in R_1 \qquad (10.18)$$

This variational principle for the increment of the displacement field is closely connected with the Newton-Raphson algorithm.

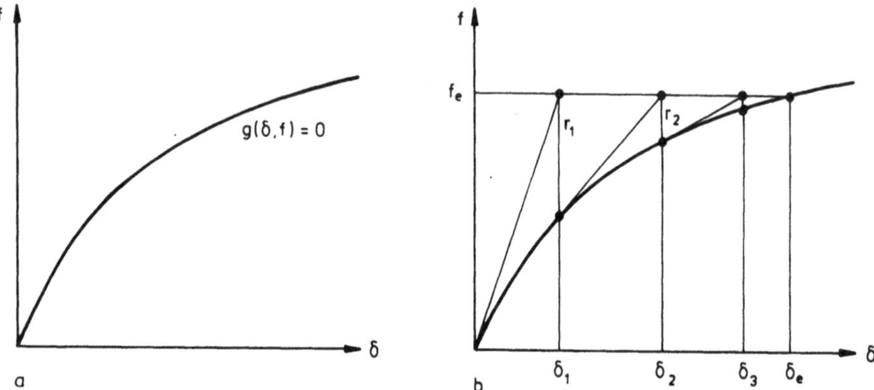

Figure 10.1

To see this let

$$k(\delta) - f = 0$$

be the discrete form (finite elements) of the first variation, see Eq. (10.17). Here $k(\delta)$ is a vector (in the linear case $k(\delta) = K\delta$) whose components are the derivatives of $F(u)$ with respect to δ_i and the components of the vector f

$$\frac{\partial}{\partial \delta_i} \left\{ \int_\Omega p \cdot u \, d\Omega - \int_{\Gamma_2} \bar{t} \cdot u \, ds \right\} = f_i$$

are the equivalent nodal forces.

To start assume for simplicity that $k(\delta)$ is a scalar-valued function, there is only one degree of freedom, δ, and let the curve in Fig. 10.1 a be the locus of all the points $\{\delta, f\}$ which satisfy the equation

$$g(\delta, f) = k(\delta) - f = 0$$

If we move along this curve the increments $d\delta$ and df must satisfy the equation

$$dg = g_{,\delta} \, d\delta + g_{,f} \, df = k'(\delta) \, d\delta - df = 0$$

According to the same logic the vector of the displacement increments, $\Delta\delta$, must, in the case of systems with more degrees of freedom, satisfy the equation

$$K_T(\delta) \, \Delta\delta - \Delta f = 0 \tag{10.19}$$

where

$$K_T = \left[\frac{\partial k_i}{\partial \delta_j}\right]$$

is the "*tangential stiffness matrix*". Eq. (10.19) is the discrete analogue of Eq. (10.18).

This tangential stiffness matrix appears also in the Newton-Raphson algorithm which, to find the zero δ of the equation

$$r(\delta) = k(\delta) - f = 0,$$

formulates the sequence

$$\delta_{i+1} = \delta_i - K_T(\delta_i)^{-1}\left(k(\delta_i) - f\right) \qquad i = 1, 2 \ldots$$

This sequence is equivalent with the iteration

$$K_T(\delta_i)\, \Delta\delta_{i+1} = r(\delta_i)$$

which coincides with Eq. (10.19) except that the load increments are now the unbalanced forces $r(\delta_i)$, see Fig. 10.1 b.

10.5 Large Displacement Analysis of Beams

The displacements are u (in horizontal direction) and w (in vertical direction). The quantities EA and EI are constant but all other quantities depend on x, see $[W_1]$ p. 142.

Def.
$$\varepsilon^N = u' + \frac{1}{2}(w')^2$$
$$\varepsilon^M = w''$$

Const.
$$N = EA\, \varepsilon^N$$
$$M = -EI\, \varepsilon^M$$

Equ.
$$-N' = p_x$$
$$-M'' - (Nw')' = p_z$$

Substituting these equations consecutively pairwise into each other, from top to bottom, we obtain the following system for the two displacements u and w alone

$$-EA\left(u' + \frac{1}{2}(w')^2\right)' = p_x$$

$$EI\, w^{IV} - \left(EA\left(u' + \frac{1}{2}(w')^2\right)w'\right)' = p_z \tag{10.20}$$

Let $v = \{u, w\}$ be the vector of the two displacements and let $\boldsymbol{D}v$ denote the left-hand side of the system (10.20). The first identity of the operator $\boldsymbol{D}$ is found if we transform the integral $(\boldsymbol{D}v, \hat{v})$.

Employing the notations

$$N = N(v) = EA\left(u' + \frac{1}{2}(w')^2\right), \quad M = M(w) = -EIw''$$

we obtain

$$\int_a^b \boldsymbol{D}v \cdot \hat{v} \, dx = \int_a^b [(\text{1st row}) \, \hat{u} + (\text{2nd row}) \, \hat{w}] \, dx$$

$$= \int_a^b [(-N'\hat{u} - (M'' + (Nw')') \, \hat{w}] \, dx$$

$$= [-N\hat{u} - (M' + Nw') \, \hat{w} + M\hat{w}']_a^b + E(v, \hat{v})$$

where

$$E(v, \hat{v}) = \int_a^b [-M\hat{w}'' + N(\hat{u}' + w'\hat{w}')] \, dx$$

$$= \int_a^b \left[\frac{M(w) \, M_w(\hat{w})}{EI} + \frac{N(v) \, N_v(\hat{v})}{EA}\right] dx$$

Hence, the first identity is

$$p: \quad v, \hat{v} \in \{C^2, C^4\} \times \{C^1, C^2\}$$

$$q: \quad G(v, \hat{v}) = \int_a^b \boldsymbol{D}v \cdot \hat{v} \, dx + [N\hat{u} + (M' + Nw') \, \hat{w} - \dot{M}\hat{w}']_a^b - E(v, \hat{v}) = 0$$

$$(10.21)$$

The integral $E(v, \hat{v})$ is the first variation of the functional

$$F(v) = \frac{1}{2} \int_a^b \left(\frac{N^2}{EA} + \frac{M^2}{EI}\right) dx$$

and, therefore, the essential condition, see section 4.12, is satisfied. However, $E(v, \hat{v})$ is not symmetric and, therefore, all of what was said above, see table 10.1, applies here as well.

The functional $R(v)$, i.e. the boundary integral in the first identity $G(v, v)$ (see section 4.12), is here the expression

$$R(v) = [Nu + (M' + Nw') \, w - Mw']_a^b$$

and its first variation the expression

$$\delta R(v, \hat{v}) = [N_v(\hat{v})\, u + (\hat{M}' + N_v(\hat{v})\, w' + N\hat{w}')\, w - \hat{M} w']_a^b$$

$$+ [N\hat{u} + (M' + Nw')\, \hat{w} - M\hat{w}']_a^b \tag{10.22}$$

where

$$N_v(\hat{v}) = \frac{d}{d\varepsilon}\, N(v + \varepsilon\hat{v})\bigg|_{\varepsilon=0} = E A\,(\hat{u}' + w'\hat{w}')$$

We obtain the form $V(v, \hat{v})$ when we add to $(-1)\, G(v, \hat{v})$ and subtract from it simultaneously the missing term of the first variation of $R(v)$, the first bracket in Eq. (10.22). Else, the form $V(v, \hat{v})$ is the expression

$$V(v, \hat{v}) = (-1)\, G(v, \hat{v}) + N(v, \hat{v}) = 0$$

where

$$N(v, \hat{v}) = [\text{1st bracket}]_a^b - [\text{1st bracket}]_a^b$$

10.6 Large Displacement Analysis of Plates

The displacements in the plane are $u = \{u_1, u_2\}$ and the deflection orthogonal to the plane is w. The material is considered to be linearly elastic and the parameters v, μ, K are considered to be constant. All other quantities are functions of $x = (x_1, x_2)$.

Def.
$$\varepsilon_{ij}^N = \frac{1}{2}\,(u_{i,j} + u_{j,i}) + \frac{1}{2}\, w_{,i} w_{,j}$$
$$i, j = 1, 2$$
$$\varepsilon_{ij}^M = w_{,ij}$$

Const.
$$N_{11} = \frac{2\mu h}{1 - v}\,(\varepsilon_{11}^N + v\varepsilon_{22}^N), \qquad N_{12} = N_{21} = 2\mu h\,\varepsilon_{12}^N$$

$$N_{22} = \frac{2\mu h}{1 - v}\,(\varepsilon_{22}^N + v\varepsilon_{11}^N)$$

$$M_{11} = -K(\varepsilon_{11}^M + v\varepsilon_{22}^M) \qquad M_{12} = M_{21} = -K(1 - v)\,\varepsilon_{12}^M$$

$$M_{22} = -K(\varepsilon_{22}^M + v\varepsilon_{11}^M)$$

Equ.
$$-N_{11,1} - N_{12,2} = p_1$$

$$-N_{21,1} - N_{22,2} = p_2$$

$$-M_{11,1} - 2M_{12,12} - M_{22,22} - (N_{11} w_{,1} + N_{12} w_{,2})_{,1}$$

$$-(N_{12} w_{,1} + N_{22} w_{,2})_{,2} = p_3$$

In the absolute notation these equations read

Def.
$$E^N(u, w) - E^N = 0_{(2 \times 2)}$$
$$E^M - E^M(w) = 0_{(2 \times 2)}$$

Const.
$$C^N[E^N] - N = 0_{(2 \times 2)}$$
$$C^M[E^M] + M = 0_{(2 \times 2)}$$

Equ.
$$-\operatorname{div} N = p_{(2)}$$
$$-\operatorname{div}^2 M - \operatorname{div}(N\nabla w) = p_{3(1)}$$

where $E^N(u, w)$ is the operator

$$E^N(u, w) = \frac{1}{2}(\nabla u + \nabla u^T) + \frac{1}{2}\nabla w \, \nabla w^T$$

and $E^M(w)$ the operator $E^M(w) = \nabla\nabla w$. The elasticity tensor C^M is the tensor C of the Kirchhoff plate in Eq. (1.28) and the elasticity tensor C^N is, up to the factor h (= thickness of the plate), identical with the tensor C of the elastic plate in Eq. (1.37).

We consider the following classes of functions:

$$\mathscr{S} = \{\Sigma = \{u, w; E^N, E^M; N, M\} \,|\, E^N, E^M, N, M \text{ symmetric}\}$$
$$\mathscr{S}^a = \{\Sigma \in \mathscr{S} \,|\, u, N \in C^1(\bar{\Omega}), w, M \in C^2(\bar{\Omega}), E^N, E^M \in C(\bar{\Omega})\}$$
$$\mathscr{S}^b = \{\Sigma \in \mathscr{S} \,|\, u \in C^1(\bar{\Omega}), w \in C^2(\bar{\Omega}), E^N, E^M, N, M \in C(\bar{\Omega})\}$$

Our aim is the derivation of work and energy principles. For such principles to exist it is sufficient that the system A satisfies the Eqs. (7.28) and (7.29) of section 7.4. We, therefore, first check for these conditions.

The elasticity tensors C^N and C^M are the tensors of the linear theory, $C[E + \varepsilon\hat{E}] = C[E] + \varepsilon C[\hat{E}]$, and they satisfy the symmetry condition $C[E] \cdot \hat{E} = C[\hat{E}] \cdot E$, hence, the integral

$$\int_\Omega (C^N[E^N] \cdot \hat{E}^N + C[E^M] \cdot \hat{E}^M) \, d\Omega$$

is the first variation of the functional

$$W(E^N, E^M) = \frac{1}{2}\int_\Omega (C^N[E^N] \cdot E^N + C^M[E^M] \cdot E^M) \, d\Omega$$

and, therewith, the first condition, Eq. (7.28), satisfied.

With the help of the auxiliary lemma

$$p: \quad w \in C^2(\bar{\Omega}), \hat{w}, N \in C^1(\bar{\Omega})$$

$$q: \quad \int_{\Omega} -\operatorname{div}(N\nabla w)\,\hat{w}\,d\Omega = -\int_{\Gamma} N\nabla w \cdot \boldsymbol{n}\,\hat{w}\,ds + \int_{\Omega} N\nabla w \cdot \nabla \hat{w}\,d\Omega$$

and the two auxiliary lemmas (7.8) and (7.10) which discuss the integrals $(-\operatorname{div} N, \hat{\boldsymbol{u}})$ and $(\operatorname{div}^2 M, \hat{w})$ we obtain, easily,

$$\int_{\Omega} \left[-\operatorname{div} N \cdot \hat{\boldsymbol{u}} - (\operatorname{div}^2 M + \operatorname{div}(N\nabla w))\hat{w} \right] d\Omega =$$

$$-\int_{\Gamma} \left(N\boldsymbol{n} \cdot \hat{\boldsymbol{u}} + (N\nabla w \cdot \boldsymbol{n} + V_n)\,\hat{w} - M_n \frac{\partial \hat{w}}{\partial n} \right) ds$$

$$-\left[\left[M_{nt}\,\hat{w} \right]\right] + \int_{\Omega} \left(N \cdot E^N_{u,w}(\hat{\boldsymbol{u}}, \hat{w}) - M \cdot E^M(\hat{w}) \right) d\Omega$$

where

$$E^N_{u,w}(\hat{\boldsymbol{u}}, w) = \frac{1}{2}(\nabla \hat{\boldsymbol{u}} + \nabla \hat{\boldsymbol{u}}^T) + \frac{1}{2}(\nabla \hat{w}\,\nabla w^T + \nabla w\,\nabla \hat{w}^T)$$

$$E^M_w(\hat{w}) = E^M(\hat{w}) = \nabla\nabla\hat{w}$$

are the Gateaux differentials of $E^N(\boldsymbol{u}, w)$ and $E^M(w)$. Hence, the second condition (7.29) is satisfied, too.

Let $\langle A(\Sigma), \hat{\Sigma} \rangle$ be the usual scalar product then the first identity of the operator A is

$$p: \quad \Sigma, \hat{\Sigma} \in \mathscr{S}^a \times \mathscr{S}^b$$

$$q: \quad G(\Sigma, \hat{\Sigma}) = \langle A(\Sigma), \hat{\Sigma} \rangle + \int_{\Gamma} \left[N\boldsymbol{n} \cdot \hat{\boldsymbol{u}} + (N\nabla w \cdot \boldsymbol{n} + V_n)\,\hat{w} - M_n \frac{\partial \hat{w}}{\partial n} \right] ds$$

$$-\left[\left[M_{nt}\,\hat{w} \right]\right] - E(\Sigma, \hat{\Sigma}) = 0$$

where

$$E(\Sigma, \hat{\Sigma}) = \int_{\Omega} \left[(E^N(\boldsymbol{u}, w) - E^N) \cdot \hat{N} + (E^M - E^M(w)) \cdot \hat{M} \right.$$

$$+ (C^N[E^N] - N) \cdot \hat{E}^N + (C^M[E^M] - M) \cdot \hat{E}^M$$

$$\left. + N \cdot E^N_{u,w}(\hat{\boldsymbol{u}}, w) - M \cdot E^M(\hat{w}) \right] d\Omega$$

is the first variation of the functional

$$F(\Sigma) = \int_{\Omega} \left[(E^N(\boldsymbol{u}, w) - E^N) \cdot N + (E^M(w) - E^M) \cdot M \right.$$

$$\left. + \frac{1}{2} C^N[E^N] \cdot E^N + \frac{1}{2} C^M[E^M] \cdot E^M \right] d\Omega$$

Here, the functional $R(\Sigma)$, the boundary integral in the first identity $G(\Sigma, \Sigma)$, is the integral

$$R(\Sigma) = \int_{\Gamma} \left[\boldsymbol{N}\boldsymbol{n} \cdot \boldsymbol{u} + (N\nabla w \cdot \boldsymbol{n} + V_n)\, w - M_n \frac{\partial w}{\partial n} \right] ds - \left[\left[M_{nt} w \right]\right]$$

Its first variation is the expression

$$\delta R(\Sigma, \hat{\Sigma}) = \int_{\Gamma} \left[\hat{\boldsymbol{N}}\boldsymbol{n} \cdot \hat{\boldsymbol{u}} + ([\hat{N}\nabla w + N\nabla \hat{w}] \cdot \boldsymbol{n} + \hat{V}_n)\, w - \hat{M}_n \frac{\partial w}{\partial n} \right] ds$$

$$- \left[\left[\hat{M}_{nt} w \right]\right] + \int_{\Gamma} \left[\boldsymbol{N}\boldsymbol{n} \cdot \hat{\boldsymbol{u}} + (N\nabla w \cdot \boldsymbol{n} + V_n)\, \hat{w} - M_n \frac{\partial \hat{w}}{\partial n} \right] ds$$

$$- \left[\left[M_{nt} \hat{w} \right]\right]$$

The form $V(\Sigma, \hat{\Sigma})$ is obtained by adding to $(-1)\, G(\Sigma, \hat{\Sigma})$ the identity

$$N(\Sigma, \hat{\Sigma}) = \int_{\Gamma} \left[\hat{\boldsymbol{N}}\boldsymbol{n} \cdot \boldsymbol{u} + \ldots - \hat{M}_n \frac{\partial w}{\partial n} \right] ds - \left[\left[\hat{M}_{nt} w \right]\right]$$

$$- \int_{\Gamma} \left[\hat{\boldsymbol{N}}\boldsymbol{n} \cdot \boldsymbol{u} + \ldots - \hat{M}_n \frac{\partial w}{\partial n} \right] ds + \left[\left[\hat{M}_{nt} w \right]\right] = 0$$

and placing the curly brackets $\{\ldots \ldots\}$, as explained in section 4.12.

10.7 The Principle "eigenwork = int. energy"

In nonlinear mechanics the identity

$$\frac{1}{2}\, G(u, u) = 0$$

cannot be interpreted as the statement that the external eigenwork is stored as internal energy. Does this mean that the principle "eigenwork = int. energy" no longer applies in nonlinear mechanics? No, it still applies. It is only that its formulation is a little bit more involved.

To obtain a mathematical expression of this principle in nonlinear mechanics we best, first, repeat the derivation of this principle for a linear problem.

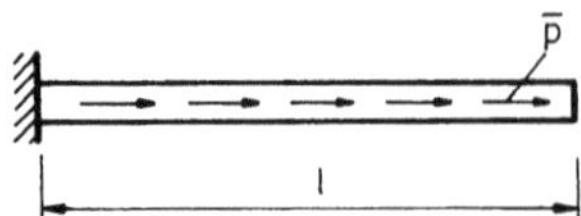

Figure 10.2

Let the bar in Fig. 10.2 be loaded with steadily increasing distributed forces $\bar{p}$

$$0 \leqslant \bar{p} \leqslant p$$

At every loading stage the displacement $\bar{u}$ satisfies the equations

$$-EA\bar{u}'' = \bar{p} \quad \bar{u}(0) = \bar{N}(l) = 0$$

Let $d\bar{u}(x)$ be the increase in displacement caused by an increase $d\bar{p}(x)$ of the external load. The engineer considers

$$dW_e = -\bar{N}(0)\,d\bar{u}(0) + \bar{N}(l)\,d\bar{u}(l) + \int_0^l \bar{p}\,d\bar{u}\,dx$$

to be the corresponding increase in external work. (For completeness we do not drop the zero terms $d\bar{u}(0)$ and $\bar{N}(l)$, see the boundary conditions).

On account of the first identity, $G(\bar{u}, d\bar{u}) = 0$, this work dW_e is equivalent with

$$dW_e = \int_0^l \frac{\bar{N}\,d\bar{N}}{EA}\,dx$$

that is equivalent with the increase in internal energy. The sum of all these single dW_e is the integral

$$\int dW_e = \int_0^l \frac{1}{EA}\left(\int_0^N \bar{N}\,d\bar{N}\right) dx = \frac{1}{2}\int_0^l \frac{N^2}{EA}\,dx$$

which, due to $G(u, u) = 0$, can be equated with an expression of external work.

$$\frac{1}{2}\int \frac{N^2}{EA}\,dx = \frac{1}{2}[Nu]_0^l + \frac{1}{2}\int_0^l pu\,dx$$

We are thus convinced that the energy stored is equal to the work done by the external loads acting through the displacements they cause.

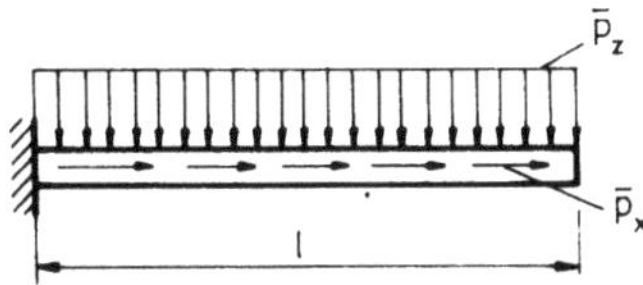

Figure 10.3

Next, let us apply the same technique to the beam in Fig. (10.3) (large displacement analysis).

Let the beam be loaded with steadily increasing distributed forces, $0 \leqslant \bar{p}_x \leqslant p_x, 0 \leqslant \bar{p}_z \leqslant p_z$. At every loading stage $\bar{p} = \{\bar{p}_x, \bar{p}_z\}$ the displacement vector $\bar{v} = \{\bar{u}, \bar{w}\}$ satisfies the equations

$$\boldsymbol{D}\bar{v} = \bar{p} \quad \text{in } 0 < x < l, \quad \bar{u}(0) = \bar{w}(0) = \bar{w}'(0) = \bar{T}(l) = \bar{M}(l) = 0$$

Let $\boldsymbol{d\bar{v}} = \{d\bar{u}, d\bar{w}\}$ be the increase in displacements corresponding to an increase $\boldsymbol{d\bar{p}} = \{d\bar{p}_x, d\bar{p}_z\}$ in forces. The corresponding increase of the external eigenwork is

$$dW_e = [\bar{N}d\bar{u} + (\bar{M}' + \bar{N}\bar{w}')\, d\bar{w} - \bar{M}d\bar{w}']_0^l + \int_0^l \boldsymbol{D}\bar{v} \cdot \boldsymbol{d\bar{v}}\, dx$$

On account of the first identity, $G(\bar{v}, \boldsymbol{d\bar{v}}) = 0$, this work is equivalent with

$$dW_e = E(\bar{v}, \boldsymbol{d\bar{v}}) = \int_0^l \left(\frac{N(\bar{v})\, N_{\bar{v}}(\boldsymbol{d\bar{v}})}{EA} + \frac{M(\bar{w})\, M_{\bar{w}}(d\bar{w})}{EI} \right) dx$$

According to section 1.6 where we studied Gateaux expansions as

$$N(\bar{v} + \boldsymbol{d\bar{v}}) = N(\bar{v}) + N_{\bar{v}}(\boldsymbol{d\bar{v}}) + \ldots = \bar{N} + d\bar{N} + \ldots$$

$$M(\bar{w} + d\bar{w}) = M(\bar{w}) + M_{\bar{w}}(d\bar{w}) + \ldots = \bar{M} + d\bar{M} + \ldots$$

we may consider the Gateaux differentials $N_{\bar{v}}(\boldsymbol{d\bar{v}})$ and $M_{\bar{w}}(d\bar{w})$ the increase in $\bar{N}$ and $\bar{M}$ resp. due to $\boldsymbol{d\bar{v}}$

$$N_{\bar{v}}(\boldsymbol{d\bar{v}}) = d\bar{N}, \quad M_{\bar{w}}(d\bar{w}) = d\bar{M}$$

With this interpretation of $N_v(\boldsymbol{d\bar{v}})$ and $M_{\bar{w}}(d\bar{w})$ we may proceed as in the linear theory, that is we consider the total work to be the integral

$$\int dW_e = \int_0^l \left(\frac{1}{EA} \left(\int_0^N \bar{N}d\bar{N} \right) + \frac{1}{EI} \left(\int_0^M \bar{M}d\bar{M} \right) \right) dx = \frac{1}{2} \int_0^l \left(\frac{N^2}{EA} + \frac{M^2}{EI} \right) dx$$

In the linear case we would turn now to the first identity, $1/2\, G(v, v) = 0$, and equate this integral, the internal energy, with an expression of the eigenwork of the external forces. But this is not possible in nonlinear mechanics. The internal energy $F(v)$ does not appear in $G(v, v)$, only its variation $E(v, v)$.

We are, thus, reminded that the first identity $G(v, \hat{v})$ is essentially an expression of virtual work. The identity states, in linear as in nonlinear mechanics, that the virtual internal work is equal to the virtual external work.

It is only in linear mechanics where $F(v) = 1/2\, E(v, v)$ that the trick with the factor $1/2$ works. But in nonlinear mechanics the eigenwork of a force N is not $1/2\, Nu$, hence, this trick must fail.

But what then is the expression of external eigenwork? Which expression equates with the internal energy $F(v)$ of the beam?

To answer this question we apply integration by parts to the integral of the internal energy and search for an expression which bears some resemblance with an expression of external work.

One such result is:

$$p: \quad v = \{u, w\} \in C^2, C^4, \ N = N(v), \ M = M(w)$$

$$q: \quad \frac{1}{2} \int \left(\frac{N^2}{EA} + \frac{M^2}{EI} \right) dx = \frac{1}{2} \left[Nu - Mw' + \left(M' + \frac{1}{2} Nw' \right) w \right]_0^l$$

$$- \frac{1}{2} \int_0^l \left(N'u + \left(M'' + \frac{1}{2} (Nw')' \right) w \right) dx$$

Another possibility is

$$p: \quad v = \{u, w\} \in C^2, C^4, \ N = N(v), \ M = M(w)$$

$$q: \quad \frac{1}{2} \int \left(\frac{N^2}{EA} + \frac{M^2}{EI} \right) dx = \frac{1}{2} \left[Nu + (Nw' + M') w - Mw' \right]_0^l$$

$$- \frac{1}{2} \int_0^l \left(N'u + (M'' + (Nw')') w + \frac{1}{2} Nw'^2 \right) dx \tag{10.23}$$

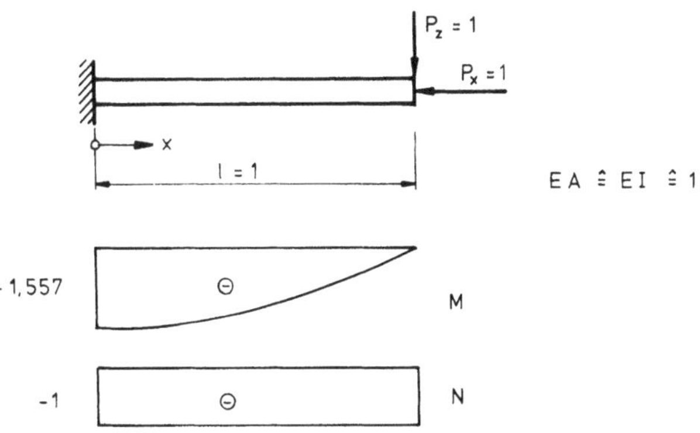

Figure 10.4

Let us check the last equation with the displacement $v = \{u, w\}$ of the beam in Fig. 10.4. The equation of the deflection w is

$$w(x) = \frac{1}{\varepsilon^3} \left[\tan\varepsilon(1 - \cos\varepsilon x) + \sin\varepsilon x - \varepsilon x \right], \qquad \varepsilon = 1 \tag{10.24}$$

and of the horizontal displacement u

$$u(x) = -\varepsilon^2 x - \frac{1}{\varepsilon^5}\left[\frac{\tan^2\varepsilon}{4}\left[\varepsilon x - \frac{1}{2}\sin 2\varepsilon x + \tan\varepsilon\left[\frac{1}{2}\sin^2\varepsilon x + \cos\varepsilon x\right]\right.\right.$$
$$\left.\left. - \sin\varepsilon x + 0.75\varepsilon x + \frac{1}{8}\sin 2\varepsilon x - \tan\varepsilon\right]\right], \qquad \varepsilon = 1 \qquad (10.25)$$

Due to the boundary conditions and $N(x) = -1$ Eq. (10.23) simplifies to

$$F(v) = \frac{1}{2}\left\{(-1)\,u(l) + 1\,w(l) - \int_0^l \frac{1}{2}(-1)\,w'^2\,dx\right\} \qquad (10.26)$$

The internal energy, the left-hand side, is

$$F(v) = \frac{1}{2}\int_0^l\left(\frac{N^2}{EA} + \frac{M^2}{EI}\right)dx = 0.9670$$

and with $u(1) = -1.1883$, $w(1) = 0.5574$ and

$$\frac{1}{2}\int_0^l w'^2\,dx = \frac{1}{2}\,0.3766$$

the "external eigenwork", the right-hand side of Eq. (10.26), becomes

$$\frac{1}{2}\left\{1.1883 + 0.5574 + \frac{1}{2}\,0.3766\right\} = 0.9670$$

The two sides in Eq. (10.26) are equal.

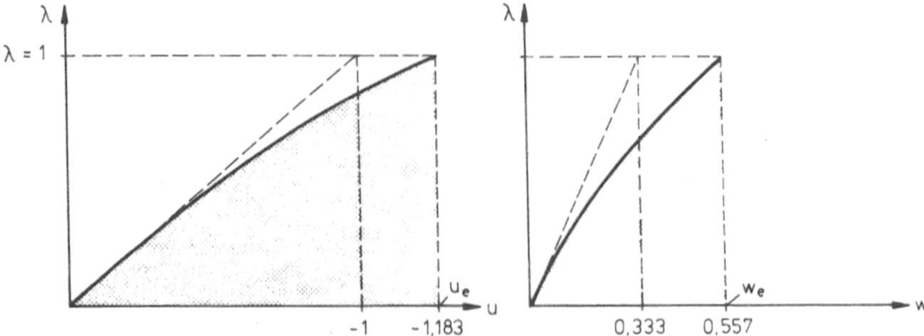

Figure 10.5

This is no surprise. We, essentially, only checked whether we did integration by parts correctly.

Hence, for an independent check, let us calculate the external eigenwork directly. Assume we really load the beam with steadily increasing end forces

$$0 \leqslant \lambda P_x \leqslant P_x, \quad 0 \leqslant \lambda P_z \leqslant P_z, \quad P_x = P_z = 1, \quad 0 \leqslant \lambda \leqslant 1$$

The displacement $u(x)$ and $w(x)$ at the loading stage λ are defined $(\varepsilon = \sqrt{\lambda})$ in Eqs. (10.24) and (10.25) and the functional connection between the load λ and the end displacements of the beam, $u(1)$ and $w(1)$, is shown in Fig. 10.5.

If the engineer is right then the area under the two curves

$$\int dW_e = \int_0^1 \lambda(u,_\lambda + w,_\lambda)\, d\lambda = \int_0^{u_e} \lambda\, d\bar{u} + \int_0^{w_e} \lambda\, d\bar{w}$$

should be equal to $F(v)$.

The increase in external eigenwork at the stage λ is

$$dW_e = \lambda P_x\, d\bar{u} + \lambda P_z\, d\bar{w} = \lambda(d\bar{u} + d\bar{w})$$

Hence, the total eigenwork, the total area, amounts to

$$\int dW_e = \int \lambda(d\bar{u} + d\bar{w}) = \int_0^1 \lambda(\bar{u},_\lambda + \bar{w},_\lambda)\, d\lambda$$

This integral is best evaluated numerically.

$$\int dW_e \simeq \int_{0+\varepsilon}^1 \lambda\left[\frac{\bar{u}(\lambda+\Delta) - \bar{u}(\lambda)}{\Delta} + \frac{\bar{w}(\lambda+\Delta) - \bar{w}(\lambda)}{\Delta}\right] d\lambda$$

Choosing $\Delta = 0.05$, $\varepsilon = 0.0001$ (at $\lambda = 0$ are $\bar{u}(\lambda)$ and $\bar{w}(\lambda)$ undefined), $N = 60$ intervals (Simpson) we obtain

$$\int dW_e = 0.9669$$

which agrees quite well with the internal energy $F(v) = 0.9670$.

All what we said above, naturally, remains valid in the large displacement analysis of plates and in the nonlinear analysis of elastic plates and bodies.

Consider, e. g., an elastic body with the internal energy so defined:

$$F(u) = \frac{1}{2} \int_\Omega E(u) \cdot C[E(u)]\, d\Omega$$

where $E(u)$ is the nonlinear strain tensor, s. Eq. (10.1) and C the standard linear elasticity tensor.

Integration by parts applied to $F(u)$ yields the following two identities.

$$p:\ \boldsymbol{u} \in C^2(\bar{\Omega}),\ S = C[E(\boldsymbol{u})]$$

$$q:\ F(\boldsymbol{u}) = \frac{1}{2}\!\int_\Gamma \tau(S,\boldsymbol{u})\cdot\boldsymbol{u}\,ds - \frac{1}{2}\int_\Gamma \nabla\boldsymbol{u}\,S\boldsymbol{n}\cdot\boldsymbol{u}\,ds$$

$$-\int_\Omega \mathrm{div}(S+\nabla\boldsymbol{u}\,S)\cdot\boldsymbol{u}\,d\Omega + \frac{1}{2}\int_\Omega \mathrm{div}(\nabla\boldsymbol{u}\,S)\cdot\boldsymbol{u}\,d\Omega$$

$$= \frac{1}{2}\int_\Gamma \tau(S,\boldsymbol{u})\cdot\boldsymbol{u}\,ds - \frac{1}{2}\int_\Omega \left(\mathrm{div}(S+\nabla\boldsymbol{u}\,S)\cdot\boldsymbol{u} + \frac{1}{2}\nabla\boldsymbol{u}\nabla\boldsymbol{u}\cdot S\right)d\Omega$$

These equations express the internal energy in terms of integrals which bear some resemblance with external work and which, hence, might be interpreted as representations of the statement "eigenwork = int. energy".

10.8 Influence Functions

In linear mechanics we express the displacement $u(x)$ of a structural element by an influence function

$$u_1(x) = \int_\Omega G_0(y,x)\,p_1(y)\,d\Omega_y \tag{10.27}$$

the L_2-scalar product between Green's function $G_0(y,x)$ and the applied load p_1.

Assume the same would be possible in nonlinear mechanics and assume

$$u_2(x) = \int_\Omega G_0(y,x)\,p_2(y)\,d\Omega_y$$

is the displacement of a second loading case, p_2.

As the scalar product is distributive ($=$ linear) the total displacement $u(x)$ is

$$u(x) = u_1(x) + u_2(x) = \int_\Omega G_0(y,x)\,(p_1+p_2)\,d\Omega_y$$

But this contradicts the fact that the governing operator, D^{NL}, is a nonlinear operator

$$D^{NL}(u_1+u_2) \neq D^{NL}u_1 + D^{NL}u_2$$

Hence, there cannot exist an influence function in the sense of Eq. (10.27) in nonlinear mechanics.

This verdict does not necessarily imply that there are no fundamental solutions in nonlinear mechanics. It only means that such solutions do not depend linearly on the magnitude of the concentrated force.

The displacements (second-order analysis) of the beam in the last section behave as

$$u(x) = \ldots \cos(\sqrt{\lambda}\,x)\ldots \qquad w(x) = \ldots \sin(\sqrt{\lambda}\,x)\ldots$$

and, therefore, do not depend linearly on the load parameter λ while the first-order displacements

$$u(x) = \lambda x \qquad w(x) = \lambda\left(\frac{1}{2}x^2 - \frac{x^3}{6}\right)$$

clearly do.

11 Finite Elements

To model the behaviour of a structure the finite element method replaces the structure by a patch of finite elements which, compared with the real structure, can undergo only a limited number of states or modes, namely all those modes whose state variables are piecewise polynomials of maximum degree, say, k.

If the system variable is the displacement u alone we speak of a displacement model, is it the elastic state $\Sigma = \{u, E, S\}$ then we speak of a mixed model. The displacement models use the energy principles of the operators D, the mixed models the energy principles of the operators A.

We first study the displacement models where according to chapter 4 we have the choice among three energy principles:

The solution $u \in C_p^{2m}$ of a regular bvp renders the basic functional $\Pi(u)$

> *stationary on C_p^m*
> *a minimum on $R_1 = C_p^m + geometric\ boundary\ conditions$*
> *a maximum on $R_2 = C_p^{2m} + static\ boundary\ conditions + field\ eq.$*

The most practical of these principles is the second, the principle of minimum potential energy, because the resulting stiffness matrix is positive definite. We shall in the following concentrate on this principle.

The reappearance of the subscript p on C^m and C^{2m} which we dropped for the sake of simplicity, see the remark on page 19, shall remind us that the energy principles are still valid on C_p^m and C_p^{2m}. The equilibrium position u renders the potential energy a minimum among all displacements in C_p^m not only among those in C^m.

The subscript p becomes important when we do finite elements. To achieve convergence the competing functions (usually) must be admissible, they must belong to the space R_1 (principle of minimum potential energy) and the lower the global smoothness condition the simpler the task to find correct trial functions.

We call functions which satisfy the necessary smoothness conditions, which belong to C_p^m, *conforming*. In the literature it happens that the finite elements themselves are called conforming. It is then understood that the global shape functions generated by an assemblage of conforming elements are conforming.

11.1 Shape Functions

Let Ω be a plate which is partitioned into triangular elements, $\Omega = \cup\ \Omega_e$, and assume we approximate a global function $u(x)$ by interpolating it at the nodes of the net.

The interpolant $u_h(x)$ is on each finite element Ω_e a function as

$$u_h^e(x) = \sum_{k=1}^{3} \delta_k^e N_k^e(x)$$

where the single numbers δ_k^e are the values of $u(x)$ at the three nodes x^k of the element and the functions N_k^e the shape functions of the element.

The global approximation $u_h(x)$ is the assemblage of the single contributions $u_h^e(x)$. If we attach the labels $1, 2, \ldots K$ to the single nodes this global function $u_h(x)$ can be written as

$$u_h(x) = \sum_{k=1}^{K} \delta_k \varphi_k(x), \qquad \delta_k = u(x^k)$$

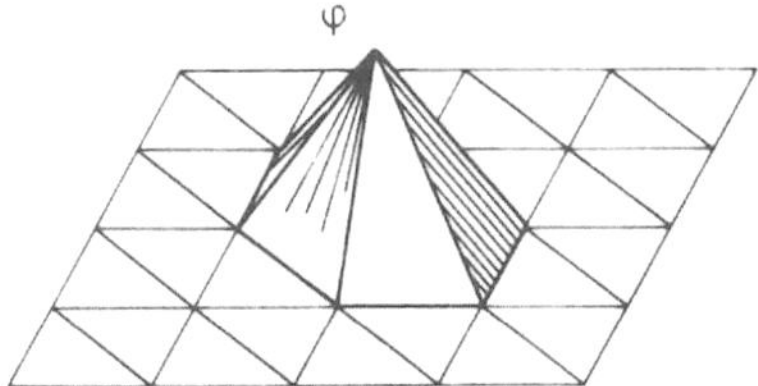

Figure 11.1

The single function φ_k, the global shape function, is the union of all those local shape functions $N_k^e(x)$ which have at x^k the value 1. That is, the global shape function is only non-zero on those elements which contain the node x^k, see Fig. 11.1.

Usually, the local shape functions $N_k^e(x)$ are polynomials and a global shape function $\varphi_k(x)$ belongs, therefore, on each element Ω_e to $C^\infty(\Omega_e)$. It is infinitely smooth locally, but not so with respect to the global domain $\Omega = \cup_e \Omega_e$.

Consider, e. g., the function in Fig. 11.1. This function belongs, locally, to $C^\infty(\Omega_e)$ but globally only to $C_p^1(\bar{\Omega})$ because the first derivatives are discontinuous across element boundaries.

These observations are motivation to introduce two classifications:

1) We say a function φ belongs to C_{loc}^n if the restrictions of φ to the elements Ω_e of the net belong to $C^n(\Omega_e)$. A notation as $\varphi \in C \cap C_{loc}^n$ then implies that φ belongs, globally, to $C(\bar{\Omega})$ and, locally, to C^n.

2) We say a function φ is a C^n-*shape function* if it belongs to $C^n(\bar{\Omega})$ and if its restriction to each element Ω_e is a polynomial.

With the help of some drawings, e. g. Fig. 11.1, we soon realize that C^0-shape functions belong to $C_p^1(\bar{\Omega})$ because they have piecewise continuous first derivatives and that C^1-shape functions belong to $C_p^2(\bar{\Omega})$, because their second derivatives are piecewise continuous.

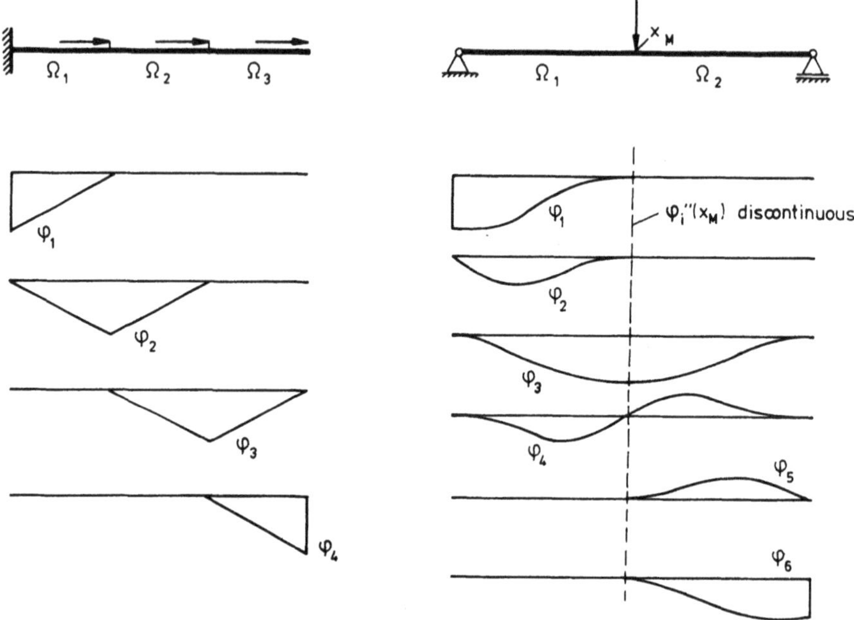

Figure 11.2

The principle of minimum potential energy requires C_p^m-functions. Hence we may conclude that C^{m-1}-shape functions are sufficient for displacement models based on the principle of minimum potential energy, see table 11.1.

principle of min. pot. energy	necessary	sufficient
$2m = 2$ bar, elastic plate etc.	C_p^1	C^0-shape functions
$2m = 4$ beam, K.-plate	C_p^2	C^1-shape functions

table 11.1

The matrix-displacement method which can be viewed as an (exact) finite element method uses e. g. C^0-shape functions (for bars) and C^1-shape functions (for beams), see Fig. 11.2. The restriction of the global functions to the single elements, the local polynomials, are just the homogeneous solutions of the single bar or beam element.

After these introductory remarks let us approximate the displacement of a simple structure, the bar in Fig. 11.3, with finite elements.

We know that the displacement of the bar renders the potential energy

$$\Pi_1(u) = \frac{1}{2} \int_0^2 \frac{N^2}{EA}\, dx - \int_0^2 pu\, dx$$

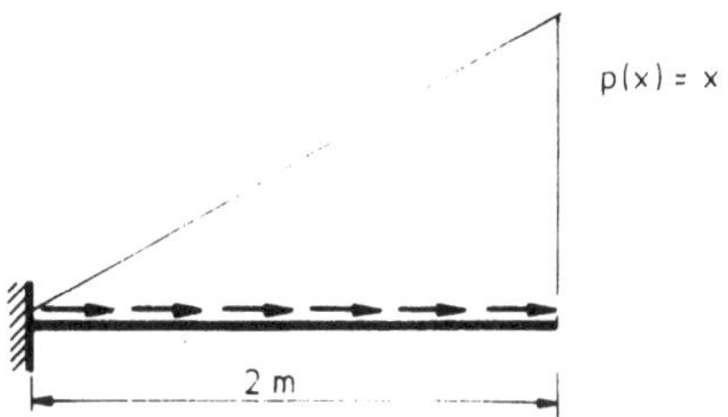

Figure 11.3

a minimum on the class

$$R_1 = \{u \in C_p^1 [0, 2]\,|\,u(0) = 0\}$$

As approximating functions we choose the three C^0-shape functions in Fig. 11.4a. All three are conforming but admissible are only φ_2 and φ_3 because only these two satisfy the boundary condition $u(0) = 0$.

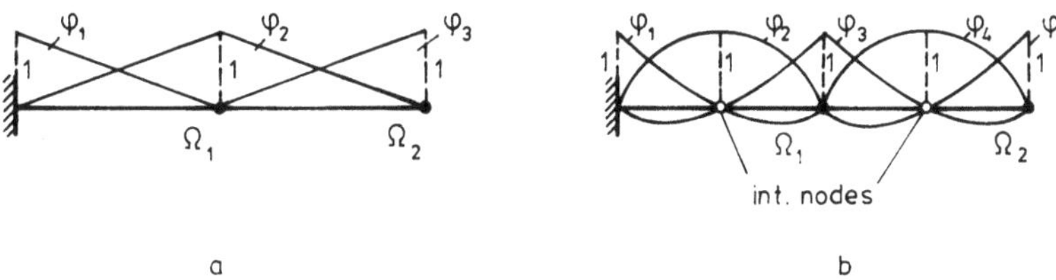

Figure 11.4

Hence, we let

$$u_h = \delta_2 \varphi_2 + \delta_3 \varphi_3$$

and we minimize the potential energy with respect to δ_2 and δ_3

$$\frac{\partial}{\partial \delta_i}\, \Pi(\delta_2 \varphi_2 + \delta_3 \varphi_3) = \frac{\partial}{\partial \delta_i}\left(\frac{1}{2}\int_0^2 \frac{N^2}{EA}\,dx - \int_0^2 pu\,dx\right) = 0,\ i = 2, 3$$

This yields the two equations

$$\begin{bmatrix} 2 & -1 \\ -1 & 1 \end{bmatrix}\begin{bmatrix} \delta_2 \\ \delta_3 \end{bmatrix} = \begin{bmatrix} 1 \\ 0.83 \end{bmatrix} \tag{11.1}$$

whose solutions, $\delta_2 = 1.83$ and $\delta_3 = 2.67$, are just the true displacements at the two nodes. That is, the finite element solution is identical with the interpolant of the true displacement with respect to φ_2 and φ_3 (this is not a general rule in finite element analysis).

This is also seen in Fig. 11.5a which displays graphically the deviation of the finite element solution from the true displacement. At two points the error $e(x) = u(x) - u_h(x)$ is zero.

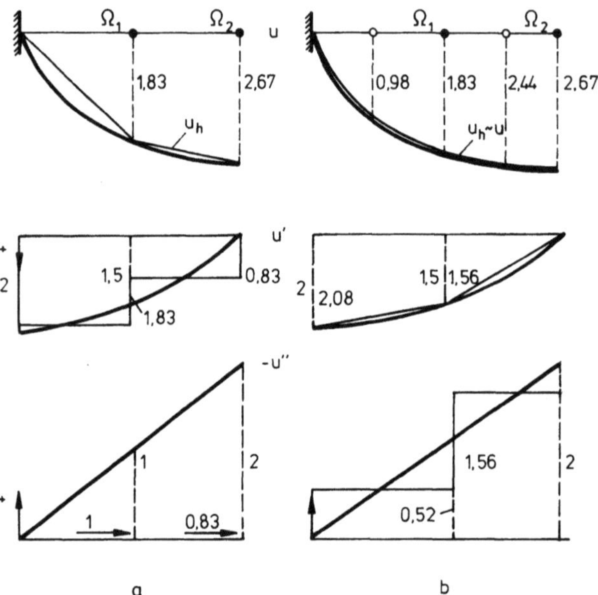

Figure 11.5

11.2 The Error in Finite Elements

To measure the size of the error function $e = u - u_h$ and its derivatives over the global domain $[0, l]$ we need some scales. The most appropriate scales to this end are the Sobolev norms of section 1.8.

The Sobolev norm of degree k of a function $u \in C^k[0, 2]$ is the number

$$\|u\|_k = \left(\int_0^2 \sum_{i=0}^k (u^{(i)})^2 \, dx\right)^{1/2} = \left(\int_0^2 (u^2 + u'^2 + u''^2 + \ldots u^{(k)2}) \, dx\right)^{1/2}$$

and the closure of $C^k[0, 2]$ with respect to this norm is the Sobolev space $H^k[0, 2]$. Note that $C^k[0, 2] \subset C_p^k[0, 2] \subset H^k[0, 2]$.

With regard to shape functions we know that the space $C_p^k[0, 2]$ is equivalent with the space $H^k[0, 2]$. If a shape function is in C_p^k then it is also in H^k and if a shape function is not in C_p^k then it is also not in H^k.

$$u \in H^k \leftrightarrow u \in C_p^k$$

If an operator D has degree $2m$ then the highest derivatives in the energy are of order m. The Sobolev space with the same index, $H^m(\Omega)$, is the space of functions with finite energy.

$$u \in H^m(\Omega) \;\rightarrow\; \frac{1}{2}\,E(u,u) < \infty$$

Every conforming shape function belongs to $C_p^m(\bar{\Omega})$, hence, to $H^m(\Omega)$ and any such function has, therefore, finite energy. The *nonconforming shape functions* have discontinuous displacement-terms. They do not belong to C_p^m and, therefore, not to $H^m(\Omega)$. Their energy is infinite. We demonstrated this in section 5.9 by expanding such a function into a Fourier series.

If we compare the finite element solution of the bar in Fig. 11.3 with the true solution, $u(x) = 6^{-1}(12x - x^3)$, then the error in the H^0-norm measures

$$\|u - u_h\|_0 = \left(\int_0^2 (u - u_h)^2 \, dx \right)^{1/2} = 1.45 \cdot 10^{-1}$$

and in the H^1-norm

$$\|u - u_h\|_1 = \left(\int_0^2 (u - u_h)^2 + (u' - u_h')^2 \, dx \right)^{1/2} = 4.82 \cdot 10^{-1}$$

The H^1-error must, by definition, be greater than the H^0-error but we also recognize that the error grows disproportionally (we should have a factor of $\sqrt{2} = 1.41\ldots$); the approximation in the first derivative is not as good as in displacements.

This is a general rule in finite element analysis: the higher the derivatives the more the two functions u and u_h drift apart. "Good approximation in displacements, bad approximation in stresses".

If we differentiate u_h too often, that is if we exceed a certain critical number $n = n_c$ then the derivatives $u_h^{(n)}$ no longer are in C_p^0, that is, u_h is not in C_p^n, $n > n_c$, and, therefore, also not in H^n, $n > n_c$. As a consequence we cannot measure the distance $\|u - u_h\|$ in Sobolev norms exceeding the critical number n_c.

In our case the critical number n_c is 1. The finite element solution belongs to C_p^1 but not to spaces C_p^n, $n > 1$, because the second derivative of u_h includes δ-functions, the concentrated loads at the nodes.

If instead of piecewise linear shape functions we use the piecewise quadratic shape functions of Fig. 11.4b then we obtain a much better approximation, see Fig. 11.5b. The error measured in the different norms decreases significantly.

$$\|u - u_h\|_0 = 7.36 \cdot 10^{-3}, \quad \|u - u_h\|_1 = 5.50 \cdot 10^{-2},$$

$$\|u - u_h\|_2 = 4.17 \cdot 10^{-1}$$

The size of these errors depends in general on two numbers which are characteristic for any finite element net: the meshwidth h (the largest single distance two points in a finite element have) and the degree k of the polynomials interpolated exactly by the shape functions of the net.

The two nets in Fig. 11.5, the piecewise linear and the piecewise quadratic, are, e.g., P_1- and P_2-complete (the shape functions of these nets interpolate all polynomials of degree $\leqslant k$ exactly).

If u_h is the finite element solution (Ritz) on a P_k-complete net and h the meshwidth then the error behaves as

$$\|u - u_h\|_s = c h^{k+1-s} \|u\|_{k+1} \quad \text{if} \quad s \geqslant 2m - (k+1)$$

where c is a generic constant which depends on k and our choice of s, see [S & F] p. 166.

The more we raise s, the more derivatives we include in our measurements, the more the convergence slows down. The rate of convergence is at its minimum in the norm, $s = n_c$, of the highest Sobolev space which contains u_h and at its maximum in the H^0-norm,

$$\|u - u_h\|_0 < c h^{k+1} \|u\|_{k+1} \tag{11.2}$$

The number $\|u\|_{k+1}$ on the right-hand side is the H^{k+1}-norm of the solution. If u is not very smooth, i.e. if u does not belong to H^{k+1} but only to a space $H^p, p < k+1$, then the index $k+1$ must be replaced by p. This means that if $k+1 > p$ then $k+1$ has no influence on the speed of the convergence.

In such a case it makes no sense to increase the degree k of the polynomials further; the finite element solution does not gain in accuracy.

When we divide the inequality (11.2) by $\|u\|_{k+1}$

$$\frac{\|u - u_h\|_0}{\|u\|_{k+1}} < c h^{k+1}$$

then we learn that the quotient on the left is bounded by some power of the meshwidth h. This means if we let $u \to 0$ and consequently $\|u\|_{k+1} \to 0$ then also the numerator tends to zero; the Ritz method is stable in the H^0-norm. "Small" displacements u have "small" Ritz approximations u_h.

Fig. 11.5 suggests that we consider the finite element solutions the exact solutions of loading cases which approximate the original distribution of forces.

The piecewise linear $(-EAu_h'' = 0$ on $\Omega_e)$ solution is the displacement when the bar is loaded with the equivalent nodal forces (the right-hand side of Eq. (11.1)), see Fig. 11.5a.

The piecewise quadratic solution $(-EAu_h'' = p_e$ on $\Omega_e)$ is the displacement when the bar is loaded with distributed piecewise constant forces $p_1 = 0.52$ and $p_2 = 1.56$, see Fig. 11.5b.

The same interpretation allows the rectangular prestressed (N) membrane in Fig. 11.6a.

$$-N\Delta w = p \quad \text{in } \Omega, \quad w = 0 \quad \text{on } \Gamma$$

If we approximate the solution with piecewise linear functions $(-N\Delta w_h = 0$ on

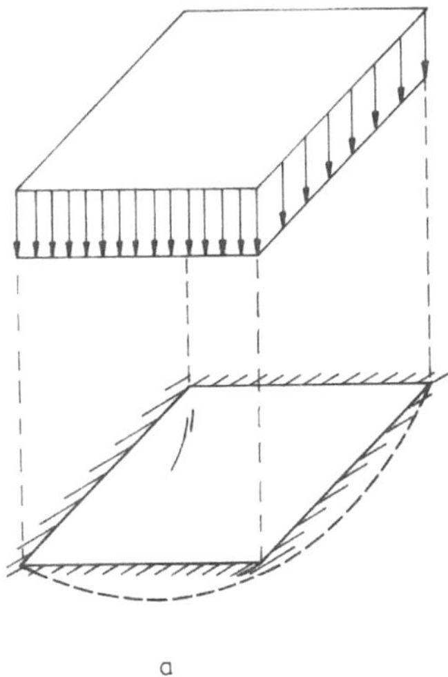
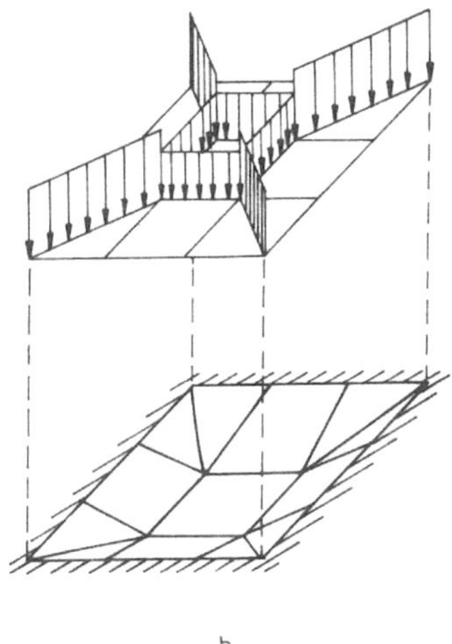

Figure 11.6

Ω_e), see Fig. 11.6b, then this effectively means that we approximate the evenly distributed pressure p with line forces acting along the interelement boundaries.

The magnitude of these forces corresponds to the jump in the normal derivatives, $[\partial w_h/\partial n]$, between two adjacent elements. So, if $G_0(y, x)$ is Green's function of the bvp the approximate solution w_h is the function

$$w_h(x) = \sum_i \int_{l_i} G_0(y, x) \, [\partial w_h/\partial n] \, ds_y$$

where $\{l_i\}$ is the set of all lines which form the finite element net. If we use C^1-shape functions instead then the line forces disappear and the finite element solution becomes the deflection of a membrane loaded with element forces, $- N\Delta \, w_h|_{\Omega_e}$, only.

$$w_h(x) = \sum_e \int_{\Omega_e} G_0(y, x) \, (- N\Delta w_h) \, d\Omega_y$$

In which case the pointwise error is

$$w(x) - w_h(x) = \sum_e \int_{\Omega_e} G_0(y, x) \, (p - N\Delta w_h) \, d\Omega_y$$

We could calculate the error if we knew Green's function $G_0(y, x)$ (but then we would need no finite elements ...).

In this situation we must replace Green's function by our engineering insight and judgement, that is we must make guesses about the distance in deflections when we know the distance in the second derivatives, or in mechanical terms, if we know the distance in forces, $p - N\Delta w_h$. This is what error analysis is all about.

Assume we approximate the deflection of a prestressed membrane

$$-N\Delta w = p \quad \text{in } \Omega \qquad w = 0 \quad \text{on } \Gamma$$

with piecewise linear polynomials. If we partition the membrane as in Fig. 11.7a then the finite element solution is the deflection of a membrane loaded with positive $(+ + +)$ and negative $(- - -)$ line forces. This, certainly, is not a good approximation because the evenly distributed pressure p does not change its sign.

The partition in Fig. 11.7b is a better choice. Now the line forces point nearly everywhere downwards, on only one line do they remain negative. This line signals a disturbance on the boundary; the tensile forces $(\partial w/\partial n)$ become infinite at the reentrant corner, see Fig. 11.7c.

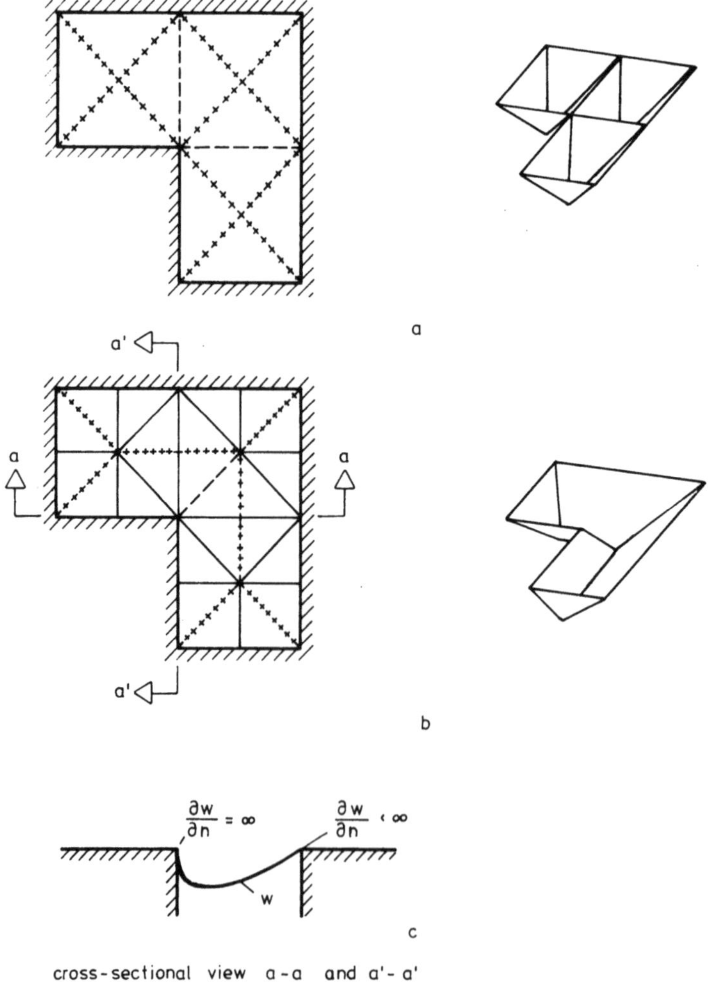

Figure 11.7

If every finite element solution of a structure is the response of the structure to external forces which approximate the true forces, how can these forces be calculated? How did we find the forces which deflect the membrane in Fig. 11.7b?

With the help of rigid-body movements!

Remember that the strain energy of pairs of functions as $\{w, r\}$ is zero, that is if the companion of w is a rigid-body movement.

$$E(w, r) = 0 \quad w \in C_p^m(\bar{\Omega}), \, r = \text{rigid-body movement}$$

Substituting the finite element solution w_h into this expression (conforming shape functions belong to C_p^m) and letting $r = 1$ we obtain

$$E(w_h, 1) = \sum_e E(w_h, 1)_{\Omega_e} = 0$$

The solution w_h belongs on each element Ω_e to $C^\infty(\Omega_e)$, hence, integration by parts (we formulate the first identity) is admissible on element level

$$E(w_h, 1)_{\Omega_e} = \int\limits_{\Omega_e} - N\Delta w_h 1 \, d\Omega + \int\limits_{\partial\Omega_e} \frac{\partial w_h}{\partial n} 1 \, ds$$

Summing all the element contributions we obtain

$$\sum_e E(w_h, 1)_{\Omega_e} = \sum_e \left\{ \int\limits_{\Omega_e} - N\Delta w_h 1 \, d\Omega + \int\limits_{\partial\Omega_e} \frac{\partial w_h}{\partial n} 1 \, ds \right\}$$

$$= \int\limits_{\Omega} - N\Delta w_h 1 \, d\Omega + \sum_i \int\limits_{l_i} \left[\frac{\partial w_h}{\partial n} \right] 1 \, ds = 0 \tag{11.3}$$

where $[w_h/\partial n]$ is the jump in the normal derivative between neighboring elements and $\{l_i\}$ the set of all lines of the finite element net.

Hence, the external loads which correspond to w_h are distributed forces $- N\Delta w_h|_{\Omega_e}$ and line forces $[\partial w_h/\partial n]$. In other words, w_h is the displacement

$$w_h = \int\limits_{\Omega} G_0(y, x) (- N\Delta w_h) \, d\Omega_y + \sum_i \int\limits_{l_i} G_0(y, x) [\partial w_h/\partial n] \, ds_y$$

If w_h is piecewise linear as in our case then $N\Delta w_h = 0$ on each element Ω_e, that is the first integral is zero in our case.

This technique, naturally, is applicable to every structure, but please note: the condition that w_h is in C_p^m, i.e. that w_h has finite energy

$$E(w_h, w_h) < \infty$$

is essential for the analysis. A nonconforming solution, $w_h \notin C_p^m$, has no finite energy, $E(w_h, w_h) = \infty$, and, hence, the argument that

$$E(w_h, w_h) = \sum_e E(w_h, w_h)_{\Omega_e}$$

fails. The jump terms $[\partial w_h/\partial n]$ obtained by pieceing together local results would not represent the true forces acting on the lines l_i. These forces are infinite if w_h is nonconforming, see section 5.9.

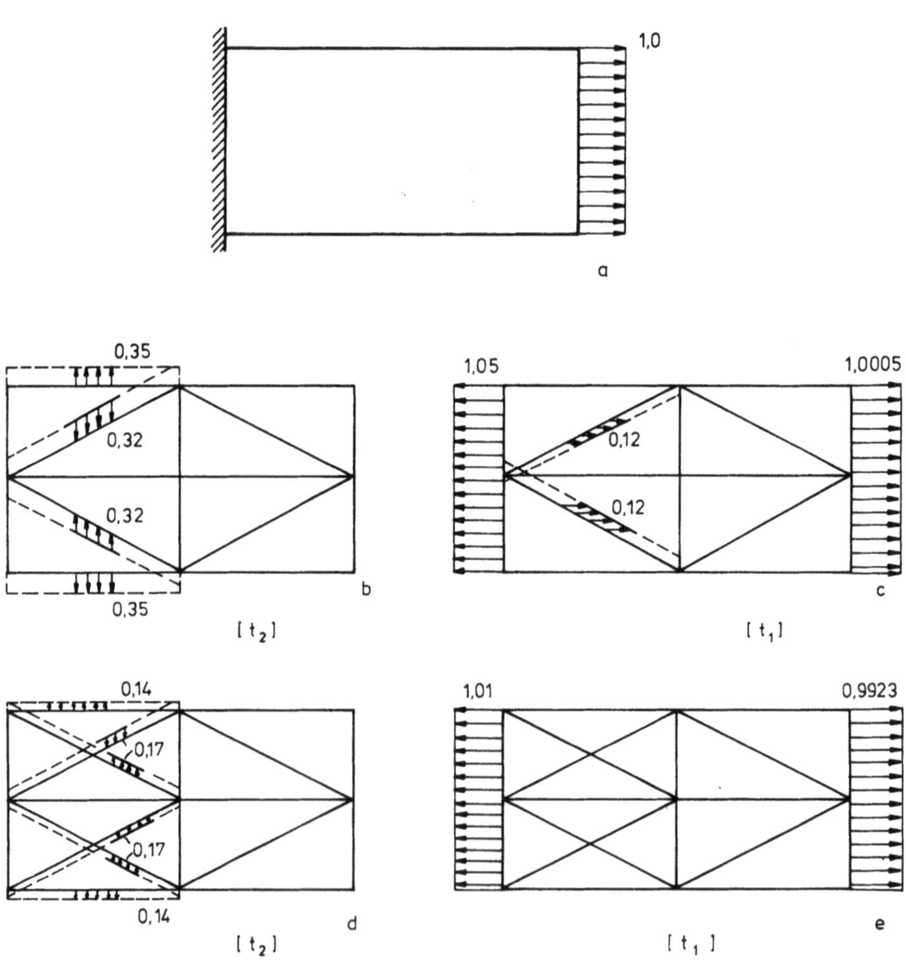

Figure 11.8

In the case of bars, elastic plates, elastic bodies and membranes, $m = 1$, the conforming shape functions are in C^0 and the jump terms concentrated forces distributed along the interelement boundaries.

In the case of fourth-order problems, $m = 2$, the conforming shape functions are in C^1 and, hence, the jump terms couples (jumps in the second derivatives) or forces (jumps in the third derivatives).

Stated more generally: the forces concentrated at the interelement boundaries depend on the global smoothness of the shape functions and the variation of the forces within the elements on the local degree of the shape functions. If φ has degree 5 then $K\Delta\Delta\varphi$ has degree 1 etc.

As a further example consider the elastic plate in Fig. 11.8a whose displacement field we approximated with the simplest possible finite element net: a net of piecewise linear C^0-shape functions.

The resulting finite element solution u_h (Ritz solution) is the exact displacement field of the plate loaded with the horizontal and vertical line forces depicted in Fig. 11.8b and c (line forces < 0.05 were neglected in the drawings).

The line forces are just the jump $[t_h] = \{[t_{h1}], [t_{h2}]\}$ of the traction vector t of the displacement field u_h between neighboring elements.

The magnitude of the forces distributed along the interior lines and on the horizontal faces of the plate serves as an error indicator, because these line forces would be zero if u_h were the exact solution.

$$\text{"error"} = \int \left(|[t_{h1}]| + |[t_{h2}]| \right) ds$$

The error decreases if we choose a finer net as in Fig. 11.8d and e.

Let us close with a remark concerning Eq. (11.3). This equation confirms that the finite element solution (and also each single conforming shape function) satisfies the equilibrium conditions. It is this equation which gives credit to our claim that every conforming finite element solution is the exact solution of a neighboring load case.

In the case of a piecewise linear approximation of a membrane Eq. (11.3) simplifies, because of $-N\Delta w_h|_{\Omega_e} = 0$, to

$$\sum_i \int_{l_i} [\partial w_h/\partial n] \, 1 \, ds = 0$$

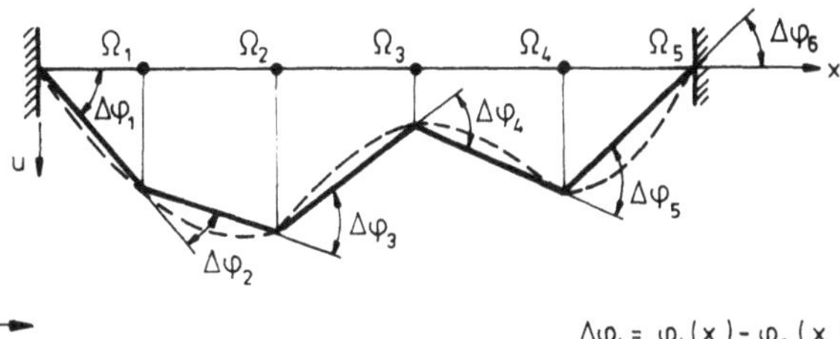

Figure 11.9

It means that the sum (integral) of the jumps in the normal derivative taken over all lines l_i is zero whenever we deflect a patch of finite elements without bending the single elements as, e.g., in Fig. 11.6.

Its one-dimensional analogue is the statement that, given a broken line as in Fig. 11.9, the sum of all the jumps in the slope is zero

$$\sum_i \tan \Delta \varphi_i = 0$$

or, in mechanical terms, the piecewise linear finite element approximation of the displacement of a bar $(-EAu'' = p)$ always satisfies the equilibrium condition

$$\sum_i P_i = 0 \quad (P_i \triangleq \tan \Delta \varphi_i)$$

11.3 Nonconforming Shape Functions

The difficulty to construct higher order shape functions in 2-D and 3-D makes the use of nonconforming shape functions, functions which violate the necessary smoothness conditions, practically unavoidable.

Which error we commit when we use nonconforming functions will best be illustrated with a simple example. Consider Fig. 11.10.

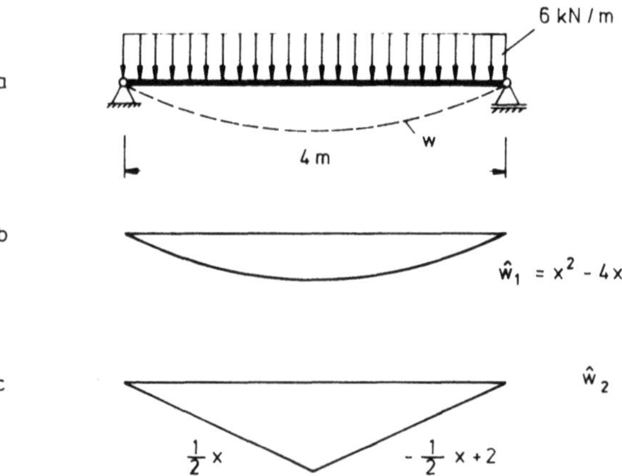

Figure 11.10

The smooth function $\hat{w}_1$ is conforming, it belongs to $C_p^2[0,1]$ but the function with the bend, the function $\hat{w}_2$ is not.

Assume we let the function $\hat{w}_1$ act as a virtual displacement on the beam in Fig. 11.10a. The first identity of the pair $\{w, \hat{w}_1\}$ is

$$G(w, \hat{w}_1) = \int_0^l p\hat{w}_1 \, dx - \int_0^l \frac{M\hat{M}_1}{EI} \, dx = -64 - (-64) = 0 \tag{11.4}$$

The result is correct. The principle of virtual displacements applies. Then we do the same with the function $\hat{w}_2$

$$G(w, \hat{w}_2) = \int_0^l p\hat{w}_2 \, dx - \int_0^l \frac{M\hat{M}_2}{EI} \, dx = 12 - 0 \stackrel{?}{=} 0 \tag{11.5}$$

The result is incorrect. The principle of virtual displacements does not apply. The reason is that $\hat{w}_2$ does not belong to $C_p^2[0,l]$; it does not meet the smoothness requirements of the first identity.

As Eq. (11.5) is also the first variation of the potential energy of the beam

$$\Pi_1(w) = \frac{1}{2}\int_0^l \frac{M^2}{EI}\,dx - \int_0^l pw\,dx \tag{11.6}$$

this negative result, Eq. (11.5), implies that the first variation of $\Pi_1(w)$ in the direction of $\hat{w}_2$ does not vanish (contrary to the first variation in the direction of $\hat{w}_1$).

It must, therefore, lead, in general, to incorrect results if we search among non-conforming functions for approximate solutions of the beam problem in Fig. 11.10a by minimizing (11.6). We usé the wrong functional, the wrong identity.

Wrong with respect to $\hat{w}_2$. The beam and the function $\hat{w}_2$ do not match. The function $\hat{w}_2$ is not an admissible virtual displacement of the beam but of the beam with a hinge at mid-span, see Fig. 11.11, at the point where $\hat{w}_2$ has its bend.

The deflections of this structure, this linkage, belong to $C^0 \cap C_{loc}^4$ and the admissible virtual displacements to $C^0 \cap C_{loc}^2$ ($\hat{w}_2$ belongs to this class).

The first identity of the linkage is obtained if we partition the interval $[0,l]$ into two parts, formulate the first identity for both intervals and add the two identities, $0 + 0 = 0$.

$$G(w,\hat{w}_2) = G(w,\hat{w}_2)_1 + G(w,\hat{w}_2)_2$$

$$= \int_0^l p\hat{w}_2\,dx - (M_l\hat{w}_l' - M_r\hat{w}_r')\left(\frac{l}{2}\right) - \int_0^l \frac{M\hat{M}}{EI}\,dx$$

$$= 12 - 12\frac{1}{2} + 12\left(-\frac{1}{2}\right) = 12 - 6 - 6 = 0 \tag{11.7}$$

The result is correct. Obviously this is the right identity. Why then, so one might ask, do we not use this identity to find the correct potential energy functional?

Remember, our rule in this respect was the following (see section 4.5): the first variation of the potential energy functional $\Pi_1(w)$ is identical with the expression of the first identity $(-1)\,G(w,\hat{w})$ at $w = w_s$, $\hat{w} \in R_{1,0}$.

But if we apply this rule to the identity (11.7) and try to find the functional $\Pi_1(w)$ then we realize that $(-1)\,G(w_s,\hat{w})$ is not the first variation of a functional $\Pi_1(w)$ because the bending moments M_l and M_r at the mid-node do not appear in the formulation of the *bvp* of the beam:

$$EIw^{IV} = p, \quad 0 < x < l, \quad w(0) = w(l) = M(0) = M(l) = 0$$

(Forget for the time being that we know the exact solution and, hence, also that $M(l/2) = 12\,\text{kNm}$). A vital piece of information is missing. Because of this the *bvp* is not regular (in the terminology of section 4.12) with respect to the identity (11.7).

If we would modify the bvp and give the previously undetermined internal bending moments M_l and M_r the status of external prescribed couples, see Fig. 11.11, that is if we would formulate the bvp as follows

$$EIw^{IV} = p, \quad 0 < x < l/2,\ l/2 < x < l, \quad M_l(l/2) = M_r(l/2) = 12\,\text{kNm}$$

$$w(0) = w(l) = M(0) = M(l) = 0, \quad w_l(l/2) = w_r(l/2), \quad V_l(l/2) = V_r(l/2)$$

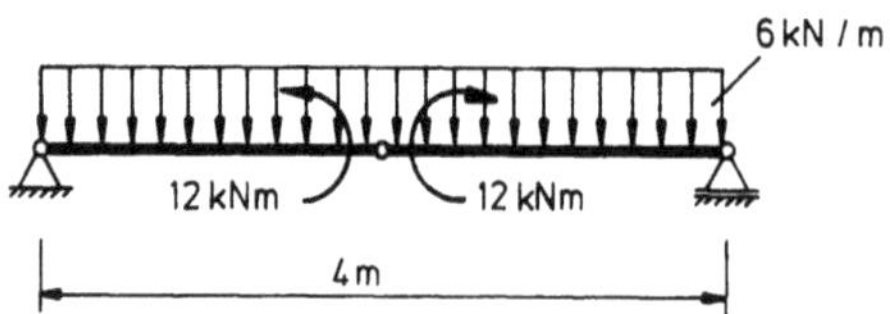

Figure 11.11

then Eq. (11.7) would be (up to a factor (-1)) the first variation of the functional

$$\Pi_1(w) = \frac{1}{2} \int_0^l \frac{M^2}{EI}\,dx - 12(w_l' - w_r')\left(\frac{l}{2}\right) - \int_0^l pw\,dx$$

the potential energy of the linkage in Fig. 11.11.

Hence, in principle nonconforming shape functions are also admissible (in connection with the correct functional). It is only that to formulate the correct functional we need more information about the solution, more than is supplied by the equations of the bvp.

In the case of the beam in Fig. 11.11 the unknown information, the value of the bending moment at mid-span could be calculated beforehand by employing equilibrium conditions. But this is no longer possible in the case of a Kirchhoff plate.

In such a case we could substitute for the unknown bending moments M_i at the interelement boundaries $\partial\Omega_i$, $i = 1, 2\ldots$, Lagrange multipliers, λ_i, $i = 1, 2\ldots$ That is, to say it in terms of our beam problem, we could replace the energy functional of the linkage by the functional

$$\Pi(w, \lambda) = \frac{1}{2} \int_0^l \frac{M^2}{EI}\,dx - \lambda(w_l' - w_r')\left(\frac{l}{2}\right) - \int_0^l pw\,dx$$

which depends on w and λ.

For an assessment of the numerical results obtainable with this method in plate bending, see [H & K].

We think the reader understands now also why nonconforming shape functions have infinite internal energy. They are the displacements of the structure when overloading causes the structure to release excessive internal actions by forming plastic hinges and therewith cause the structure to become a linkage.

Or stated otherwise: if we approximate the displacement field of a structure with nonconforming shape functions then this actually means that we load the structure with infinite forces distributed along the interelement boundaries as, e. g., in the case of the beam in Fig. 11.12. The interpolate w_n is the deflection caused by pairs of infinite couples located at the bends.

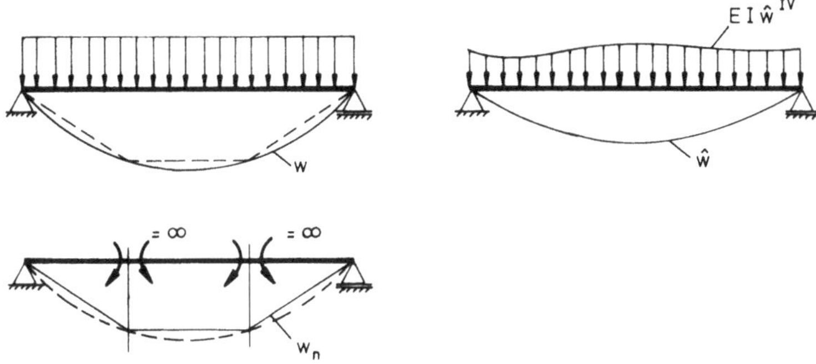

Figure 11.12

It is obvious that the interpolate w_n with increasing n, with increasing number of points where w_n and w match tends to w. But does this hold true for the higher derivatives, the moments and the loads? How do we compare a continuous quantity, p, with discrete (infinite) quantities? How do we measure the infinite couples?—By making use of Betti's principle.

The load p can be considered a functional

$$P(\hat{w}) = \int_0^l p\,\hat{w}\,dx$$

on the set

$$S = \{\hat{w} \in C^4\,[0, l]\,|\,\hat{w}(0) = \hat{w}(l) = \hat{M}(0) = \hat{M}(l) = 0\}$$

of all virtual displacements which are in C^4 (we need this high regularity to perform integration by parts). That is, with every $\hat{w}$ we can associate a number, $P(\hat{w})$, the work done by p acting through $\hat{w}$.

Because w satisfies $EIw^{IV} = p$, because Betti's principle ($=$ 4-times integration by parts) applies and because the boundary conditions are homogeneous this functional can also be calculated by letting the load $EI\hat{w}^{IV}$ act through w.

$$P(\hat{w}) = \int_0^l p\,\hat{w}\,dx = \int_0^l EIw^{IV}\,\hat{w}\,dx = \int_0^l w\,EI\hat{w}^{IV}\,dx$$

This observation is the answer to our problem.

We say a sequence $\{w_n\}$ approximates w if

$$\lim_{n \to \infty} \int_0^l w_n EI \hat{w}^{IV}\, dx = \int_0^l w EI \hat{w}^{IV}\, dx = \int_0^l p\hat{w}\, dx \qquad \forall\, \hat{w} \in S \tag{11.8}$$

that is if w_n, in the limit, satisfies Betti's principle the way the real w does.

Let Ω_e, $e = 1, 2 \ldots n$, the partition of the beam corresponding to w_n then, for one n, the left-hand side reads

$$\int_0^l w_n EI \hat{w}^{IV}\, dx = \sum_e \int_{\Omega_e} w_n EI \hat{w}^{IV}\, dx = -\sum_e \int_{\Omega_e} w_n' EI \hat{w}'''\, dx$$

$$= \sum_e -[w_n' EI \hat{w}'']_{\Gamma_e} = \sum_e \Delta w_n'(x_e)\, \hat{M}(x_e)$$

and, hence, the condition (11.8) means that in the limit the work done by the infinite many couples $\hat{M}(x_e)$ rotating through the infinitesimal small $\Delta w_n'(x_e)$ is just the work done by the continuous load p acting through w.

$$\lim_{n \to \infty} \sum_{e=1}^n \Delta w_n'(x_e)\, \hat{M}(x_e) = \int_0^l p\hat{w}\, dx \qquad \forall\, \hat{w} \in S$$

The definition (11.8) is now also motivation to approximate the function w the following way:

Let

$$w_n = \sum_{i=1}^n \delta_i \varphi_i$$

a sum of (not necessarily) nonconforming shape functions. We say w_n is an approximation of w with respect to the subset

$$S_n = \{\psi_1, \psi_2, \ldots \psi_n | \psi_i \in S\}$$

if

$$\int_0^l w_n EI \psi_i^{IV}\, dx = \int_0^l p\psi_i\, dx \qquad i = 1, 2 \ldots n$$

that is if

$$K\delta = f$$

where

$$K_{ij} = \int_0^l \varphi_i EI \psi_j^{IV}\, dx \qquad f_i = \int_0^l p\psi_i\, dx$$

The advantage this "Betti-approach" has is that it requires the trial functions, the shape functions φ_i, to be only in $L_2[0, l]$. But instead the test functions ψ_i must now be in C^4 (so nothing is gained in the end). Furthermore, because the test and trial functions are no longer identical, the stiffness matrix is no longer symmetric.

11.4 The Patch Test

The first test devised to check the usefulness of nonconforming shape functions was the patch test by Irons and Razzaque, [I & R].

This test makes use of the observation that with decreasing element size the strain within the single element will approximately be constant. The patch test checks whether the finite element net reproduces such states correctly.

What is meant by this is best seen if we study the test itself: Let Ω_S be a subset of the net Ω and P_m a polynomial of degree m. With this polynomial we formulate the bvp

$$Du = 0 \quad \text{in} \ \Omega_S, \quad u|_{\partial\Omega_S} = P_m \tag{11.9}$$

whose solution is, naturally, just the polynomial P_m itself because P_m is a displacement with constant strains and, hence, a homogeneous solution of $Du = 0$.

This bvp is a d-bvp, see section 4.6, and, therefore, the associated functional $\Pi_1(u)$ simply

$$\Pi_1(u) = \frac{1}{2} E(u, u)$$

We know, furthermore, that the solution $u_s = P_m$ renders this functional a minimum on

$$R_1 = \{u \in C_p^m(\bar{\Omega}_S) | u|_{\partial\Omega_S} = P_m\}$$

Now, let R_1^* be the set of all nonconforming shape functions φ_i which satisfy the geometric boundary conditions. If the function which minimizes $\Pi_1(u)$ on R_1^* is just P_m then we say that the finite elements have passed the test with respect to this particular patch Ω_S and this polynomial P_m.

Note that for reasons of completeness we require the shape functions φ_i of finite element nets, be they conforming or nonconforming, to interpolate the polynomials up to the degree m, the degree of the energy, exactly. Hence, the solution P_m is possible in R_1^*.

As an example consider the four (partially) nonconforming functions φ_i in Fig. 11.13 (a shape function of a bar is conforming if it is in C^0, that is if it is continuous).

If we pose the bvp

$$-EAu'' = 0, \quad 0 < x < l, \quad u(0) = 0, u(2) = 1$$

where the boundary values are the values of the first-order polynomial $P_1 = 0.5\,x$ and minimize the potential energy

$$\Pi_1(u) = \frac{1}{2}\int_0^2 \frac{N^2}{EA}\,dx$$

then we obtain the solution

$$u_h(x) = 0\,\varphi_1 + 0\,\varphi_2 + 1\,\varphi_3 + 1\,\varphi_4$$

but not P_1 even though P_1 is in R_1^*

$$P_1 = 0\,\varphi_1 + 0.5\,\varphi_2 + 0.5\,\varphi_3 + 1\,\varphi_4$$

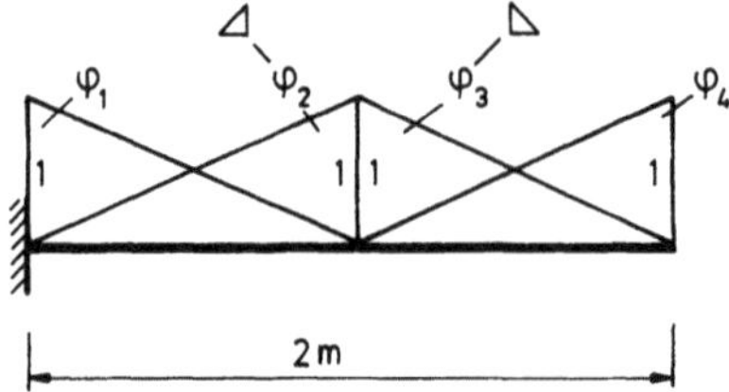

Figure 11.13

The reason for this incorrect result is that we used the wrong identity.

The correct first identity for pairs of functions, $\{u, \hat{\varphi}_i\}$, from $C_p^2 \times (C_p^0 \cap C_{\text{loc}}^2)$ and, therewith, for $u = P_1$ and $\hat{\varphi}_i \in R_{1,0}^*$ is

$$(-1)\,G(u, \hat{\varphi}_i) = -\int_0^l -EA\,u''\,\hat{\varphi}_i\,dx - [N\hat{\varphi}_i]_0^l - N\!\left(\frac{l}{2}\right)(\hat{\varphi}_{il} - \hat{\varphi}_{ir})\left(\frac{l}{2}\right)$$

$$+ \int_0^l \frac{N\hat{N}}{EA}\,dx = -N\!\left(\frac{l}{2}\right)(\hat{\varphi}_{il} - \hat{\varphi}_{ir})\left(\frac{l}{2}\right) + \int_0^l \frac{N\hat{N}}{EA}\,dx = 0$$

$$(11.10)$$

If we would let $u = \sum \delta_i \varphi_i$ and if we would choose consecutively for $\hat{\varphi}_i$ the test functions φ_1, φ_2, φ_3, φ_4 then this identity would render the correct coefficients δ_i, namely 0, 0.5, 0.5, 1.

The error we commit is that we neglect in the expression (11.10) the contribution of the work done at the internal boundaries. We consider the expression without these terms

$$(-1)\,G(u, \hat{\varphi}_i) = \int_0^l \frac{N\hat{N}}{EA}\,dx \overset{?}{=} 0$$

This would be no error if the work at the internal boundaries

$$N\left(\frac{l}{2}\right)(\hat{\varphi}_{il} - \hat{\varphi}_{ir})\left(\frac{l}{2}\right) \tag{11.11}$$

would be zero for all functions $\hat{\varphi}_i$.

This is Strang's and Fix' version of the patch test, see [S&F] p. 177: the non-conforming shape functions are admissible if and only if the work done at the internal boundaries vanishes for all pairs $\{P_m, \hat{\varphi}_i\}$, $\hat{\varphi}_i \in R_1^*$.

If this is guaranteed then there is no difference between the correct and the incorrect identity, and then even the incorrect identity renders the correct result.

In the case of bars there are, naturally, no nonconforming functions φ_i which satisfy Eq. (11.11). If φ_i satisfies this equation then it is continuous, i. e. conforming.

In the case of elastic plates and bodies the correct first identity for pairs as $\{P_1, \hat{\varphi}\}$ where $\hat{\varphi} = \{\hat{\varphi}_i\}$ is a vector-valued function of nonconforming C_p^0-shape functions (the functions $\hat{\varphi}_i$ are only piecewise continuous) is

$$G(P_1, \hat{\varphi}) = \int_\Gamma \tau(P_1) \cdot \hat{\varphi}\, ds + \sum_i \int_{\partial\Omega_i^{\text{int}}} \tau(P_1) \cdot \hat{\varphi}\, ds - E(P_1, \hat{\varphi}) = 0$$

Here $\partial\Omega_i^{\text{int}}$ denotes that part of an element boundary $\partial\Omega_i$ which lies in the interior, in Ω.

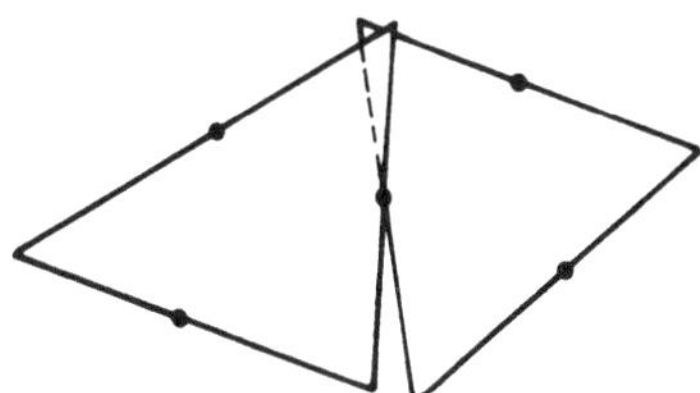

Figure 11.14

Instead of integrating over each internal boundary twice the integration can be done once if we sum the contributions of neighboring elements. This would yield (in a somewhat symbolic notation)

$$\sum_i \int_{\partial\Omega_i^{\text{int}}} \tau(P_1) \cdot \hat{\varphi}\, ds = \sum_i \int_{\partial\Omega_i^{\text{int}}} \tau(P_1)\,(\hat{\varphi}_l - \hat{\varphi}_r)\, ds \tag{11.12}$$

Because $\tau(u) = \tau(P_1)$ is a constant vector (constant strains!) the condition that Eq. (11.12) is zero is equivalent with the condition that the integral of the jumps in displacements of the nonconforming global shape function must disappear when we integrate over the boundaries $\partial\Omega_i$ of the elements Ω_i.

The so-called Veubeke-element with constant strains ($=$ linear functions φ) is an element which connects with its neighbors at the three midnodes, see Fig. 11.14.

The function φ_1 obtained on assigning the value $\mathbf{1} = \{1, 1\}$ to node 1 and to all other nodes the value $\mathbf{0} = \{0, 0\}$ is, e.g., an admissible nonconforming function because the integral of the jump in displacement along the boundary of every element Ω_i vanishes.

In the case of a Kirchhoff plate the nonconforming shape functions are, usually, C^0-shape functions and the correct first identity for pairs of functions as $\{P_2, \hat{\varphi}\}$, $\hat{\varphi}$ is a C^0-function, is

$$G(P_2, \hat{\varphi}) = -\int_\Gamma M_n(P_2) \frac{\partial \hat{\varphi}}{\partial n}\, ds - \sum_i \int_{\partial \Omega^{\text{int}}} M_n(P_2) \left(\frac{\partial \hat{\varphi}_l}{\partial n} - \frac{\partial \hat{\varphi}_r}{\partial n} \right) ds$$

$$- E(P_2, \hat{\varphi}) = 0$$

For a nonconforming global shape function to be admissible we must require that the integral of the jumps of the normal derivative along the boundary $\partial \Omega_i$ of each element Ω_i is zero.

In practice the patch test is performed with the help of the computer. One simply solves the *bvp* (11.9) for different patches and compares the solution with the exact solution P_m.

Though the patch test seems very plausible Stummel, [S2], [S3], could demonstrate that it is neither necessary nor sufficient. In particular is it not enough to require that the internal work, see Eq. (11.11), caused by the discontinuities vanishes with respect to all polynomials P_m. Instead this must be true (in the limit, when the mesh-parameter h tends to zero) with respect to all C_0^∞-functions.

But Stummel could also demonstrate that many of the best known nonconforming elements pass this generalized patch test and, therefore, converge to the correct solution.

11.5 Hybrid Energy Principles

The finite element method considered so far was based on the principle of minimum potential energy derived in chapter 4 for functions $u \in C_p^m(\bar{\Omega})$.

These global smoothness conditions put strong constraints on the competing functions and, hence, the question: could we not formulate energy principles which require less smooth functions, which are not so demanding?

Such energy principles exist indeed. They are called hybrid energy principles.

The theory of these hybrid energy principles will be developed, systematically, in the following.

The basic idea is to consider the finite element net a composite structure as in chapter 3.

As our model problem we choose the bar in Fig. 11.15 which is partitioned into two finite elements, Ω_1 and Ω_2.

Figure 11.15

First we try, as in chapter 3, by adding zeros, $0 + 0 = 0$, to formulate a first identity for the whole system.

$$p: \ u, \hat{u} \in C^2_{\text{loc}} \times C^1_{\text{loc}}$$

$$q: \ G(u, \hat{u}) = G(u, \hat{u})_1 + G(u, \hat{u})_2 = \int_0^{x_M} - EA u'' \hat{u} \, dx + [N\hat{u}]_0^{x_M}$$

$$- \int_0^{x_M} \frac{N\hat{N}}{EA} \, dx + \int_{x_M}^l - EA u'' \hat{u} \, dx + [N\hat{u}]_{x_M}^l - \int_{x_M}^l \frac{N\hat{N}}{EA} \, dx = 0 \tag{11.13}$$

This expression contains, side by side, the identity of the first element, Ω_1, and the second element, Ω_2.

To establish a connection between these two isolated expressions we have two choices:

a) We let the function u be continuous at the interface, $u_l = u_r$

b) We let the axial force N be continuous at the interface, $N_l = N_r$.

The functions u which satisfy condition a) form the class C_d and the functions which satisfy condition b) the class C_f.

We first consider in the following what happens with the expression (11.13) on the class C_d.

If the pair $\{u, \hat{u}\}$ belongs to $(C_d \cap C^2_{\text{loc}}) \times (C_d \cap C^1_{\text{loc}})$ we obtain the expression

$$G(u, \hat{u}) = \int_0^l - EA u'' \hat{u} \, dx + [N\hat{u}]_0^l$$

$$+ (N_l - N_r)(x_M)\, \hat{u}(x_M) - \int_0^l \frac{N\hat{N}}{EA} \, dx = 0 \tag{11.14}$$

In terms of section 4.12 the abstract structure of this Eq. (11.14) is

$$G(u, \hat{u}) = (Du, \hat{u}) + [r_e(u), \hat{u}]_0^l + q_i(u)(x_M)\, \hat{u}(x_M) - E(u, \hat{u}) = 0$$

where

$$r_e(u) = N(u) \qquad q_i(u)(x) = \big(N_l(u) - N_r(u)\big)(x)$$

are external and internal boundary operators.

For energy principles to exist it is sufficient that the *bvp* of the composite bar

$$- EA u'' = p, \quad 0 < x < x_M, x_M < x < l$$

$$u(0) = 0, \quad u_l(x_M) = u_r(x_M), \quad N_l(x_M) = N_r(x_M), \quad N(l) = 0 \tag{11.15}$$

is regular with regard to this identity (a *bvp* is regular if of two conjugated boundary terms exactly one is prescribed on every part of the boundary).

Of the two conjugated terms in $[Nu]_0^l$ the first, N, is prescribed at the right end of the bar, at $x = l$, and the second, u, at the left end, at $x = 0$, and in the expression $q(u)\,u$ the first component, $q(u) = N_l - N_r$, is prescribed at x_M, the internal boundary; the other is not.

Hence, the *bvp* is regular and, therefore, energy principles exist.

The basic functional is the expression

$$\Pi(u) = \frac{1}{2} \int_0^l \frac{N^2}{EA}\,dx - \int_0^l pu\,dx + N(0)\,u(0)$$

which on the class

$$R_1 = \{u \in C_d \cap C^1_{\mathrm{loc}} | u(0) = 0\}$$

becomes the functional of the principle of minimum potential energy.

$$\Pi_1(u) = \frac{1}{2} \int_0^l \frac{N^2}{EA}\,dx - \int_0^l pu\,dx$$

and on the class

$$R_2^{\mathrm{mod}} = \{u \in C_d \cap C^2_{\mathrm{loc}} | -EAu'' = p,\ N(l) = 0\}$$

the functional of the principle of stationary value of complementary energy.

$$\Pi_2^{\mathrm{mod}}(u) = \frac{1}{2} \int_0^l \frac{N^2}{EA}\,dx + (N_l - N_r)\,(x_M)\,u(x_M)$$

The conforming shape functions associated with the class $C_d \cap C^1_{\mathrm{loc}}$ are, naturally, just the C^0-shape functions and what we derived so far is, with the exception of the functional Π_2^{mod}, not new.

This new functional, Π_2^{mod}, differs from the functional Π_2 of chapter 4 by an additional term which takes into account the jump of the axial force at the interface; the functions in R_2^{mod} do not necessarily satisfy the condition $(N_l - N_r) = 0$ at x_M.

As a consequence of this the displacement of the bar is not a maximum of the basic functional on R_2^{mod} but only a stationary point.

These were the results for the class $C_d \cap C^n_{\mathrm{loc}}$.

Let us formulate the expression (11.13) now on $C_f \cap C^n_{\mathrm{loc}}$.

$$p:\ u, \hat{u} \in (C_f \cap C^2_{\mathrm{loc}}) \times (C_f \cap C^1_{\mathrm{loc}})$$

$$q:\ G(u, \hat{u}) = \int_0^l -EAu''\hat{u}\,dx + [N\hat{u}]_0^l + N(x_M)\,(\hat{u}_l - \hat{u}_r)\,(x_M)$$

$$-\int_0^l \frac{N\hat{N}}{EA}\,dx = 0$$

The formal structure of this equation is

$$G(u, \hat{u}) = (Du, \hat{u}) + [r(u), \hat{u}]_0^l + r(u)\,(x_M)\,s(\hat{u})\,(x_M) - E(u, \hat{u}) = 0$$

where

$$r(u) = N(u) \qquad s(u)\,(x) = (u_l - u_r)\,(x)$$

The *bvp* (11.15) is regular with respect to this identity because at the internal boundary the quantity

$$s(u)\,(x_M) = (u_l - u_r)\,(x_M) = 0$$

is prescribed and on the external boundary u or N. Hence, energy principles exist.

The basic functional is the expression

$$\Pi^{\mathrm{mod}}(u) = \frac{1}{2} \int_0^l \frac{N^2}{EA}\,dx - \int_0^l pu\,dx + N(0)\,u(0) - N(x_M)\,(u_l - u_r)\,(x_M)$$

which on the class

$$R_1^{\mathrm{mod}} = \{u \in C_f \cap C_{\mathrm{loc}}^1 | u(0) = 0\}$$

becomes the functional of the principle of stationary value of potential energy

$$\Pi_1^{\mathrm{mod}}(u) = \frac{1}{2} \int_0^l \frac{N^2}{EA}\,dx - \int_0^l pu\,dx - N(x_M)\,(u_l - u_r)\,(x_M)$$

and on the class

$$R_2 = \{u \in C_f \cap C_{\mathrm{loc}}^2 | N(l) = 0\}$$

the functional of the principle of maximum complementary energy

$$\Pi_2(u) = -\frac{1}{2} \int_0^l \frac{N^2}{EA}\,dx$$

What we did here with a bar can be done with every structural element.

The hybrid energy principles are obtained if we write down the first identity for each single finite element and then choose an appropriate continuity across interelement boundaries.

We should only hint at one specific difference between operators of second and fourth degree.

In the case of operators of second degree we have continuity in one boundary term and in the case of operators of fourth degree continuity in two boundary terms.

The reason for this is that operators of degree 4 possess at every internal boundary four boundary terms, 2 displacements and 2 forces, and, therefore, continuity can be (must be) established by matching two neighboring quantities.

In the case of a Kirchhoff plate we could think of a plate which possesses a global continuity in the following pairs of functions

$$w, \frac{\partial w}{\partial n}; \quad M_n, V_n(F_k); \quad w, M_n; \quad \frac{\partial w}{\partial n}, V_n(F_k)$$

The most obvious choice would, naturally, be the first model, which represents continuity in displacement terms (C_d). The C^1-shape functions represent such models. The second model corresponds to the previous class C_f. The third and fourth model are too exotic to deserve much interest.

Summarizing our results we may state:

The functionals of the modefied and non-modefied energy principles with respect to $C_d \cap C_{\text{loc}}^n$ and $C_f \cap C_{\text{loc}}^n$ functions are:

	$C_d \cap C_{\text{loc}}^n$	$C_f \cap C_{\text{loc}}^n$	
regular:	$\Pi = \Pi^{\text{reg}}$	$\Pi^{\text{mod}} = \Pi^{\text{reg}} + \Delta d \cdot f$	*hybrid*
regular:	$\Pi_1 = \Pi_1^{\text{reg}}$	$\Pi_1^{\text{mod}} = \Pi_1^{\text{reg}} + \Delta d \cdot f$	*hybrid*
hybrid:	$\Pi_2^{\text{mod}} = \Pi_2^{\text{reg}} + \Delta f \cdot d$	$\Pi_2 = \Pi_2^{\text{reg}}$	*regular*

The functionals Π^{reg} are the functionals of chapter 4, the functionals on the classes $C_p^m(\bar{\Omega})$ or $C_p^{2m}(\bar{\Omega})$. So, it may happen that the modified functional coincides with the original functional, that there is no difference. If, compared with the original functional, additional terms appear then we term the functional Π^{mod}.

These additional terms represent the work of discontinuous forces or displacements

$$\Delta f = \textit{jump of a force-term}, \quad d = \textit{displacement}$$
$$\Delta d = \textit{jump in displacement}, \quad f = \textit{force term}$$

The hybrid functionals, Π^{mod}, can be modified further by replacing the terms d and f which accompany the jump terms by Lagrange multipliers.

11.6 Hybrid Energy Principles for Operators A

Our introduction to hybrid energy principles has been very formal. For practical applications we propose a simpler procedure:

Assume the finite element net consists of N elements. We then simply write the form $V(u, \hat{u})$ N-times side by side.

If we substitute into this expression the pair

$$\{u_s, \hat{u}\} \quad \hat{u} \in C_d \cap C_{\text{loc}}^n \quad \text{or} \quad \{u_s, \hat{u}\} \quad \hat{u} \in C_f \cap C_{\text{loc}}^n$$

where u_s is the solution of the *bvp* then we, automatically, obtain the first variation of the modified (or non-modified) functional on the classes $C_d \cap C_{loc}^n$ or $C_f \cap C_{loc}^n$.

We illustrate this approach with the elastic plate in Fig. 11.16 which we assume to be partitioned into N elements.

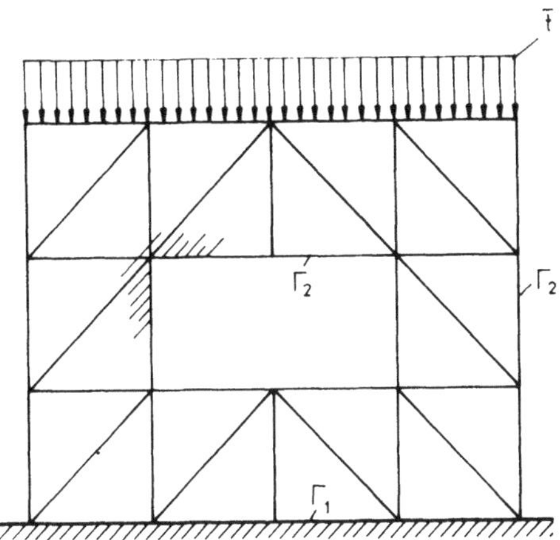

Figure 11.16

The *bvp* of the plate formulates as

$$A(\Sigma) = \begin{bmatrix} 0 \\ 0 \\ p \end{bmatrix} \quad \begin{array}{l} u|_{\Gamma_1} = \bar{u} \\ Sn|_{\Gamma_2} = \bar{t} \end{array} \tag{11.16}$$

For simplicity we assume that the solution $\Sigma = \{u, E, S\}$ belongs to $C^\infty(\bar{\Omega})$ (with each component).

Let $\mathcal{S}_{loc}$ be the class of all elements $\Sigma = \{u, E, S\}$ which belong on each finite element Ω_i to $C^\infty(\bar{\Omega}_i)$. Formulating the expression $V(\Sigma, \hat{\Sigma})$ N times we obtain

$$V(\Sigma, \hat{\Sigma}) = \sum_{i=1}^{N} \left[\left\{ -\langle A(\Sigma), \hat{\Sigma} \rangle - \int_{\partial\Omega_i} Sn \cdot \hat{u}\, ds + \int_{\partial\Omega_i} u \cdot \hat{S}n\, ds \right\} \right.$$
$$\left. - \int_{\partial\Omega_i} u \cdot \hat{S}n\, ds + E(\Sigma, \hat{\Sigma})_{\Omega_i} \right]$$

where the integrals in a bracket $[\ldots\ldots]$ are to be taken over the element Ω_i and its boundary $\partial\Omega_i$.

No connection yet exists between the single brackets, the single elements. To provide for such connections we can choose between two interface conditions

$$u^{(i)} - u^{(j)}|_{\partial\Omega_i \cap \partial\Omega_j} = 0, \quad (Sn)^{(i)} + (Sn)^{(j)}|_{\partial\Omega_i \cap \partial\Omega_j} = 0 \tag{11.17}$$

Continuity in displacements on interelement boundaries or continuity in tractions.

Let $\mathscr{S}_d$ and $\mathscr{S}_f$ denote the set of all elastic states Σ which satisfy the first and the second condition, resp.

Energy principles are obtained if we restrict the form $V(\Sigma, \hat{\Sigma})$ to the classes $\mathscr{S}_d \cap \mathscr{S}_{\text{loc}}$ and $\mathscr{S}_f \cap \mathscr{S}_{\text{loc}}$. We start with the first class.

The form $V(\Sigma, \hat{\Sigma})$ becomes, if we substitute the data of the bvp into it and let $\hat{\Sigma}$ an elastic state from $\mathscr{S}_d \cap \mathscr{S}_{\text{loc}}$,

$$\delta \Pi(\Sigma, \hat{\Sigma}) = \int_{\Gamma_1} (\bar{u} - u) \cdot \hat{S}n \, ds - \int_{\Gamma_1} Sn \cdot \hat{u} \, ds - \int_{\Gamma_2} \bar{t} \cdot \hat{u} \, ds + E(\Sigma, \hat{\Sigma})$$

$$- \int_{\Omega} p \cdot \hat{u} \, d\Omega = 0$$

Here we used the auxiliary lemma

$$p: \Sigma = \Sigma_s, \hat{\Sigma} \in \mathscr{S}_d \cap \mathscr{S}_{\text{loc}}$$

$$q: \sum_i \int_{\partial \Omega_i} Sn \cdot \hat{u} \, ds = \int_{\Gamma_1} Sn \cdot \hat{u} \, ds + \int_{\Gamma_2} \bar{t} \cdot \hat{u} \, ds$$

The form $\delta \Pi(\Sigma, \hat{\Sigma})$ is the first variation of the basic functional

$$\Pi(\Sigma) = \int_{\Gamma_1} (\bar{u} - u) \cdot Sn \, ds - \int_{\Gamma_2} \bar{t} \cdot u \, ds - \int_{\Omega} p \cdot u \, d\Omega + \frac{1}{2} E(\Sigma, \Sigma)$$

whose restriction to R_1 is the functional

$$\Pi_1(\Sigma) = \frac{1}{2} E(\Sigma, \Sigma) - \int_{\Omega} p \cdot u \, d\Omega - \int_{\Gamma_2} \bar{t} \cdot u \, ds$$

the functional of the principle of minimum potential energy.

The principle of stationary value of complementary energy is obtained if we restrict $\Pi(\Sigma)$ to the class

$$R_2^{\text{mod}} = \{\Sigma \in \mathscr{S}_d \cap \mathscr{S}_{\text{loc}} | C^{-1}[S] - E = 0, \, -\text{div} \, S = p, \, Sn = \bar{t} \text{ on } \Gamma_2\}$$

Here, $\Pi(\Sigma)$ has the form

$$\Pi_2^{\text{mod}}(S, u) = -\frac{1}{2} \int_{\Omega} C^{-1}[S] \cdot S \, d\Omega + \int_{\Gamma_1} \bar{u} \cdot Sn \, ds$$

$$+ \sum_{i=1}^{N} \int_{\partial \Omega_i^{\text{int}}} u \cdot Sn \, ds \tag{11.18}$$

where $\partial\Omega_i^{\text{int}}$ indicates that integration is done only over the element boundaries $\partial\Omega_i$ which lie inside the domain, which do not coincide with Γ_1 or Γ_2. The internal boundary function u plays the role of a Lagrange multiplier.

The result agrees, up to the usual factor (-1), with Eq. (24) in [P&T].

We turn now to the class $\mathscr{S}_f \cap \mathscr{S}_{\text{loc}}$. The form $V(\Sigma, \Sigma)$ becomes, if we substitute the data of the bvp into it, and let $\hat{\Sigma}$ be an element of this class

$$\delta\Pi(\Sigma,\hat{\Sigma}) = \int_{\Gamma_1} \bar{u}\cdot\hat{S}n\,ds + \int_{\Gamma_2} (Sn-\bar{t})\cdot\hat{u}\,ds + \int_{\Gamma_2} u\cdot\hat{S}n\,ds - \int_{\Omega} p\cdot\hat{u}\,d\Omega$$

$$+ E(\Sigma,\hat{\Sigma}) - \sum_{i=1}^{N} \left\{ \int_{\partial\Omega_i} Sn\cdot\hat{u}\,ds + \int_{\partial\Omega_i} u\cdot\hat{S}n\,ds \right\}$$

Here, we used the auxiliary lemma

$$p:\ \Sigma = \Sigma_s,\ \hat{\Sigma} \in \mathscr{S}_f \cap \mathscr{S}_{\text{loc}}$$

$$q:\ \sum_i \int_{\partial\Omega_i} u\cdot\hat{S}n\,ds = \int_{\Gamma_1} \bar{u}\cdot\hat{S}n\,ds + \int_{\Gamma_2} u\cdot\hat{S}n\,ds$$

The form $\delta\Pi(\Sigma,\hat{\Sigma})$ is the first variation of the functional

$$\Pi(\Sigma) = \int_{\Gamma_1} \bar{u}\cdot Sn\,ds + \int_{\Gamma_2} (Sn-\bar{t})\cdot u\,ds$$

$$+ \frac{1}{2} E(\Sigma,\Sigma) - \int_{\Omega} p\cdot u\,d\Omega - \sum_i \int_{\partial\Omega_i} u\cdot Sn\,ds$$

which on the class

$$R_1^{\text{mod}} = \{\Sigma \in \mathscr{S}_f \cap \mathscr{S}_{\text{loc}} | E(u) - E = 0,\ C[E] - S = 0,\ u = \bar{u}\ \text{on}\ \Gamma_1\}$$

becomes

$$\Pi_1^{\text{mod}}(u,S) = -\int_{\Gamma_2} \bar{t}\cdot u\,ds - \sum_{i=1}^{N} \int_{\partial\Omega_i^{\text{int}}} u\cdot Sn\,ds$$

$$- \int_{\Omega} p\cdot u\,d\Omega - \frac{1}{2}\int_{\Omega} C[E(u)]\cdot E(u)\,d\Omega$$

This is the functional of the principle of stationary value of potential energy, see [P&T] Eq. (44).

The traction vector Sn on the internal boundaries plays the role of a Lagrange multiplier.

Little by little we subtract
Faith and fallacy from fact,
The illusory from the true
And then starve upon the residue.

Samuel Hoffenstein

THE END

12 References

[A 1]   H. Antes, "Über Fehler und Möglichkeiten ihrer Abschätzung bei numerischen Berechnungen von Schalentragwerken", Mitteilungen aus dem Institut für Mechanik, Inst. f. Mechanik der Ruhr-Universität Bochum 1980

[A & W]   D. N. Arnold, W. L. Wendland, "On the asymptotic convergence of collocation methods". Math. Comp. 41, pp. 349–381, 1983

[A 2]   I. Avramidis, "Übertragungs- und Steifigkeitsmatrizen für den elastisch gebetteten Zug- und Druckstab nach der Theorie II. Ordnung", Die Bautechnik, 3, p. 99–104, 4, pp. 140–143 (1982)

[B 1]   I. Babuška, "The Finite Element Method with Lagrange Multipliers", numer. Math., 20, 179–192 (1973)

[B 2]   C. A. Brebbia, The Boundary Element Method for Engineers, Pentech Press, London, Plymouth 1978

[B 3]   C. A. Brebbia (Ed.), Progress in Boundary Element Methods, Volume 1, Pentech Press, London, Plymouth 1981

[B 4]   F. Brezzi, "On the Existence, Uniqueness and Approximation of Saddle-Point Problems Arising from Lagrangian Multipliers", Revue Française d'Automatique Informatique et Recherche Opérationelle Numer. Anal. R 2, 129–151, 1974

[B & B]   P. K. Banerjee, R. Butterfield, Boundary Element Methods in Engineering Science, McGraw Hill, London, New York, 1981

[B & C]   M. Bernadou, P. G. Ciarlet, "Sur l'ellipticité du modèle linéaire de coque de W. T. Koiter", in: Computing Methods in Applied Sciences and Engineering, (R. Glowinski and J. L. Lions, Ed.) pp. 89–136, Lecture Notes in Economics and Mathematical Systems, Vol. 134, Springer-Verlag, Berlin

[B & T & W]   C. A. Brebbia, J. C. F. Telles, L. C. Wrobel, Boundary Element Techniques, Springer-Verlag, Heidelberg, New York, Tokyo, 1983

[C 1]   D. E. Carlson, Linear Thermoelasticity, in [F 2]

[C 2]   A. Castigliano, "Nuova teoria intorna dell'equilibrio dei sistemi elastici", Atti della Academia delle scienze Torino 10 (1875)

[C 3]   A. Castigliano, Theorie de l'équilibre des systèmes elastiques, Turin 1879

[C 4]   A. Chajes, Principles of Structural Stability Theory, Civil Engineering and Engineering Mechanics Series, Prentice-Hall, Inc. Englewood Cliffs, New Jersey 1974

[C 5]   P. G. Ciarlet, The Finite Element Method for Elliptic Problems, North-Holland Amsterdam 1978

[C 6]   L. Collatz, Numerische Behandlung von Differentialgleichungen, Springer-Verlag, Berlin/Göttingen/Heidelberg 1951

[C & O]   G. F. Carey, J. T. Oden, Finite Elements: A second course, Vol. II Prentice Hall, Inc. Englewood Cliffs, New Jersey 1983

[C & S]   G. Czerwenka, W. Schnell, Einführung in die Rechenmethoden des Leichtbaus, BI-Hochschultaschenbücher 124/124a und 125/125a, Bibliographisches Institut Mannheim 1967

[F 1]   G. Fichera, Existence Theorems in Elasticity, in [F 2]

[F 2]   S. Flügge (Ed.), Encyclopedia of Physics, Volume VI a/2, Editor: C. Truesdell, Springer-Verlag, Berlin/Heidelberg/New York 1972

[F3]   W. Flügge (Ed.), Handbook of Engineering Mechanics, McGraw Hill Book Company, New York, Toronto, London 1962

[F4]   Y. C. Fung, Foundations of Solid Mechanics, Prentice Hall, Inc. Englewood Cliffs, New Jersey 1965

[G1]   K. Girkmann, Flächentragwerke, Springer-Verlag Wien, 1963, 6th edition

[G2]   M. E. Gurtin, The Linear Theory of Elasticity, in [F2]

[H1]   F. Hartmann, Elastische Potentiale in Gebieten mit Ecken, Dissertation, University of Dortmund 1980

[H2]   F. Hartmann, "The derivation of stiffness matrices from integral equations", Appl. Math. Modelling, 1981, Vol. 5, October 1981, pp. 355–365

[H3]   F. Hartmann, "The Somigliana identity on piecewise smooth surfaces", Journal of Elasticity, Vol. 11, No. 4. October 1981, pp. 403–423

[H4]   F. Hartmann, "The physical nature of elastic layers", Journal of Elasticity, Vol. 12, No. 1, January 1982, pp. 19–29

[H5]   F. Hartmann, "Elastic potentials on piecewise smooth surfaces", Journal of Elasticity, Vol. 12, No. 1, January 1982, pp. 31–50

[H6]   F. Hartmann, Elastostatics, in [B3]

[H7]   F. Hartmann, Die mathematischen Grundlagen der Statik, Habilitation-thesis, University of Dortmund 1982

[H8]   F. Hartmann, "Castigliano's Theorem and its Limits", Zeitschrift für angewandte Mathematik und Mechanik, (ZAMM), Vol. 62, No. 12, pp. 645–650, 1982

[H9]   F. Hartmann, "Castigliano and Sobolev", Zeitschrift für angewandte Mathematik und Mechanik, (ZAMM), Vol. 65, No. 2, pp. 121–125, 1985

[H10]   F. Hartmann, "Castigliano's Theorem and stiffness matrices", Ingenieurarchiv, Vol. 52, 1984, pp. 182–187

[H11]   F. Hartmann, "Conforming and Nonconforming Boundary Elements" Proceedings of the 6th Int. Conf. on Boundary Element Methods Ed. C. A. Brebbia, CML-Publication (1984)

[H12]   S. Hoffenstein, in: M. Kline, Mathematics, the loss of certainty, Oxford University Press, New York 1980, p. 241

[H&K]   J. W. Harvey and S. Kelsey, "Triangular plate bending elements with enforced compatibility", AIAA J.9, 1023–1026

[H&K&P]   F. Hartmann, C. Katz and B. Protopsaltis, "Boundary elements and symmetry", Ingenieurarchiv, to appear

[I&R]   B. M. Irons and A. Razzaque, "Experience with the patch test for convergence of finite elements", in: The mathematical foundations of the finite element method with applications to partial differential equations, Ed. A. K. Aziz, Academic Press, New York, London 1972

[J&S]   M. A. Jaswon and G. T. Symm, Integral Equation Methods in Potential Theory and Elastostatics, Academic Press, 1979

[K1]   W. T. Koiter, On the foundations of the linear theory of thin elastic shells I, II, pp. 169–182/183–198, Proceedings of the Kon. Ned. Akad. v. Wetenschappen, Series B, Vol. LXXIII, 1970

[K2]   V. D. Kupradze (Ed.), Three-dimensional Problems of the Mathematical Theory of Elasticity and Thermoelasticity, North-Holland Publishing Co., 1979

[L1]   R. S. Lehman, "Developments near an analytic corner of solutions of elliptic partial differential equations", J. Math. Mech. 8, 727–760

[L&T]   M. Lawo and G. Thierauf, Stabtragwerke, Matrizenmethoden der Statik und Dynamik, Friedr. Vieweg & Sohn, Braunschweig/Wiesbaden 1980

[M1]   J. Mason, Variational, Incremental and Energy Methods in Solid Mechanics and Shell Theory, Elsevier scientific publishing company, Amsterdam/Oxford/New York 1980

[M&O]   F. A. Mirza and M. D. Olson, "The mixed finite element method in plane elasticity", Int. J. Num. Meth. Eng., Vol. 15, 237–289 (1980)

[N1]   P. M. Naghdi, The Theory of Shells and Plates, in [F2]

[N2]   P. M. Naghdi, Foundations of elastic shell theory, Progress in solid mech. Vol. 4 (Ed. I. N. Sneddon and R. Hill), pp. 1–90. North Holland Amsterdam = Tech.

Rep. No. 15, Contract Nonr-222 (69), Univ. of Calif., Berkeley (Jan. 1962)

[N3]   J. L. Nowinski, Applications of Functional Analysis in Engineering, Plenum Press, New York and London 1981

[O1]   J. T. Oden, "The classical variational principles of mechanics", in: Energy Methods in Finite Element Analysis, Ed. R. Glowinski, E. Y. Rodin, O. C. Zienkiewicz, John Wiley & Sons, New York 1979

[O2]   J. T. Oden, Applied Functional Analysis, Prentice-Hall Civil Engineering and Engineering Mechanic Series, Englewood Cliffs, N. J., 1972

[O3]   J. T. Oden, Finite Elements of Nonlinear Continua, McGraw Hill, New York 1972

[O&R]   J. T. Oden, J. N. Reddy, An introduction to the mathematical theory of finite elements, John Wiley & Sons, New York 1976

[P1]   H. Parkus, Thermal Stresses, in [F3]

[P2]   C. Petersen, Statik und Stabilität der Baukonstruktionen, Friedr. Vieweg & Sohn, Braunschweig/Wiesbaden 1980

[P&T]   T. H. H. Pian, Pin Tong, "Basis of finite element methods for solid continua", Int. J. Num. Meth. in Eng., Vol. 1, 3–28 (1969)

[R1]   K. Rektorys, Variational Methods in Mathematics, Science and Engineering, 2nd edition, D. Reidel Publishing Company, Dordrecht—Holland/Boston—USA/London/England 1980

[S1]   M. Stern, "A general boundary integral formulation for the numerical solution of plate bending problems", Int. J. Solids Structures, 15, pp. 769–782, 1979

[S&F]   G. Strang and G. J. Fix, An analysis of the Finite Element Method, Prentice-Hall Inc., Englewood Cliffs, N. J. 1973

[S2]   F. Stummel, "The generalized patch test", SIAM J. Numerical Analysis, Vol. 16, No. 3, June 1979, pp. 449–471

[S3]   F. Stummel, "The limitations of the patch test", Int. J. Numer. Methods Eng., 15, 177–188, 1980

[T1]   D. Talaslidis, Inkrementelle Schalentheorie und numerische Behandlung diskretisierter Eigenwertprobleme der Strukturmechanik auf der Grundlage verallgemeinerter Arbeitsaussagen, Thesis, University of Bochum, West Germany, 1978

[T&W]   D. Talaslidis and W. Wunderlich, "Static and dynamic analysis of Kirchhoff shells based on a mixed finite element formulation", Computers & Structures, Vol. 10, pp. 239–249 (1979)

[T2]   W. Thomson (Lord Kelvin), "On the equations of equilibrium of an elastic solid", Cambr. Dubl. Math. J. 3, 87–89 (1848)

[T3]   E. Tonti, "On the mathematical structure of a large class of physical theories", Rend. Accad. Naz. Lincei, Class. Sci. fis. mat. nat. 52 (1972), pp. 48–56

[V1]   P. Villaggio, Qualitative methods in elasticity, Noordhoff Int. Publishing, Leyden 1977

[W1]   K. Washizu, Variational Methods in Elasticity and Plasticity, 2nd edition, Pergamon Press, Oxford, New York, 1975

[W2]   F. van der Weeën, "Application of the boundary integral equation method to Reissner's plate model", Int. J. for Num. Methods, Vol. 18, pp. 1–10 (1982)

[W3]   W. L. Wendland, "On Galerkin collocation methods for integral equations of elliptic boundary value problems", in: Numerical Treatment of Integral Equations, Ed: J. Albrecht and L. Collatz, ISNM 53, Birkhäuser Verlag, Basel, 1980, pp. 244–275

[W4]   H. Weyl, "David Hilbert and His Mathematical Work", Bulletin of the American Mathematical Society 50, pp. 612–654 (1944)

[W5]   W. Wunderlich, "Grundlagen und Anwendung eines verallgemeinerten Variationsverfahrens", in: Finite Elemente in der Statik, Ed. K. E. Buck, D. W. Scharpf, E. Stein, W. Wunderlich, Verlag von Wilhelm Ernst & Sohn, Berlin, München, Düsseldorf 1973

[W6]   W. Wunderlich, "Mixed models for plates and shells: principles—elements—examples", in: Hybrid and mixed finite element methods, ed. by S. N. Atluri, R. H. Gallagher and D. C. Zienkiewicz, John Wiley and Sons 1983, 215–240

The figures 1.21 [G 1], 1.22 [W 2], 5.26 (source unknown), 6.8 [G 1], 8.1 [C 5], 9.6 [C & S], and 11.1 [O 3] are reproduced by courtesy of Springer-Verlag Wien, John Wiley & Sons Ltd., North-Holland Publishing Company, Bibliographisches Institut and McGraw Hill Book Company.

13 Subject Index

Engineering
with
Computers

An International Journal for Computer-aided Mechanical and Structural Engineering

Editors: **Ted Belytschko, Steven J. Fenves**

Now there is a journal that publishes papers which creatively integrate analytical methods and numerical models with techniques of software engineering, including databases, geometric modeling, and computer graphics. The express purpose of **Engineering with Computers** is to present papers which describe this integration of analysis and computer-aided design and which lead toward and underlying theory of computer-aided engineering.

Civil, structural, mechanical, manufacturing, and aerospace engineers can keep up-to-date with the state-of-the-art activity in:
- user-friendly integrated design and analysis methods
- integration of CAD and analysis systems
- database management techniques for engineering applications
- computer graphics and geometric modeling for enhanced design and analysis
- symbolic and other high-level languages
- knowledge-based expert systems for design
- finite element and boundary element methods pertinent to CAD
- practical optimization techniques
- educational and training programs, procedures, and philosophies

Subscription Information is available from your bookseller or directly from Springer-Verlag, Journal Promotion Dept., P.O.Box 105 280, D-6900 Heidelberg, FRG

Springer-Verlag
Berlin
Heidelberg
New York
Tokyo

MIX
Papier aus verantwortungsvollen Quellen
Paper from responsible sources
FSC® C105338

If you have any concerns about our products,
you can contact us on
ProductSafety@springernature.com

In case Publisher is established outside the EU,
the EU authorized representative is:
Springer Nature Customer Service Center GmbH
Europaplatz 3, 69115 Heidelberg, Germany

Printed by Libri Plureos GmbH
in Hamburg, Germany